Phytoecdysteroids

Ugiloy Yusufovna Yusupova ·
Nurmurod Sheralievich Ramazonov ·
Vladimir Nikolaevich Syrov ·
Shomansur Shosaidovich Sagdullaev

Phytoecdysteroids

Properties, Biological Activity and Applications

Springer

Ugiloy Yusufovna Yusupova
Yunusov Institute of the Chemistry of Plant Substances
Tashkent, Uzbekistan

Nurmurod Sheralievich Ramazonov
Yunusov Institute of the Chemistry of Plant Substances
Tashkent, Uzbekistan

Vladimir Nikolaevich Syrov
Yunusov Institute of the Chemistry of Plant Substances
Tashkent, Uzbekistan

Shomansur Shosaidovich Sagdullaev
Yunusov Institute of the Chemistry of Plant Substances
Tashkent, Uzbekistan

The project was funded through a grant from the Republic of Uzbekistan State Foundation [grant number FZ-2020100797].

ISBN 978-981-16-6710-7 ISBN 978-981-16-6711-4 (eBook)
https://doi.org/10.1007/978-981-16-6711-4

This Springer imprint is published by the registered company Springer Nature Singapore Pte Ltd.
The registered company address is: 152 Beach Road, #21-01/04 Gateway East, Singapore 189721, Singapore

Preface

Ecdysteroids are one of the most widespread steroid compounds in nature. They were found in more than 90% of the world's animal species, named after arthropods, the species number of which reaches 1 million, as well as in some other invertebrates. In 1966, K. Nakanishi and his colleagues discovered ecdysteroids for the first time in the plant *Podocarpus nakaii*, which is very popular in oriental medicine. The presence of ecdysteroids was later discovered in many types of plants, including flowering ones.

The role of these compounds in organisms of lower animals, where as hormones the y regulate a number of extremely important vital processes, is well known. For assessing the role of ecdysteroids in the vital activity of plants, one should bear in mind the close ecological interconnection and interdependence between the world of plants and insects. Unlike plants, most invertebrates do not have an enzyme apparatus for the complete synthesis of steroids. Plants supply a necessary material for building cell membranes and insect hormones in the form of sterols.

Ecdysteroids secreted from plants (phytoecdysteroids) being in most cases hormones for arthropods, they are not for warm-blooded ones. However, they were able to regulate many physiological processes in their bodies. In this case, phytoecdysteroids do not have toxic or any other side effects. It was found that beneficial shifts occurring under the influence of phytoecdysteroids in the mammalian organisms are accompanied by a noticeable improvement in their functional status, which is manifested by activation of the central nervous system, followed by increased efficiency, and increased adaptive capabilities of the body to environmental stressors.

Ecdysteroids in both animal and plants world play an extremely important role in the regulation of vital processes in their organisms, despite the fact that they are far from each other in evolutionary terms. This circumstance led to a significant interest in scientific research and important results were obtained that have both fundamental theoretical and practical significance.

In recent years, in the field of chemistry and pharmacology of plant steroids, most of the work is devoted specifically to phytoecdysteroids, as the most original and promising group of compounds with metabolic activity. Many research institutions are conducting active research in this direction. The largest number of works in the field of ecdysteroid chemistry is devoted to their study both in terms of establishing the structures of newly isolated compounds and in terms of identifying promising plants producing ecdysteroids for use in medicine, cosmetology, the food industry, and agriculture (Scientists—V. V. Volodin, Institute of Biology, Komi Scientific Center of the Ural Branch of the Russian Academy of Sciences, Syktyvkar; B. I. Tuleuov—International Phytochemistry Scientific and Production Holding, Karaganda, Republic of Kazakhstan). Of great interest are the works on the synthesis of ecdysteroids and their derivatives, leading to the production of new substances with more prolonged action (works of Prof. V.N. Odinokov—Institute of Petrochemistry and Catalysis of the Russian Academy of Sciences, Ufa; and Prof. I. V. Zavarzin—Institute of Organic Chemistry named after Acad. N. D. Zelinsky Russian Academy of Sciences, Moscow).

Numerous studies have established that phytoecdysteroids are quite common secondary metabolites of plants, more than 500 compounds of this series have already been identified. Based on them, more than 300 ecdysteroid-containing substances have been created on the basis of which biologically active additives and preparations are developed for use in various fields of human activity.

For many years at the Institute of the Chemistry of Plant Substances named after Academician S. Yu. The Yunusov, Academy of Sciences of the Republic of Uzbekistan, also conducts multifaceted comprehensive studies of compounds of this class, isolated from plants of Uzbekistan and neighboring countries. This monograph is devoted to the analysis and generalization of the data obtained.

The authors are grateful to the initiators of these works at the Institute, carried out both in the chemical-pharmacological and technological aspects, to Profs. N. K. Abubakirov, Z. Saatov, A. G. Kurmukov, T. T. Shakirov, N. D. Abdullaev A. U. Mamatkhanov, as well as all employees of the laboratory of glycoside chemistry, pharmacological and experimental-technological laboratories involved in the implementation of these studies.

Tashkent, Uzbekistan — Ugiloy Yusufovna Yusupova
Tashkent, Uzbekistan — Nurmurod Sheralievich Ramazonov
Tashkent, Uzbekistan — Vladimir Nikolaevich Syrov
Tashkent, Uzbekistan — Shomansur Shosaidovich Sagdullaev

Acknowledgements The authors would like to thank the ICPS (Uzbek Academy of Sciences) for providing all necessary equipment, chemicals, and animals for the study.

We are indebted to Prof. Bakhtiyor Rasulev and Dr. Durbek Usmanov for their comments which enabled us to correct errors and clarify the presentation.

We would like to thank the staff at Springer, in particular Shin'ichi Koizumi and Vidyaa Shri Krishna Kumar, for their help and support.

Contents

Chapter 1
Phytoecdysteroids

Abstract Ecdysteroids, molting hormones of insects, can exert several mild, non-hormonal bioactivities in mammals, including humans. Thus, to date, ecdysteroids have been found in plants belonging to more than 100 families from the fern divisions—*Polypodiphyta*, gymnosperms—*Pinophyta,* and angiosperms—*Magnoliophyta.* Over 500 phytoecdysteroids analogs have been identified so far and it has been speculated that there are over 1,000 possible structures in nature. The essential characteristics of ecdysteroids are a *cis*-(5β-H) junction of rings A and B, a 7-*en*-6-one chromophore, and a *trans*-(14α-OH) junction of rings C and D. These are synthesized from mevalonic acid and cholesterol. The order of introduction of the last hydroxyl groups into the sterol core and side-chain during the biosynthesis of ecdysteroids in insects and plants is also different. These have been isolated from all parts of plants in much higher amounts than those present in insects. In families, along with ecdysteroid-positive, there may be ecdysteroid-negative genera, and even ecdysteroidpositive and ecdysteroid-negative species can be present even in genera. More than 120 phytoecdysteroids isolated from *Ajuga, Rhaponticum*, *Silene* and are listed in this chapter.

Keywords Phytoecdysteroids · Biosynthesis · Distribution · Insect · Plant

1.1 Characteristics of Phytoecdysteroids

The term "ecdysteroids" comes from the Greek word ecdysis (molting) and combines a group of polyhydroxylated sterols with a structure similar to molting hormone and arthropod metamorphosis (Karlson 1976).

Ecdysteroids are a class of steroids of natural origin, structurally close to the hormones of molting arthropods and possessing high biological activity (Golovatskaya 2004; Bajguz et al. 2015).

This definition is quite satisfactory for the first acquaintance with ecdysteroids in order to understand their place and significance among other natural biologically active compounds. However, the current level of knowledge requires a more rigorous use of the term and definition of ecdysteroids, which was pointed out by ecdysone discoverer Carlson P. at the international meeting on ecdysones, which was held

U. Y. Yusupova et al., *Phytoecdysteroids*,
https://doi.org/10.1007/978-981-16-6711-4_1

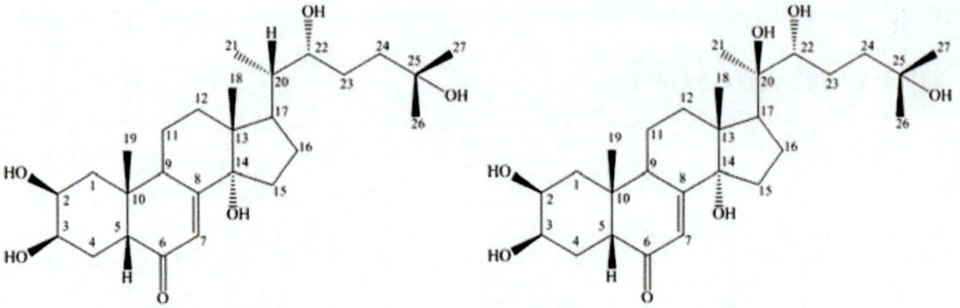

α-ecdysone (1) ecdysterone (2)

Fig. 1.1 The first isolated ecdysteroids from various sources

in 1994 in Ceske Budejovice in the Czech Republic. For example, he pointed out the inadmissibility of using the terms "ecdysteroids" and "ecdysones" as synonyms. The latter term is erroneous, since "ecdysone" is the trivial name for a particular compound—the true hormone of molting insects (Karlson 1994).

For the first time, the crystalline molting hormone was isolated in 1954 (Butenandt and Karlson 1954). The structure of α-ecdysone (1) (Fig. 1.1) was established in 1965 using chemical and instrumental methods and involved X-ray diffraction analysis. It was shown that α-ecdysone (1) is (22R)—2β, 3β, 14α, 22,25-pentahydroxy-5β-cholest-7-en-6-one (Karlson et al. 1963; Reggi et al. 1975).

When ecdysterone (2) (Fig. 1.1) and other ecdysone-like compounds were discovered after ecdysone, the general term "ecdysteroids" was proposed for them in 1976 by analogy with the names of other steroids, for example, "corticosteroids" (Karlson and Koolman 1976). Regarding the definition of "ecdysteroids," at the suggestion of R. Lafont and D. Horn, a distinction was made between true ecdysteroids and ecdysteroid-like compounds (Lafont and Horn 1989).

True ecdysteroids include all steroids having a cis-joint of rings A and B, a 7-en-6-on group and a 14α-hydroxyl group, regardless of their activity in the molt hormone test. Ecdysteroid-like compounds may not have all of the above criteria, they may lack either 14-α-OH, or 7-ene, or they may have one additional double bond at position 4 or 14. According to Prof. Lafont R., who expressed the international meeting on phytoedysteroids in Syktyvkar, this definition is no longer quite satisfactory, since it excludes some compounds biogenetically associated with ecdysone and even having the activity of the molting hormone (Lafont and Horn 1989).

Phytoecdysteroids are known to be fairly widespread secondary plant metabolites. It turned out that these compounds, having in general structural similarity to the α-ecdysone molecule (1) (the main hormone of molting invertebrates), are characterized by much greater variability of chemical structures than zoo-ecdysteroids. To date, more than 500 different structures of phytoecdysteroids have been identified. It is practically impossible to establish absolute boundaries for this group of compounds, since they form a kind of continuous series with other plant steroids with a similar chemical structure: brassinosteroids, vitanolides, steroid glycosides, etc. Ecdysteroids are very diverse in their structure. So, the total number of carbon atoms in a structure can be 27, 28, 29, or 30 for compounds with a complete side chain, or 24, 21, and 19 for compounds having a side chain breaking in the positions C_{24}–C_{25}, C_{20-22}, and C_{17}–C_{20} (Bajguz et al. 2015).

Ecdysteroids may contain:

(1) 3β-OH, 3α-OH or 3-oxo groups;
(2) OH groups in the steroid nucleus can be localized at positions C_1, C_2, C_5, C_{11}, C_{19};
(3) OH groups at positions C_{20}, C_{22}, C_{23}, C_{24}, C_{25}, and C_{26-27} for compounds having a side chain;
(4) additional unsaturated bonds Δ 9–11, Δ 24–25, Δ 25–26, or Δ 24–28;
(5) 5- or 6-membered lactone rings.

Ecdysteroids can form conjugates at positions C_2, C_3, C_{20}, C_{22}, or C_{25} with inorganic acids (phosphates), organic acids (acetates, cinnamates, benzoates, palmitates, p-coumarrates), sugars (galactosides, glucosides), or acetone (2,3- and 20,22-acetonides).

Since a combination of any individual modifications can take place, theoretically one can expect the existence of more than 1000 different structures. It can be assumed that the phytoecdysteroids identified to date represent only a small fraction of those compounds that exist in nature (Barbier 1978; Chaubey 2018).

1.2 Biosynthesis of Phytoecdysteroids

The ability to form the unequal level of content and the structural diversity of ecdysteroids in plants of various species can be understood on the basis of modern ideas about the characteristics of the biosynthesis of phytoecdysteroids and the mechanisms of its regulation.

In insects, cholesterol (C_{27}-sterol) is the main metabolic precursor of ecdysteroids. Since insects are not able to synthesize the de novo steroid core, they receive cholesterol with food. Predatory types of sterols are obtained directly from animal-born food, and phytophages are forced to use plant sterols, which are homologs of cholesterol and differ from it by the presence of additional methyl (campesterol) and ethyl (stigmasterol and sitosterol) substituents at the 24th carbon atom (respectively C_{28}- and C_{29}-sterols). With the help of enzymatic dealkylation reactions, most arthropods can convert C_{28}- and C_{29}-phytosterols into cholesterol.

It is interesting to note that only in a few species of insects, for example, bees (Apis mellifera, Hymenoptera) that are not able to dealkylate phytosterols, the hormonal system is adapted to a plant diet, as a result of which molester hormone turned out to be makisterone A, the only true insect molting hormone containing a methyl radical at C_{24} and biogenetically associated not with cholesterol, but with campesterol. In addition, 24-epimakisterone A was detected in at least one insect species, while makisterone C was found in the other, having 29 carbon atoms in the molecule (Lafont, personal communication).

Although the study of the biosynthesis of ecdysteroids in plants has only a relatively recent history and many of its stages, according to the figurative expression of professor Lafont R., are still a "black box", a general biosynthetic pathway has already appeared in the literature in the 90 s (Lafont et al. 1991; Lafont 1998; Dinan et al. 2009).

Many of the modifications found in phytoecdysteroids are also found in other classes of plant triterpenoids (e.g., brassinosteroids, sterols). This fact raises the possibility that enzymes of the biosynthetic pathways may be common to the plants producing these various classes of metabolites. A lack of specificity in these enzymes may cause the generation of many diverse metabolites or biosynthetic products without massive genetic redundancy of the corresponding enzymes involved (Kreis and Muller-Uri 2018). The understanding of the biosynthetic pathway(s) for phytoecdysteroids is limited. A summary of the knowledge of the biosynthesis has been reviewed by several authors (Rees 1995; Lafont 1997). Studies by Adler and co-workers have demonstrated that active biosynthesis of phytoecdysteroids takes place in developing tissue and that these compounds are transported to other organs (Grebenok and Adler 1991; Tomas et al. 1992). Phytoecdysteroids are synthesized in plants from mevalonic acid via cholesterol and/or lanosterol (Fig. 1.2a, b) (Lafont et al. 2011).

So, plants are able to synthesize de novo ecdysteroids from their precursors (e.g., mevalonic acid) through the formation of sterols as the closest metabolic precursors, which was demonstrated in a number of experiments using labeled cholesterol and mevalonate shortly after the discovery of phytoecdysteroids in the late 60s and early 70s. The use of precursors containing labeled atoms in the molecule at strictly defined positions made it possible to study the sequence of some stages of ecdysteroid biosynthesis in a number of plant species (Sauer et al. 1968; Erdemoglu et al. 2003; Hikino et al. 1970). It turned out that along with the common stages insects and plants have significant differences in the implementation and sequence of some reactions. It is believed that the formation of the Δ7 bond is one of the earliest stages of the biosynthesis of ecdysteroids in both insects and plants. However, as an exception, we can give an example of the fact that, for example, in the fern *Cheilantes tenuifolia*, the introduction of the Δ7 bond is most likely to occur at the last stages. This is indicated by the presence in this species of plants of cheilants containing all the elements of the ecdysone structure, except for the double bond between C_7 and C_8 (Akhrem et al. 1973).

The formation of a cis rings' junction between A and B, on which the biological activity of ecdysteroids depends, also occurs in the early stages of biosynthesis, the mechanism of formation of which is somewhat different in insects and plants (Lloyd-Jones et al. 1973; Das et al. 2020). The mechanism of formation of the Δ7-6-keto group in both insects and plants is still unclear.

Hydroxylation at C_{14} in insects most likely occurs before the introduction of the 6-keto group and/or the cis-A / B articulation (Rees 1995; Das et al. 2020). Whereas in plants this stage follows after the formation of 5β-H and Δ7-6-keto grouping (Lafont 1998). The order of introduction of the last hydroxyl groups into the sterol core and side-chain during the biosynthesis of ecdysteroids in insects and plants is also different. In insects, hydroxylation is carried out in the sequence C_{25}, C_{22}, C_2 (Rees 1995), while in plants it is most likely in the sequence C_2, C_{22}, C_{25}, although this conclusion may be premature due to limited data (Lafont 1998). The experimental data presented fit into the following proposed scheme for the biosynthesis of ecdysteroids in plants (Fig. 1.2). Unlike zooecdysteroids, for which the detection of C_{28}-ecdysteroid (Makisterone A) turned out to be the exception, phytoecdysteroids are characterized by a significant structural diversity and their large number of molecules contains 28 and 29 carbon atoms. This fact suggests that many plants use C_{28} and C_{29} sterols for the synthesis of ecdysteroids without prior dealkylation. Although we did not find any information in the literature on the inclusion of labeled C_{28} and C_{29} sterols in ecdysteroids, it is quite logical to supplement them with the biosynthesis of ecdysteroids, in which alkyl sterols can also be metabolic precursors of ecdysteroids. Moreover, it can be assumed that in plants producing C_{28} and C_{29} ecdysteroids, ordinary C_{27} ecdysteroids (for example, 20-hydroxyecdysone) may be of secondary origin through dealkylation of C_{28} and C_{29} ecdysteroids.

In recent years, studies have appeared on the mechanism of biosynthesis of ecdysteroids and its regulation in the annual plant *Spinacia oleracea* (*Chenopodiaceae*) using in vivo labeled precursors (Grebenok and Adler 1991; Grebenok et al. 1991). Based on the analysis of the distribution of ecdysteroids in plant organs from the time of seed germination, it was established that the biosynthesis of ecdysteroids begins at the stage of ontogenesis associated with the development of leaves, and the place of biosynthesis is the leaves preceding the apical, into which the synthesized ecdysteroids are then transported. It was found that in biosynthetically active leaves, the formation of Δ7-sterol of latosterol (the most likely precursor of ecdysone), ecdysone, ecdysone-3-phosphate, and 20-hydroxyecdysone occurs. Ecdysone and ecdysone-3-phosphate are endogenous precursors of 20-hydroxyecdysone. Cytochrome-P450-ecdysone-20-monoxygenase, which catalyzes the hydroxylation of ecdysone to 20-hydroxyecdysone, was isolated from biosynthetic *Spinach* leaves (Grebenok et al. 1996). Ecdysone and ecdysone—3-phosphate may also fulfill a transport function, delivering the ecdysone molecule to the accumulating organs—the developing leaves of the apex of the stem and roots (Grebenok et al. 1994; Okamoto et al. 2018). An extremely important result for understanding the mechanism of regulation of ecdysteroid biosynthesis was the discovery of the effect of its inhibition by the end products of biosynthesis—ecdysone and 20-hydroxyecdysone polyphosphates, according to the type of negative feedback (Lafont et al. 2011) (Fig. 1.2). The fact that ecdysteroid polyphosphates were found in the species Zea mays (Devarenne et al. 1995), which does not contain free ecdysteroids and, therefore, was not previously considered a producer of ecdysteroids, suggests (Lafont 1998) that genes encoding enzymes involved in the biosynthesis of ecdysteroids is present in many (if not all!) plant species, but their expression occurs only in some. Then, a different level of accumulation of free ecdysteroids (from very high for some species to the complete absence of free ecdysteroids in others, such as Zea mays) may be a consequence of the regulatory mechanism described above involving ecdysteroids polyphosphates.

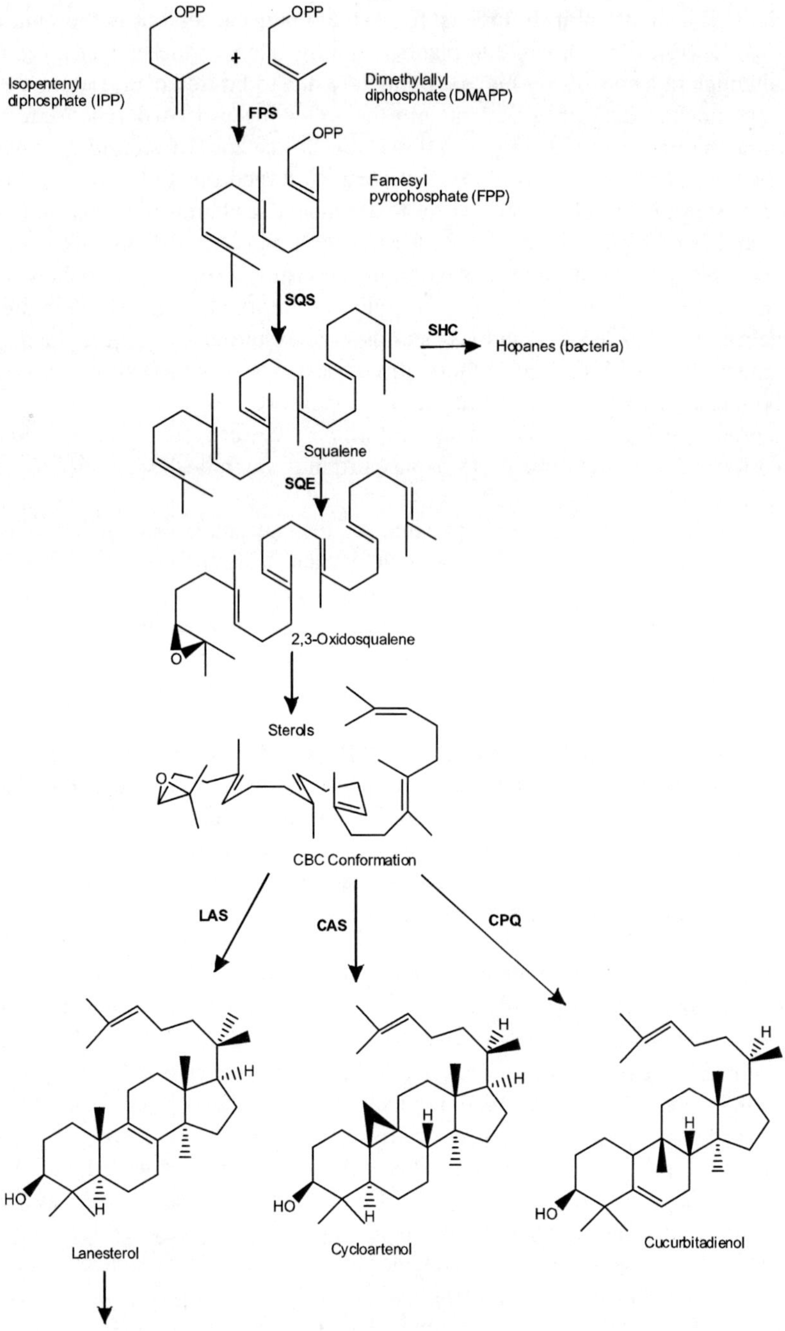
OPP
OPP
Isopentenyl diphosphate (IPP)
Dimethylallyl diphosphate (DMAPP)
FPS
OPP
Farnesyl pyrophosphate (FPP)
SQS
SHC
Hopanes (bacteria)
Squalene
SQE
2,3-Oxidosqualene
Sterols
CBC Conformation
LAS
CAS
CPQ
Lanesterol
Cycloartenol
Cucurbitadienol

◂**Fig. 1.2** The biosynthetic route to sterols. Sterols are synthesized via the mevalonic acid (MVA) pathway. The enzymes that catalyze the various steps are indicated in boxes*. **b** The biosynthetic route to sterols. Sterols are synthesized via the mevalonic acid (MVA) pathway. The enzymes that catalyze the various steps are indicated in boxes* **Enzyme abbreviations:* FPS—farnesyl pyrophosphate synthase; SQS—squalene synthase; SQE—squalene monooxygenase or epoxidase; SHC—squalene-hopene cyclase; LAS—lanosterol synthase; CAS—cycloartenol synthase; CPQ—cucurbitadienol synthase

The regulatory mechanism of phytoecdysteroids biosynthesis described above allows us to consider data on the distribution of ecdysteroids in the plant world from a new perspective.

1.3 Distribution of Ecdysteroids in the Plant World

In 1966, at a symposium on natural compounds, Nakanishi K. from the University of Tohoku (Japan) announced the discovery of a compound with the activity of the insect molting hormone in the plant *Podocarpus nakaii* (Nakanishi et al. 1966).

He by himself described the history of this sensational discovery in a letter to the famous Australian biochemist Denis Horn, who later published this letter (Spindler et al. 2009).

We considered it important to familiarize our readers and especially young researchers with the history of this discovery. Nakanishi writes about his interest in medicinal plants of South-West Asia with antitumor properties. During one trip to the mountain region of central Taiwan Island in 1963, he collected leaves of a *Podocarpus* called locals ‘ai pai-yu-chin’ and used it as an antitumor agent. Although the extracts of this plant turned out to be inactive when tested for cytotoxicity, in the future he managed to isolate a mixture of terpenoid compounds that turned out to be new, the structures of which were determined, but not immediately (Takemoto et al. 1967).

In August 1966, presenting the abstract of the report to the Pacific Scientific Congress in Tokyo, the researchers still attributed the 29-nor-triterpenoid formula to the main component. However, already at the opening of the congress, Masado Koride, graduate student of Nakanishi, managed to establish the structure of this compound, which was called ponasterone A (3) (Nakanishi et al. 1966).

In order to verify whether the isolated compound is new, the researchers studied the literature and found that the structure of ponasterone A (3) differs by the position of only one hydroxyl group from ecdysone—the insect molting hormone, the structure of which was established by X-ray analysis. Nakanishi writes that they twice turned out to be very close to their discovery. Due to differences in the signals of methyl groups of the side chain in the NMR spectrum, Corida did not see any similarities between these compounds until its complete structure was established (Nakanishi et al. 1966).

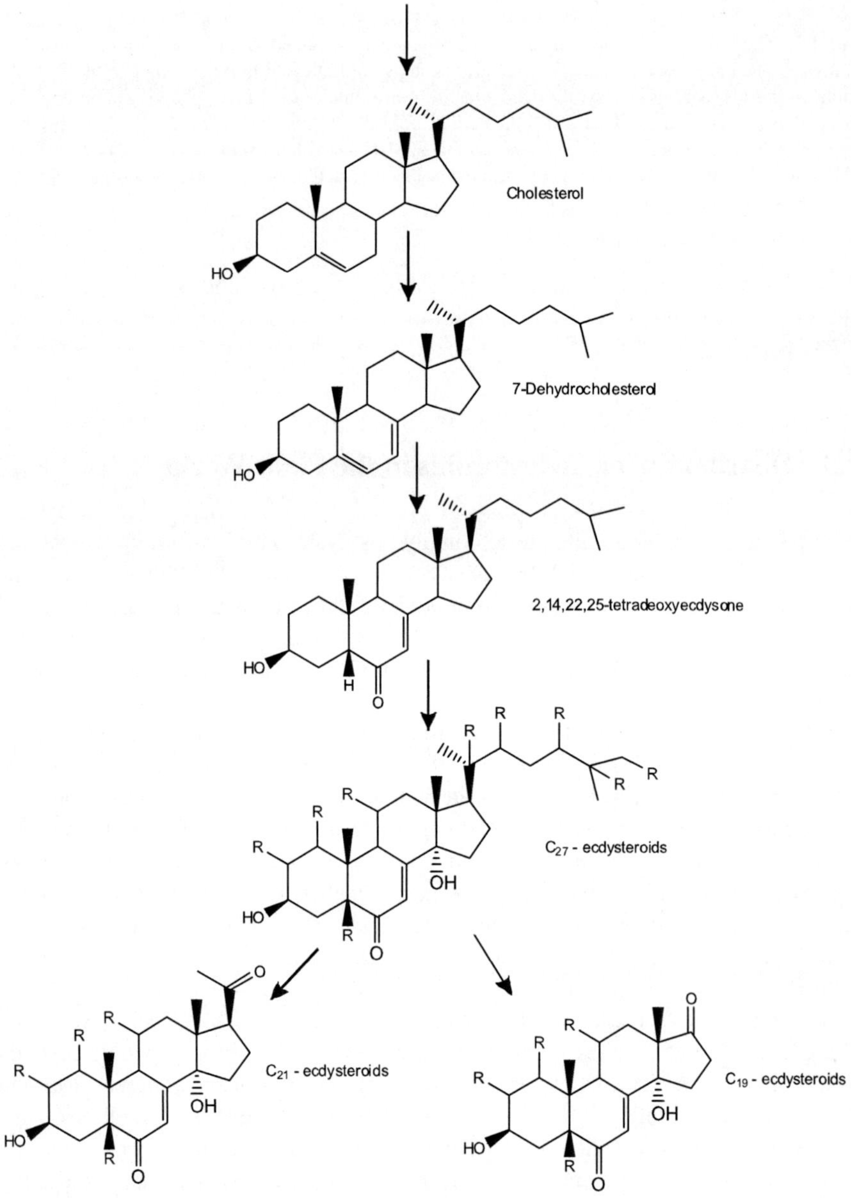

Fig. 1.2 (continued)

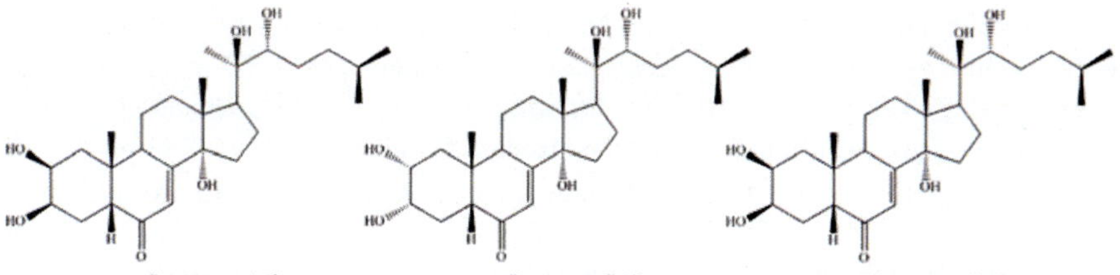

Fig. 1.3 Substances were isolated from another Australian species *Podocarpus nakaii*

When Butenandt A., one of the discoverers of ecdysone, visited Japan, Nakanishi attended both of his lectures on the isolation and structure elucidation of ecdysone. But even then, after realizing the proximity of the structures of ponasterone A (3) and α-ecdysone (1), they could not even assume that ponasterone A (Fig. 1.3) may have the activity of the insect molting hormone.

However, a sample of ponasterone A (3) was nevertheless sent to Australia by Horn D. and it was soon found that the substance, when tested on *Calliphora* larvae, is really active as a natural molting hormone. During a sensational statement by Nakanishi at a symposium on natural compounds, another Japanese researcher Takemoto T., in a private discussion with Nakanishi, told him that he and his co-workers isolated 20-hydroxyecdysone (2) and its structural isomer, called inocosterone, from common in Japan weed *Achiranthes radix* (Takemoto et al. 1967).

Following the discovery by Nakanishi of the ponasterones A (3), B (4), and C (5) (Fig. 1.3), 20-hydroxyecdysone (2) was isolated from another Australian species *Podocarpus nakaii* (Troczka et al. 2012; Galbraith and Horn 1969), which turned out to be identical to the molt hormone isolated from crustaceans.

Soon after, in Czechoslovakia, 20-hydroxyecdysone (2) was isolated from the roots of the common millipede (*Polipodium vulgare*) (Jizba et al. 1967).

These researchers reported that a compound with the same melting point was isolated back in 1933 by Volmar and Ribbe. However, due to its high solubility in water, the authors considered this compound to be a glycoside of no scientific interest, and did not even try to determine its structure. Thus, the detection of ecdysteroids in plants could occur much earlier than they were isolated from insects and crustaceans.

These events triggered a massive study of plants for the presence of compounds with the activity of the insect molting hormone. On the one hand, plants turned out to be important natural sources of these hormones, since their content in some plants is many times higher than that in insects. On the other hand, almost immediately after the discovery of phytoecdysteroids, their physiological activity for mammals was shown, manifested in non-specific enhancement of protein synthesis (Otaka et al. 1969; Cheng et al. 2013), which opened the prospect of using ecdysteroids in medicine. In the late 60s and early 70s, the flora of Japan was especially carefully examined. It was found that among 1056 studied plant species belonging to 738 genera and 186 families, extracts of plants belonging to 77 families had the activity of molting hormone (Imai et al. 1969a). On the territory of the former Soviet Union, the scientific schools of academician N. K. Abubakirov made a great contribution to the identification of new ecdysteroid-containing plant species (Dinan et al. 2009; Saatov et al. 1993), and Prof. Yu. D. Kholodova (Ukraine) (Murtazina et al. 2014).

Currently, screening studies are actively carried out in the botanical garden of Tomsk State University (Zibareva 1991), the Institute of Biology of the Komi Scientific Center, Ural Branch of the Russian Academy of Sciences (Syktyvkar) (Volodin et al. 2002), and in other research institutions.

According to the results of intensive studies of plants on the content of ecdysteroids conducted in the 70s, it was believed that ecdysteroids are characteristic mainly of fern and gymnosperms, and they are less common in flowering plants. For example, of the 64 studied species of ferns growing in New Zealand, the activity of molting hormone showed the extracts of 24 species when tested on housefly larvae (Russell and Fenemore 1970).

Among gymnosperms, compounds with the activity of the hormone of insect molting were found in 73 species (Qi et al. 2011). Among 639 studied species of angiosperms of various systematic groups, only 23 species showed the activity of the molt hormone (Russell and Fenemore 1971). On the basis of these data, as well as information that ferns and gymnosperms are practically not attacked by insects, it was suggested that ecdysteroids determine the protective mechanism of stability of ferns and gymnosperms against phytophages.

However, the results of plants' studies in the 90s and nowadays show that ecdysteroids are also widespread among flowering plants (Lafont 1998; Dinan 2000; Thiem et al. 2017). According to the literature, the largest number of ecdysteroid-containing species was found in the families: amarantha—*Amaranthaceae* (23 species belonging to 9 genera), *Asteraceae* (23 species, mainly belonging to the genera *Rhaponticum* and *Serratula*, clove—*Caryophyllaceae* (34 species, mainly from of the genera *Silene* and *Lychnis*), Marevae—*Chenopodiaceae* (13 species from 8 genera), Labioata—*Labiatae* (14 species, 13 of which are from the *Ajuga* genus), Ranunculaceae—*Ranunculaceae* (30 species, of which 11 species from the hellebore genus—*Helleborus*, are characterized by increased ecdysteroids, and the remaining 19 belong to the genera of wind Itza—*Anemone* and cross—*Pulsatilla* low-ecdysteroids).

Thus, to date, ecdysteroids have been found in plants belonging to more than 100 families from the fern divisions—*Polypodiphyta*, gymnosperms—*Pinophyta* and angiosperms—*Magnoliophyta*. Until the 80s, it was believed that fungi, algae, and mosses do not contain ecdysteroids. However, this point of view changed when ecdysteroid-like compounds—pinnasterols and paxillosterone, respectively, (Fukuzawa et al. 1981) were discovered in the red seaweed *Laurencia pinnata* and the fungus *Paxillus panuoides*, and true ecdysteroids were isolated from the fruiting bodies of the fungus *Polyporus umbellatus* (Ohsawa et al. 1992). Recent data indicate that the ability to biosynthesize ecdysteroids arose (or arose repeatedly) already in the early stages of plant evolution (Akhrem et al. 1973).

As a result of a chemical study of the ecdysteroid components of the roots and stems of *Cyathula officnalis*, two stereoisomers of cyasterone, 28-epicytaster (β-Me) (6), mp 248–250 °C and 25-epi-28-epi-cyasterone (α-Me) (7) (Fig. 1.4) were isolated, along with the well-known cyasterone. The structure of 28-epi-cyasterone (6) and

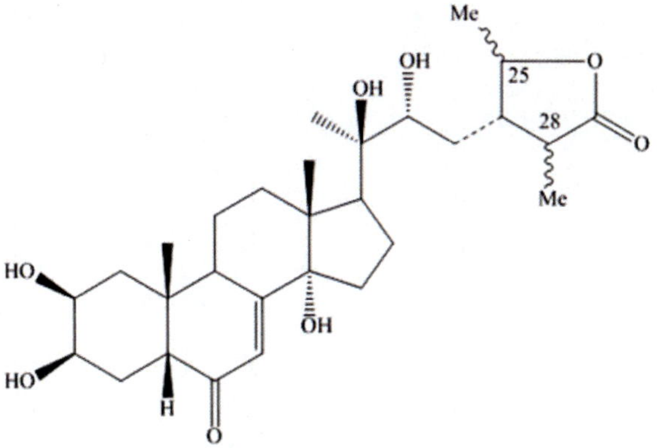

28-epi-cyasterone (β–Me) (6)
25-epi-28-epi-cyasterone (α–Me) (7)

Fig. 1.4 Ecdysteroids isolated from *Cyathula officnalis*

25-epi-28-epi-cyasterone (7) was established on the basis of spectral analysis data. The 24S, 25S, 28S-configuration of 25-epi-28-epi-cyasterone (7) and the 24S, 25S-configuration for 28-epi-cyasterone (6) were confirmed by X-ray diffraction analysis (Okuzumi et al. 2005).

25S-inocosterone was isolated from *Achyranthes bidentata* by various chromatographic methods and identified by spectral and chemical methods (9) (Fig. 1.5), so pl. 242–244 °C, [α] 20D + 78.3° (c 0.07; MeOH), as well as ecdysterone (2) (Zhu et al. 2004).

Due to the fact that modern representatives of the plant kingdom, in which ecdysteroids are found, are located in taxa both close and far apart in phylogenetic relation to each other, a strong view has developed in the literature that the distribution of ecdysteroids is not associated with a phylogenetic classification. The following arguments are in favor of this statement: if we talk about taxa of high rank, then in the *Magnoliophyta* family the ecdysteroids were found in all subclasses, 40 orders and more than 80 families of different levels of advancement; in families, along with

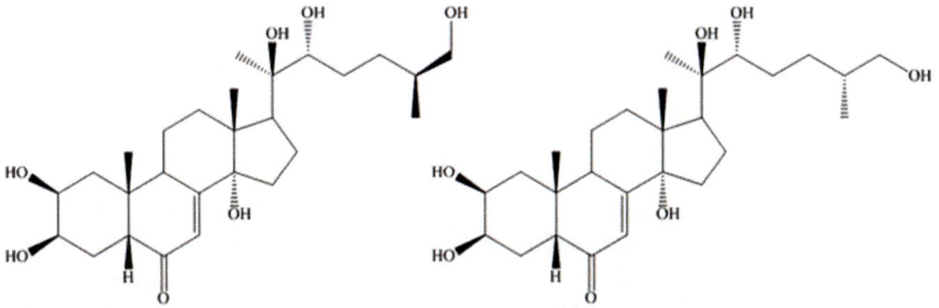

Inokosterone (8) 25S-Inokosterone (9)

Fig. 1.5 Plant ecdysteroids *Achyranthes bidentata*

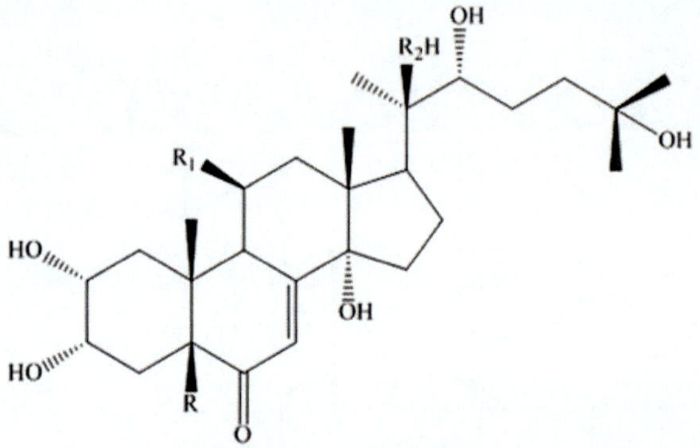

Fig. 1.6 Plant ecdysteroids *Rhaponticum carthamoides*

Lesterone (R=R_2=H; R_1=OH) (10)
20-acetate-rapisterone D=(R=OH; R_1=H; R_2=Ac) (11)

ecdysteroid-positive, there may be ecdysteroid-negative genera and, moreover, even ecdysteroid-positive and ecdysteroid-negative species can be present even in genera.

On the distribution of ecdysteroids in the plant world, the genus *Helleborus*, containing the only ecdysteroid-positive species *H. Niger*, the family *Lamiaceae* and *Ranunculaceae*, each containing one ecdysteroid-positive genus, survivor *Ajuga*, is given as an illustration of the last statement, the family *Caryophyllaceae,* containing ecdysteroid-positive (*Lychnis, Silene, and Dianthus)* and ecdysteroid-negative (soap dish—*Saponaria* and *Cucubalis*) genera (Yusupova et al. 2019, 2020a, b; Okuzumi et al. 2005; Ramazanov 2005).

From the seeds of *Rhaponticum carthamoides*, along with the well-known ecdysteroids, new ecdysteroids, stesterone and 20-acetate-rapisterone D were isolated (11) (Fig. 1.6). For the first time, isomeric C-28 ecdysteroids—makisterone A (12) and 24-epi-makisterone A (13) were isolated and identified from a plant of the genus *Rhaponticum.* From the *Chenopodium* album plant containing minor amounts of ecdysteroids, 2-deoxyecdysone was isolated and identified using HPLC (52). New sources of ecdysteroids were found and methods for their isolation, as well as taxa from various parts of *Taxus cuspidata, Stemmacantha uniflora, Matteuccia strupthiopteris,* were developed (Lukyantseva et al. 2001).

However, in subsequent years, the genus *Helleborus* was supplemented by ten positive species (Hardman and Benjamin 1976). The studies of the Altai flora in the 1980s brought sufficient clarity to the distribution of ecdysteroid-containing species in the *Caryophyllaceae* family: ecdysteroid-positive genera *Lychnis* and *Silene* belong to the same tribe *Lychnideae,* in the other two studied tribes *Alsineae* non-positive genus (Dinan 1995).

From the roots of *Rh. carthamoides,* along with well-known compounds, new ecdysteroid derivatives, leuzeasterone (15), [α] 20D + 51.6° (c 0.41; MeOH), and (24Z)-29-hydroxy-24 (28)-dehydromakisterone were isolated (16) (Fig. 1.7), [α] 20D + 32.5° (c 0.24; MeOH). The structure of leuzeasterone (15) and (24Z) -29-hydroxy-24 (28)-dehydromakisterone (16) was established using IR, ^{1}H NMR, ^{13}C NMR and mass spectra (Vokáč et al. 2002).

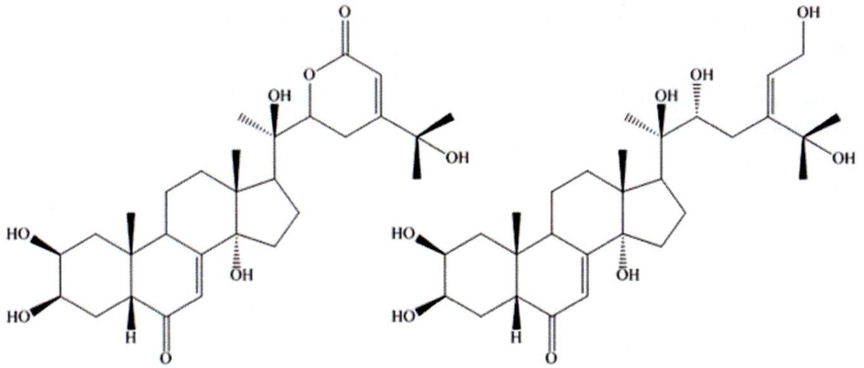

Leuzasterone (15) (24Z)-29-hydroxy-24(28)-degidromakisterone (16)

Fig. 1.7 The structure of new ecdysteroid derivatives

The concentration trend of ecdysteroid-positive species in certain genera and tribes is characteristic of the *Chenopodiaceae* family (Dinan 1995, 1998). Significant amounts of ecdysteroids were found in representatives of the tribes *Chenopodiaceae* and *Atripliceae*. Representatives of the tribes *Caryophyllaceae* and *Chenopodiaceae* contain relatively low amounts of ecdysteroids, and no ecdysteroid-containing species were found in the tribe *Salicornieae*. It turned out that in the tribe *Beteae* from the *Procumbences* section, two *Beta* genera contain ecdysteroids, and the rest are ecdysteroid-negative. For the genetically controversial genus (*Beta*), beetroot detection of ecdysteroids in the species *B. patellaris* and *B. webbiana* served as additional biochemical characters to separate them together with the species *B. procumbens* as a separate genus *Patellifolia* from the rest of the genus *Beta* (Nachtigall et al. 2016).

Analysis of the distribution of ecdysteroids in the *Amaranthaceae* family showed that most ecdysteroid-containing species belong to the 6 genera of the tribe *Amarantheae* (Takhtadzhyan 1987; Das et al. 2020). In the *Asteraceae* family, ecdysteroids were found in representatives of 5 genera (all from the Cardueae tribe) in representatives of the Cardueae tribe, with most of the producers belonging to the subtribe Centaureinae (Rao 1967). More recently, ecdysteroids were found in many representatives of the genera Pulsatilla and Anemone (Ranunculaceae)—both genera are included in the tribe Anemoneae (Takhtadzhyan 1987; Zhang et al. 2019). In some cases, the rank of a taxon combining ecdysteroid-containing species may vary depending on the size of a particular family.

The genus *Helleborus* mentioned earlier is so isolated in the *Ranunculaceae* family that it even stands out as a separate subfamily (Zhang et al. 2019; Meng et al. 2001). In most families, the rank of a taxon that unites ecdysteroid-containing species in the families of the flowering plant department is the tribe.

1.4 Structure and Sources of Phytoecdysteroids

Sources of ecdysteroids from which they were first isolated structures and gross formulas are given in Table 1.1.

Table 1.1 Ecdysteroids isolated from plants of the genera *Ajuga, Rhaponticum,* and *Silene*

№	Ecdysteroids	Name of plants	References
1	**$C_{27}H_{44}O_6$** **α-ecdysone** 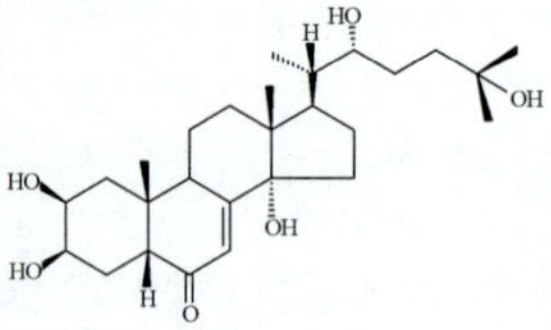	***A. turkestanica***	Saatov et al. (1999)
		A. australis	Baltaev (2000)
		Rh. integrifolium	Ramazonov et al. (2007)
		S. antirrhina	Meng et al. (2001)
		S. armeria	Zibareva (2000)
		S. asterias	Zibareva (2000)
		S. bashkirorum	Zibareva (2000)
		S. bellidifolia	Zibareva (2000)
		S. boryi	Zibareva (2000)
		S. brachypoda	Zibareva (2000)
2	**$C_{27}H_{44}O_7$** **Ecdysterone (20-hydroxyecdysone)** 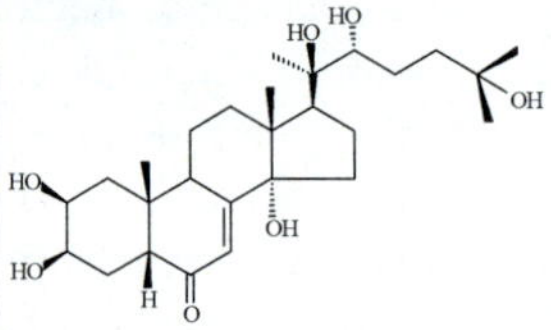	***A. decumbens***	Imai et al. (1969b)
		A. incisa	Imai et al. (1969b)
		A. iva	Bakrim et al. (2014)
		A. japonica	De-fu et al. (2002)
		A. multiflora	Kim et al. (2005)
		A. reptans	Kim et al. (2005), Matsuoka et al. (1969)
		A. nipponensis	Imai et al. (1969b)
		A. remota	Camps and Coll (1993)
		A. turkestanica	Matsuoka et al. (1969)
		A. chamaepitus	Camps et al. (1984)
		A. australis	Kubo et al. (1983)
		A. ciliata	Kubo et al. (1983)
		A. laxmanni	Kubo et al. (1983)
		A. linearifolia	Kubo et al. (1983)
		A. bracteosa	Kubo et al. (1983)
		Rh. carthamoides	Buděšínsky et al. (2008)
		Rh. integrifolium	Baltaev (1979)
		S. acaulis	Zibareva (1999, 2000)
		S. alba	Zibareva (2000)
		S. alpestris	Zibareva (2000)
		S. tomentella	Ramazanov et al. (1995)

(continued)

Table 1.1 (continued)

№	Ecdysteroids	Name of plants	References
3	**$C_{27}H_{44}O_6$** **Ponasterone A** 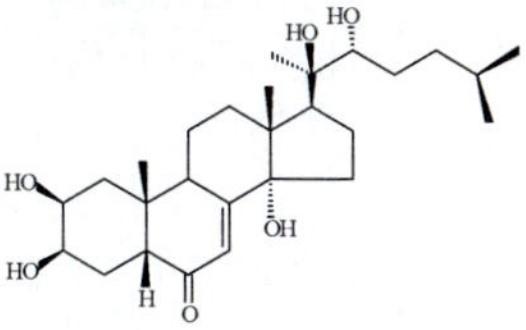	***Podocarpus nakaii***	Nakanishi et al. (1966)
4	**$C_{27}H_{44}O_6$** **Ponasterone B** 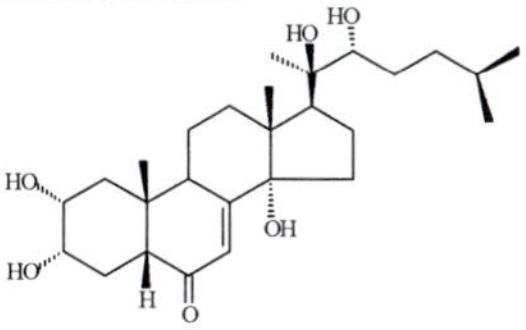	***P. nakaii***	Nakanishi et al. (1966)
5	**$C_{27}H_{44}O_7$** **Ponasterone C** 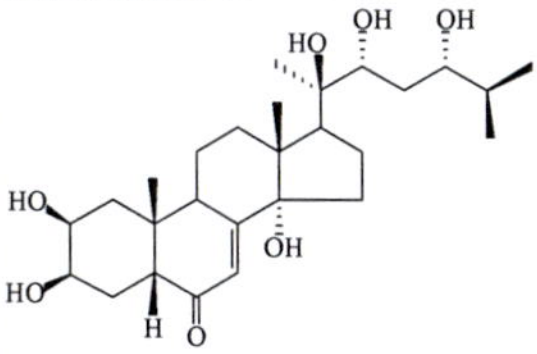	***P. nakaii***	Nakanishi et al. (1966)
6	**$C_{29}H_{44}O_8$** **28-epi-cyasterone** 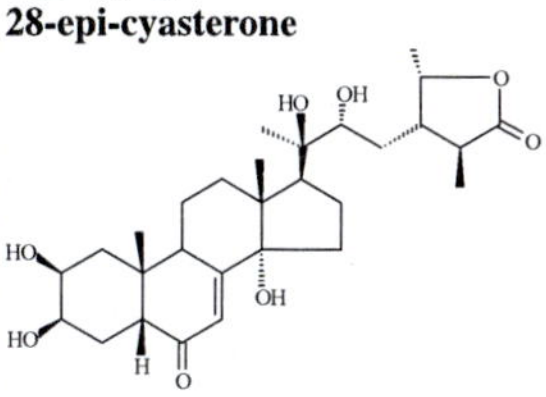	***C. officnalis***	Okuzumi et al. (2005)
7	**$C_{29}H_{44}O_8$** **25-epi-28-epi-cyasterone** 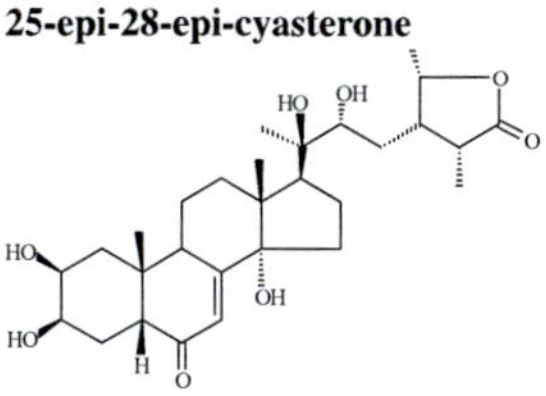	***C. officnalis***	Okuzumi et al. (2005)

(continued)

Table 1.1 (continued)

№	Ecdysteroids	Name of plants	References
8	**$C_{27}H_{44}O_7$** **Inokosterone** 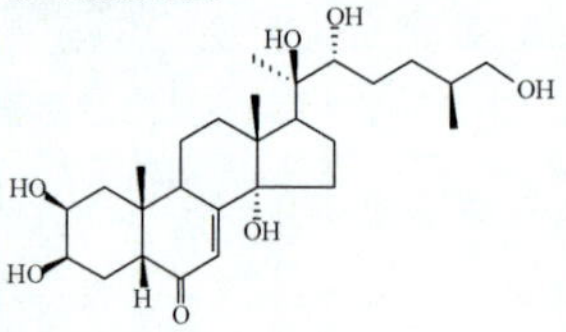	***S. disticha***	Meng et al. (2001)
		S. divaricata	Zibareva (2000)
		S. echinata	Zibareva (2000), Meng et al. (2001)
		S. fabaria	Zibareva (2000)
		S. fabariodes	Zibareva (2000)
		S. firma	Zibareva (2000)
		S. flavescens	Zibareva (1999)
9	**$C_{27}H_{44}O_7$** **Inokosterone-25S** 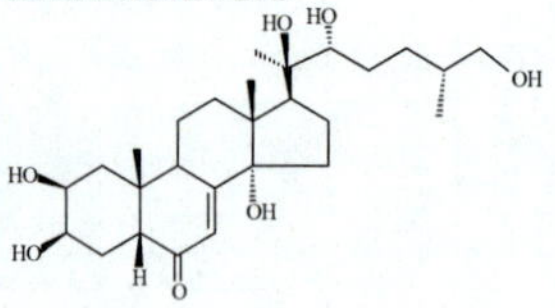	***A. bidentata***	Zhu et al. (2004)
10	**$C_{27}H_{44}O_8$** **Lesterone** 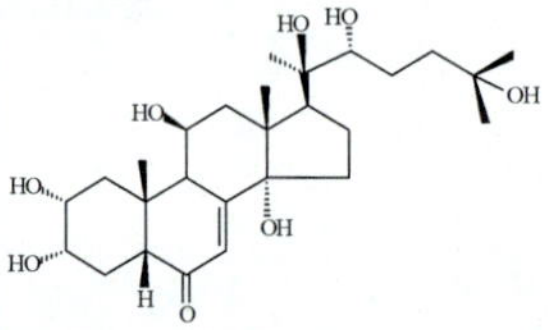	***Rh. carthamoides***	Borovikova et al. (1999)
11	**$C_{29}H_{46}O_8$** **20-Acetate-rapisterone D** 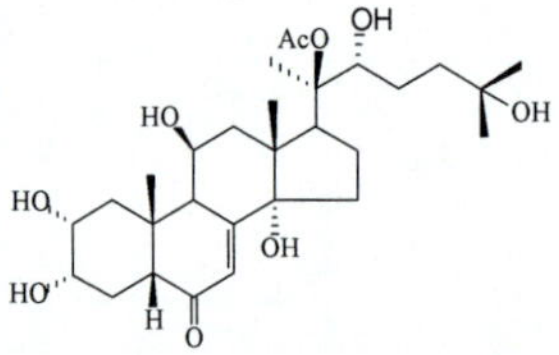	***Rh. carthamoides***	Borovikova et al. (1999)
12	**$C_{28}H_{46}O_7$** **Makisterone A** 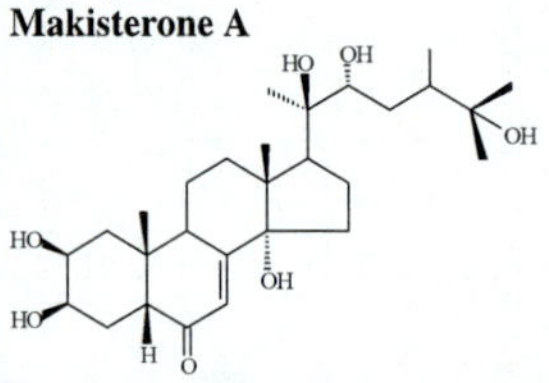	***P. macrophyllus***	Imai et al. (1968)

(continued)

Table 1.1 (continued)

№	Ecdysteroids	Name of plants	References
13	**$C_{28}H_{46}O_7$** **24-epi-Makisterone A** 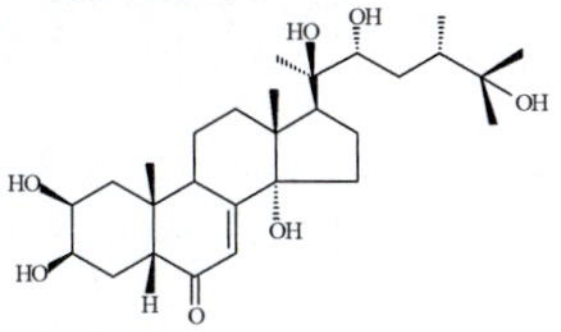	***Rh. carthamoides***	Buděšínsky et al. (2008)
14	**$C_{27}H_{44}O_5$** **2-Deoxyecdysone** 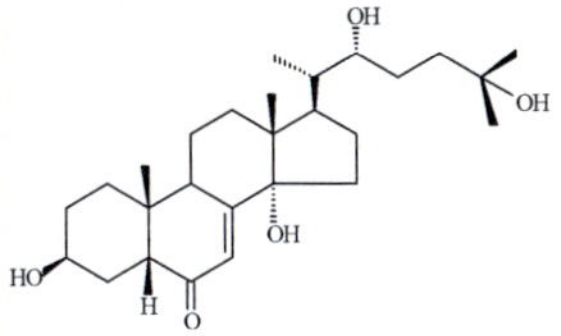	***S. nutans***	Girault et al. (1990)
15	**$C_{29}H_{43}O_8$** **Leuzeasterone** 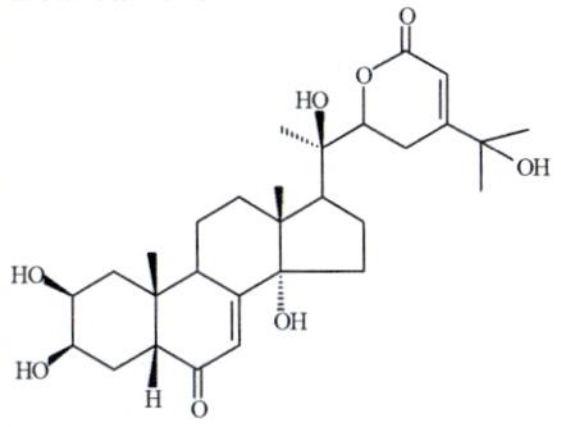	***Rh. carthamoides***	Vokáč et al. (2002)
16	**$C_{29}H_{46}O_8$** **(24Z)–29-hydroxy-24 (28)-dehydromakisterone** 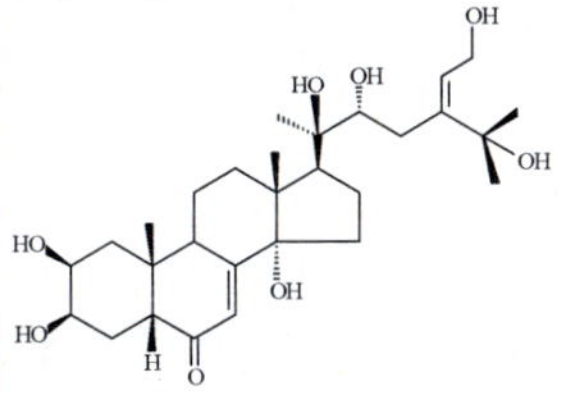	***Rh. carthamoides***	Buděšínsky et al. (2008)
17	**$C_{28}H_{46}O_7$** **Makisterone B** 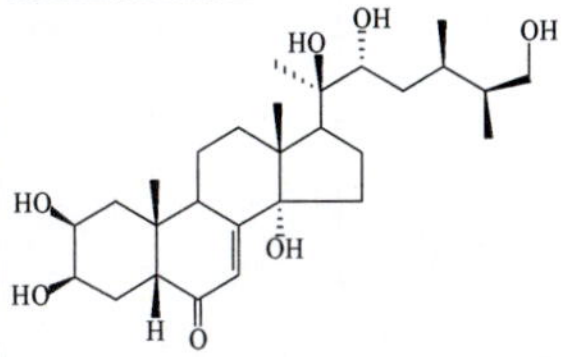	***P. macrophyllus***	Imai et al. (1968)

(continued)

Table 1.1 (continued)

№	Ecdysteroids	Name of plants	References
18	**$C_{29}H_{48}O_7$** **Makisterone C** 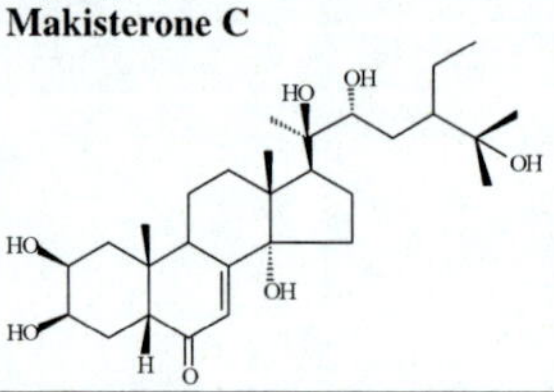	***Rh. carthamoides***	Buděšínsky et al. (2008)
		P. macrophyllus	Imai et al. (1968)
19	**$C_{29}H_{48}O_7$** **Makisterone D** 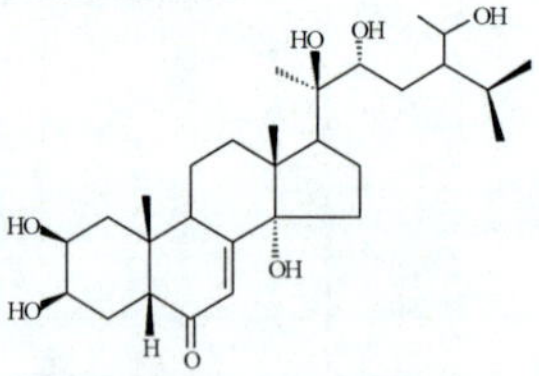	***P. macrophyllus***	Imai et al. (1968)
20	**$C_{28}H_{44}O_8$** **24(28)-Dehydroxy-Makisterone A** 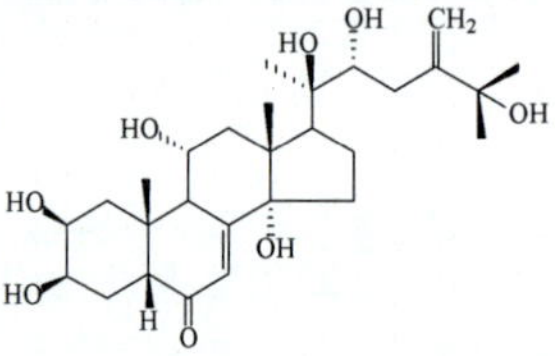	***S. frivaldszkyana***	Zibareva (1999, 2000)
		S. fruticosa	Zibareva (2000)
		S. fuscata	Zibareva (2000)
		S. gallica	Zibareva (2000)
		S. gallica	Zibareva (2000)
		S. gigantean	Zibareva (2000)
		S. goulimyi	Zibareva (2000)
		S. graminifolia	Saatov et al. (1993), Akhrem et al. (1973)
		S. gyntensis	Saatov et al. (1993)
		S. jenissensis	Saatov et al. (1993), Akhrem et al. (1973)
		S. hifacensis	Zibareva (2000)
		S. inaperta	Zibareva (2000)
		S. italica	Zibareva (1999, 2000)
		S. laeta	Zibareva (1999, 2000)
		S. latifolia	Saatov et al. (1993)
21	**$C_{27}H_{42}O_5$** **Silenosterone** 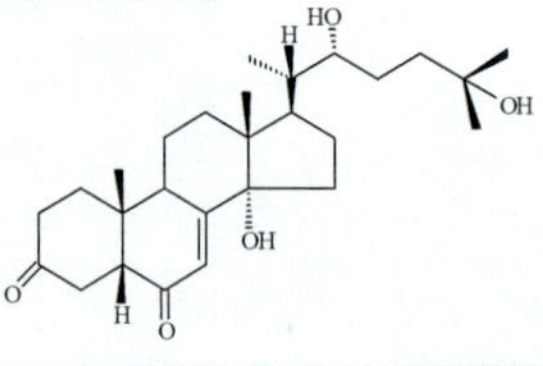	***S. scabrifolia***	Saatov et al. (1979)

(continued)

Table 1.1 (continued)

№	Ecdysteroids	Name of plants	References
22	**$C_{27}H_{44}O_5$** **2-Deoxy-α-ecdysone** 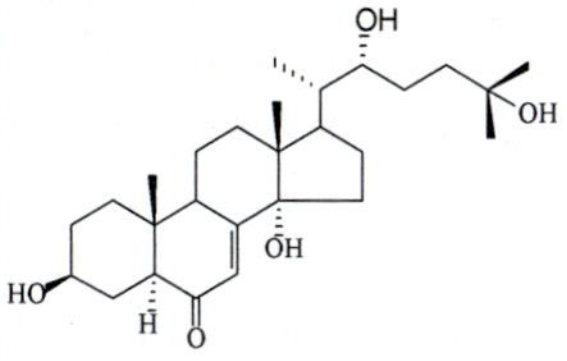	***S. nutans***	Girault et al. (1990)
23	**$C_{27}H_{44}O_6$** **2-Deoxyecdysterone** 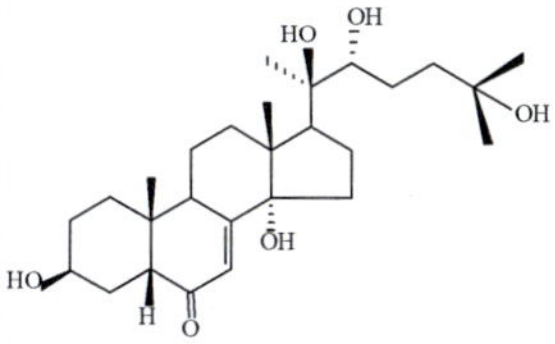	***S. tomentella***	Ramazanov et al. (1995)
24	**$C_{33}H_{54}O_{12}$** **Sileneoside A** 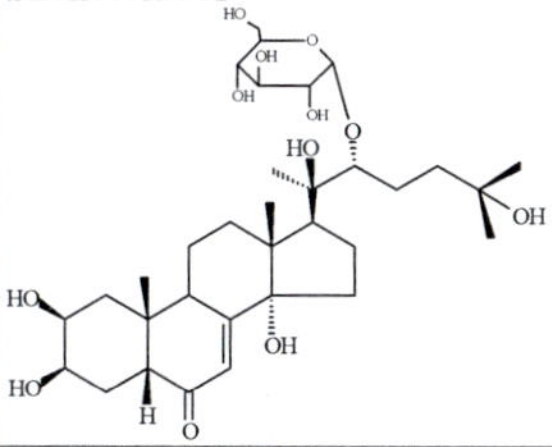	***S. brahuica***	Saatov et al. (1981)
		S. viridiflora	Mamadalieva et al. (2003, 2004)
		S. vulgaris	Zibareva (2000)
		S. waldsteinii	Zibareva (2000)
25	**$C_{39}H_{64}O_{17}$** **Sileneoside B** 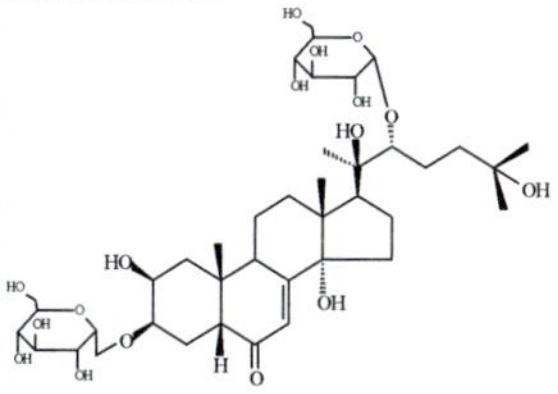	***S. brahuica***	Saatov et al. (1982)
26	**$C_{33}H_{54}O_{13}$** **Sileneoside C** 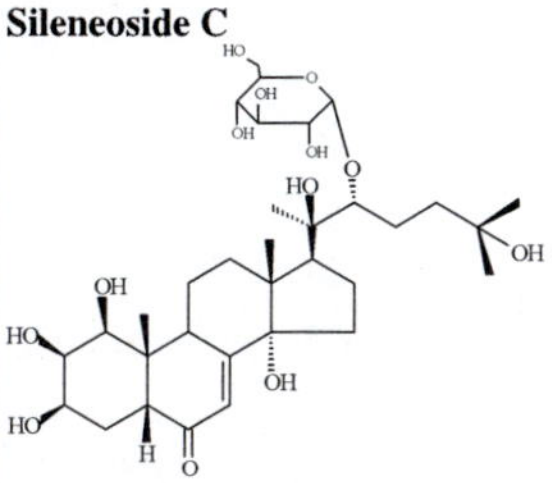	***S. brahuica***	Saatov et al. (1982)

(continued)

Table 1.1 (continued)

№	Ecdysteroids	Name of plants	References
27	**$C_{33}H_{54}O_{12}$** **Sileneoside D** 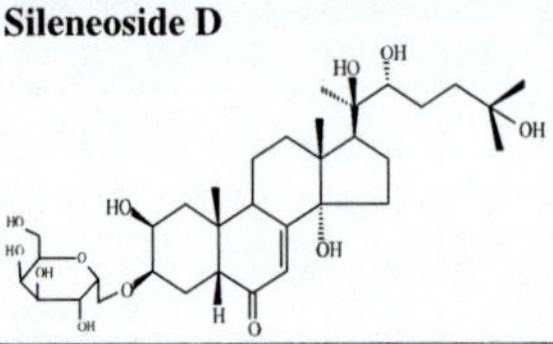	***S. brahuica***	Saatov et al. (1984)
		S. tomentella	Dzhukharova et al. (1993)
		S. turgida	Zibareva (2000)
		S. vallesia	Zibareva (1999)
28	**$C_{27}H_{46}O_6$** **Cheilanton A** 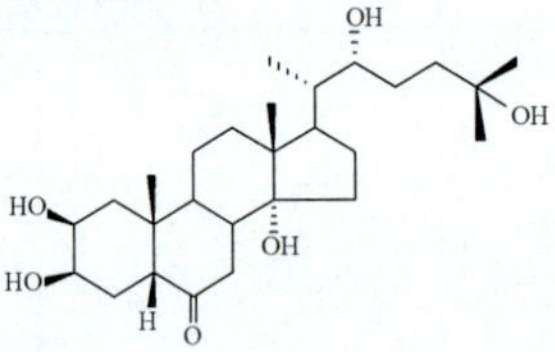	***Cheilanthes tenuifolia***	Faux et al. (1970), Rashkes et al. (1971)
29	**$C_{27}H_{44}O_5$** **Cheilanton B** 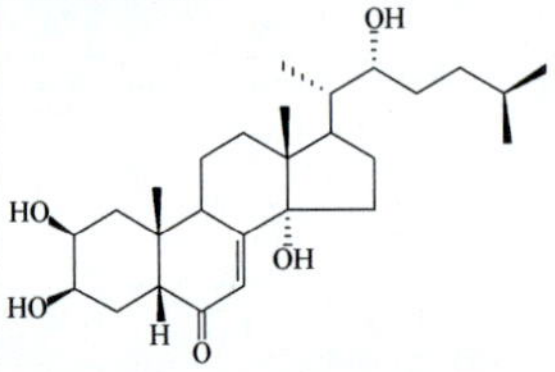	***Ch. tenuifolia***	Faux et al. (1970), Rashkes et al. (1971)
30	**$C_{34}H_{48}O_8$** **22-O-Benzoate-5α-ecdysterone** 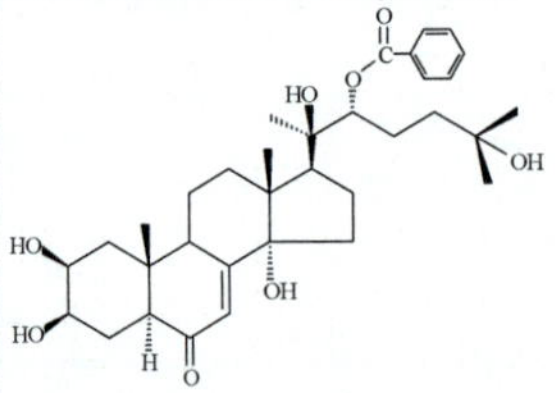	***S. scabrifolia***	Saatov et al. (1987)
31	**$C_{36}H_{50}O_9$** **22-Benzoate viticosterone E** 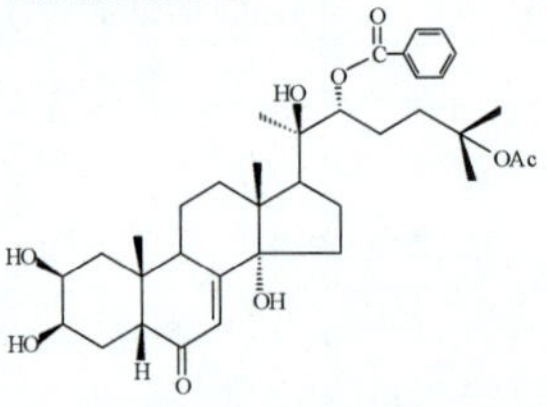	***S. wallichiana***	Saatov et al. (1988)

(continued)

Table 1.1 (continued)

№	Ecdysteroids	Name of plants	References
32	$C_{27}H_{44}O_8$ **Turkesterone** 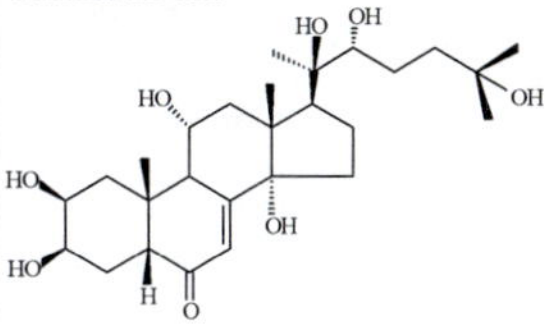	***A. turkestanica***	Usmanov et al. (1971)
		Rh. carthamoides	Buděšínsky et al. (2008)
		S. linicola	Mamadalieva et al. (2002)
		S. longicalcina	Zibareva (1999, 2000)
		S. mellifera	Zibareva (1999, 2000)
		S. micropetala	Zibareva (1999)
		S. multiflora	Zibareva (1999)
		S. nemoralis	Zibareva (2000)
		S. noctiflora	Zibareva (2000)
		S. nutans	Girault et al. (1990)
		S. otites	Girault et al. (1990), Zibareva (1999, 2000)
		S. paradoxa	Zibareva (1999, 2000)
33	$C_{35}H_{56}O_{14}$ **Sileneoside H** 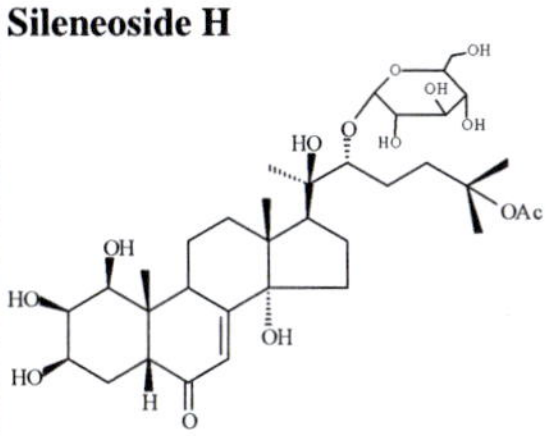	***S. brachuica***	Sadykov et al. (2000)
		S. bupleroides	Zibareva (2000)
		S. burchelli	Zibareva (1999)
		S. campanula	Zibareva (2000)
		S. campanulata	Zibareva (1999)
		S. watson	Akhrem et al. (1973)
		S. caramanica	Zibareva (1999)
		S. catholica	Zibareva (1999 2000)
34	$C_{39}H_{64}O_{17}$ **Sileneoside G** 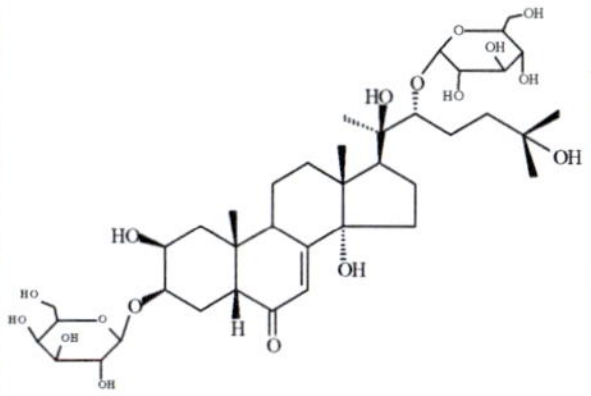	***S. brachuica***	Sadykov and Saatov (1998)
35	$C_{31}H_{46}O_9$ **22-Acetate-cyasterone** 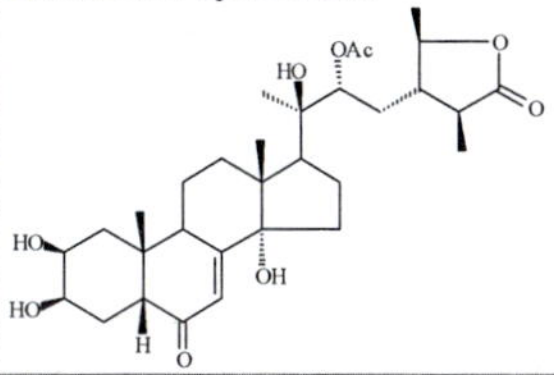	***A. turkestanica***	Borovikova et al. (1999), Usmanov et al. (1978)
		A. australis	Saatov et al. (1999), Kim et al. (2005)
		A. bracteosa	Kim et al. (2005)

(continued)

Table 1.1 (continued)

№	Ecdysteroids	Name of plants	References
36	$C_{29}H_{40}O_8$ **Ajugalactone** 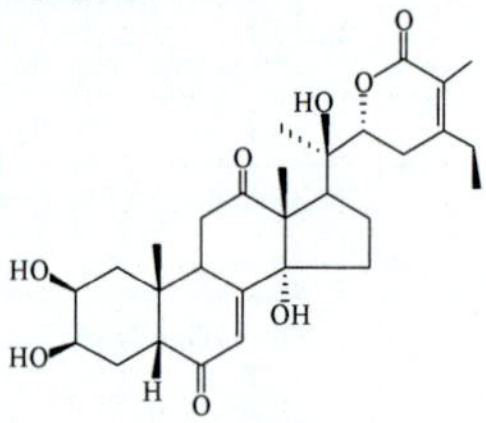	***A. turkestanica***	Saatov et al. (1977)
		A. decumbens	Camps and Coll (1993), Koreeda et al. (1970)
		A. chamaepitus	Camps et al. (1984), Baltaev (2000)
		A. reptans	Lafont and Wilson (1996)
		A. bracteosa	Alekseeva et al. (1998), Koreeda et al. (1970)
		A. remota	Kubo et al. (1983)
		A. laxmanni	Kubo et al. (1981)
		A. linearifolia	Alekseeva (1999)
		A. genevensis	Alekseeva (1999)
37	$C_{29}H_{46}O_7$ **Ajugasterone B** 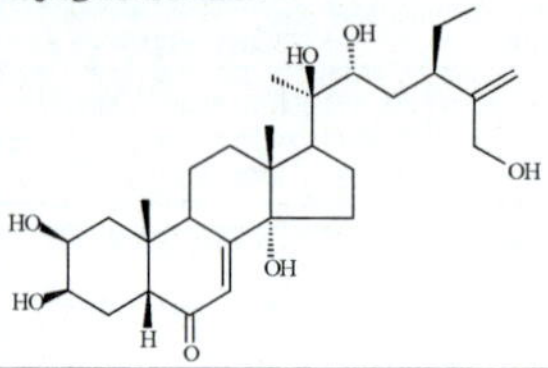	***A. incisa***	Imai et al. ((1969a, b), Camps and Coll (1993)
		A. turkestanica	Usmanov et al. (1977)
		A. bracteosa	Camps and Coll (1993)
		A. reptans	Alekseeva (1999)
38	$C_{29}H_{44}O_8$ **Cyasterone** 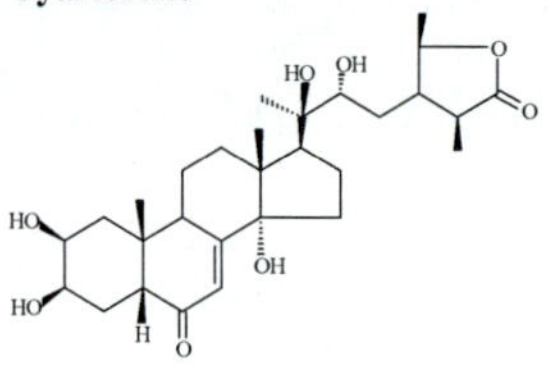	***A. decumbens***	Imai et al. (1969a)
		A. incisa Maxim	Imai et al. (1969a)
		A. japonica	Ikan and Ravid (1971)
		A. australis	Saatov et al. (1977)
		A. miltiflora	Kim et al. (2005)
		A. reptans	Kim et al. (2005)
		A. turkestanica	Saatov et al. (1977)
		A. chia	Ikan and Ravid (1971)
		A. remota	Kubo et al. (1981)
		A. iva	Bakrim et al. (2014)
39	$C_{28}H_{44}O_{10}$ **Carthamasterone A** 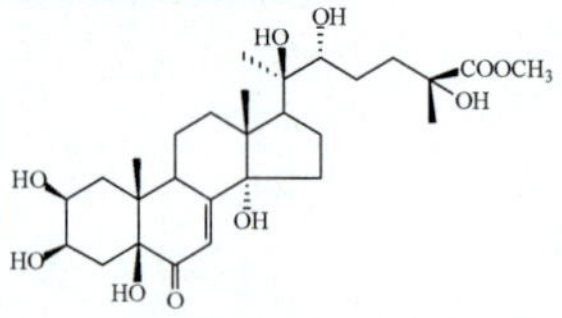	***Rh. carthamoides***	Ramazanov et al. (1997)

(continued)

Table 1.1 (continued)

№	Ecdysteroids	Name of plants	References
40	**$C_{27}H_{42}O_9$** **Carthamasterone B** 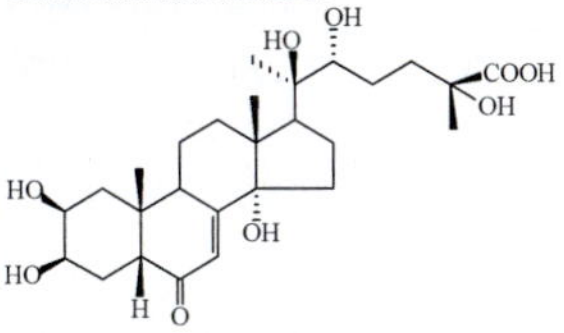	***Rh. carthamoides***	Ramazanov et al. (1997)
41	**$C_{27}H_{44}O_8$** **Polypodine B** 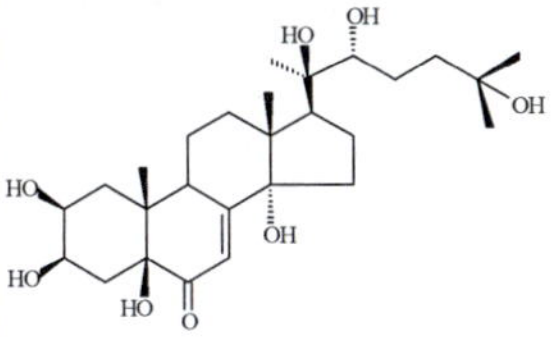	***A. reptans***	Calcagno et al. (1995)
		A. incia	Matsuoka et al. (1969)
		A. ciliata	Kubo et al. (1983)
		A. nipponensis	Camps and Coll (1993)
		A. bracteosa	Kim et al. (2005)
		Rh. carthamoides	Ramazanov et al. (1997)
		S. altaica	Zibareva (2000)
		S. tomentella	Ramazanov et al. (1995)
42	**$C_{34}H_{48}O_9$** **22-O-Benzoate-polypodine B** 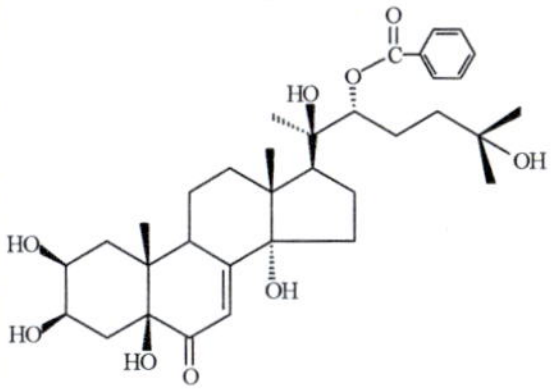	***Rh. carthamoides***	Sadykov et al. (1997)
43	**$C_{29}H_{46}O_8$** **Viticosterone E** 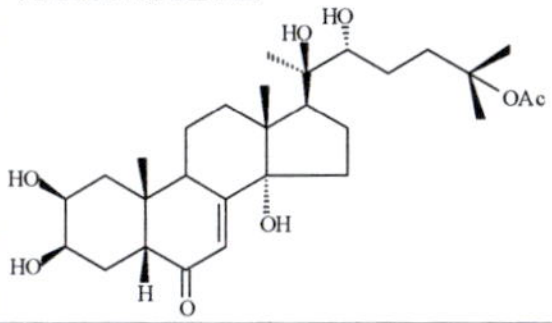	***S. tomentella***	Ramazanov et al. (1995)
		S. turgida	Zibareva (1999, 2000)
		S. vallesia	Zibareva (1999, 2000)
44	**$C_{29}H_{46}O_9$** **25-Acetate-integristeroneA** 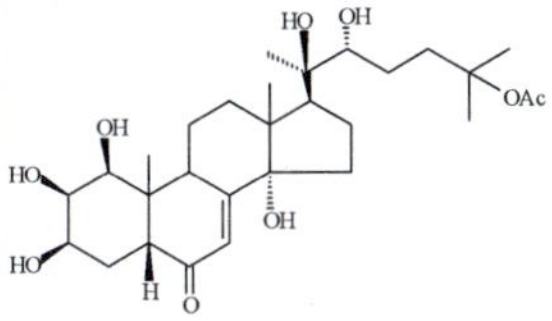	***S. brachuica***	Saatov et al. (1993), Sadikov and Saatov (1999)
		S. bupleroides	Suksamrarn et al. (2002)
		S. burchelli	Zibareva (2000)
		S. campanula	Zibareva (1999)
		S. campanulata	Zibareva (2000)
		S. watson	Zibareva (1999)
		S. caramanica	Zibareva (1999), Fukuzawa et al. (1981)
		S. catholica	Zibareva (1999, 2000)

(continued)

Table 1.1 (continued)

№	Ecdysteroids	Name of plants	References
45	**$C_{27}H_{44}O_8$** **Integristerone A** 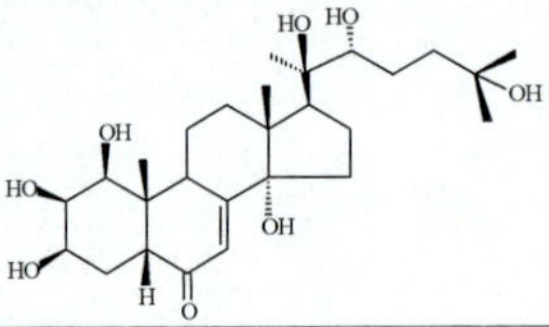	***S. claviformis***	Sadykov et al. (2001)
		S. nutans	Girault et al. (1990)
		S. otites	Girault et al. (1990)
46	**$C_{36}H_{50}O_7$** **Tomentesterone A** 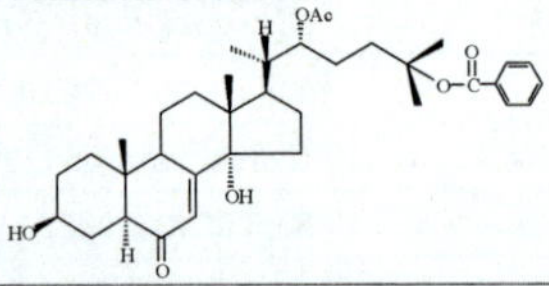	***S. tomentella***	Ramazanov et al. (1995)
		S. turgida	Zibareva (1999, 2000)
		S. vallesia	Zibareva (1999, 2000)
47	**$C_{34}H_{48}O_6$** **Tomentesterone B** 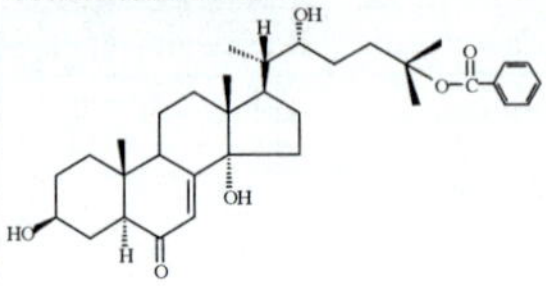	***S. tomentella***	Ramazanov et al. (1995)
48	**$C_{34}H_{48}O_8$** **22-O-Benzoate-ecdysterone** 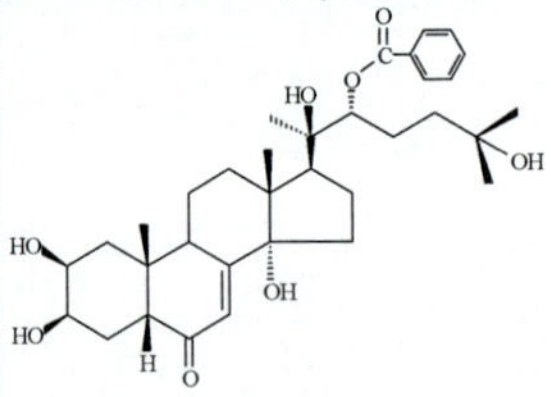	***S. scabrifolia***	Saatov et al. (1986)
49	**$C_{34}H_{48}O_7$** **3-O-Benzoate2-deoxy-ecdysterone** 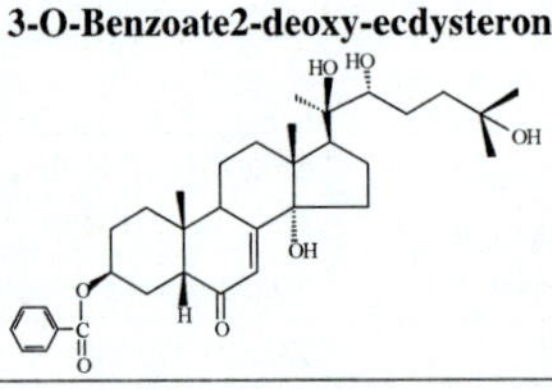	***S. wallichiana***	Mamadalieva et al. (2000)
		S. zawadskii	Zibareva (2000)
50	**$C_{29}H_{46}O_6$** **22-O-Acetate 2-deoxy-α-ecdysone** 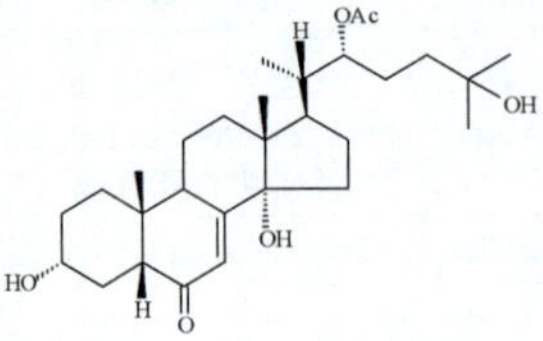	***S. otites***	Girault et al. (1990)

(continued)

Table 1.1 (continued)

№	Ecdysteroids	Name of plants	References
51	**$C_{29}H_{48}O_7$** **Amarasterone A** 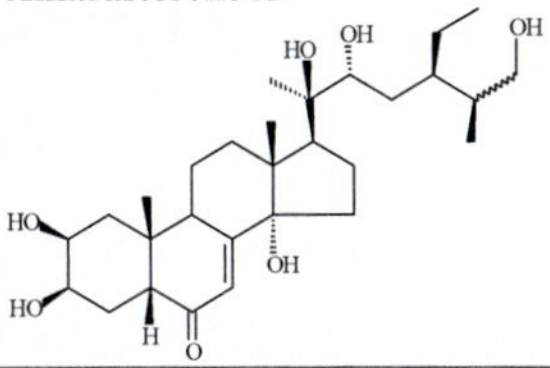	***Ajuga australis***	Kim et al. (2005)
		Rhponticum Carthamoides	Buděšínsky et al. (2008)
52	**$C_{29}H_{46}O_8$** **2-Acetate-20-hydroxyecdysone** 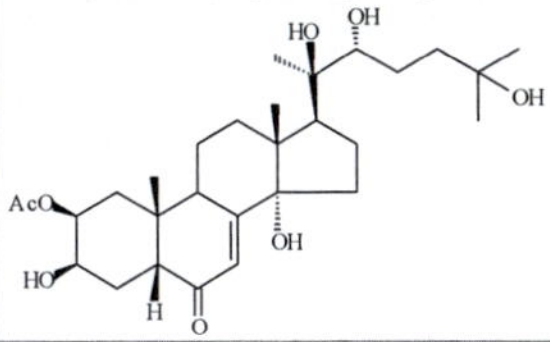	***A. reptans***	Kim et al. (2005)
		A. multiflera	Calcagno et al. (1995)
53	**$C_{29}H_{46}O_8$** **3-Acetate-** **20- hydroxyecdysone** 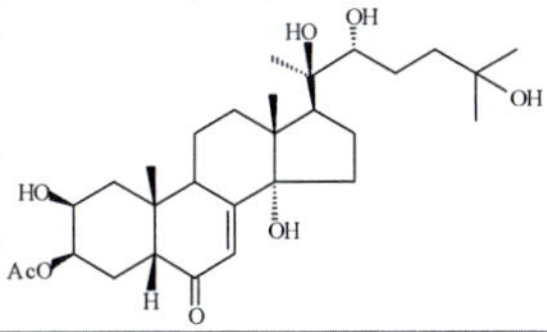	***A. multiflora***	Kim et al. (2005)
		A. reptans	Calcagno et al. (1995)
54	**$C_{29}H_{46}O_8$** **22-Acetate-20-hydroxyecdysone** 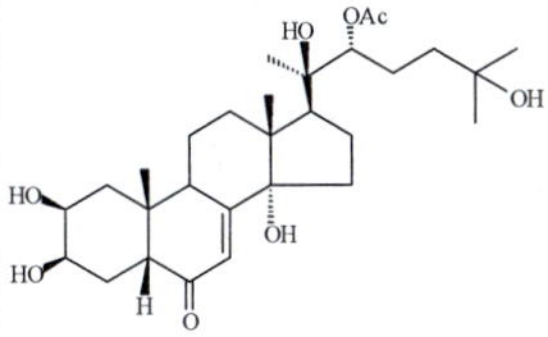	***Silene otites (Caryophyllacea)***	Báthori et al. (1997)
55	**$C_{29}H_{46}O_6$** **3-Acetate-2-deoxy-α-ecdysone** 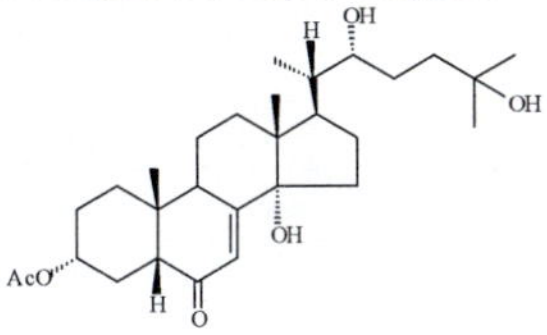	***S. scabrifolia***	Saatov et al. (1987)

(continued)

Table 1.1 (continued)

№	Ecdysteroids	Name of plants	References
56	**$C_{29}H_{46}O_7$** **3-Acetate-2-deoxyecdysterone** 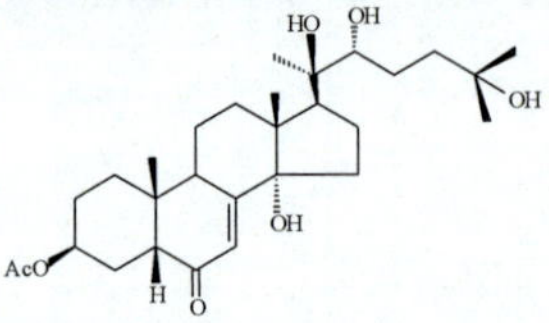	***S. praemixta***	Saatov et al. (1985)
57	**$C_{29}H_{46}O_7$** **22-O-Acetate-2-deoxyecdysterone** 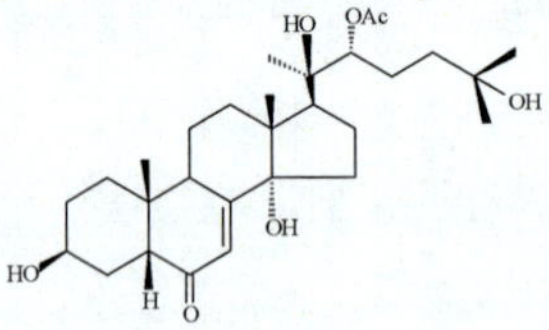	***S. otites***	Girault et al. (1990)
58	**$C_{29}H_{46}O_7$** **25-O-Acetate-2-deoxyecdysterone** 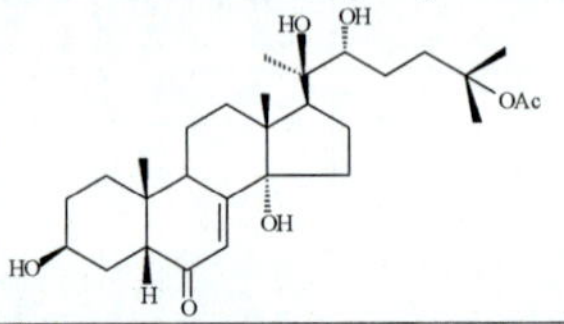	***S. wallichiana***	Mamadalieva et al. (2000)
		S. zawadskii	Zibareva (2000)
59	**$C_{32}H_{52}O_7$** **20,22-O-Acetal-izovaleriane-aldegid ecdysterone** 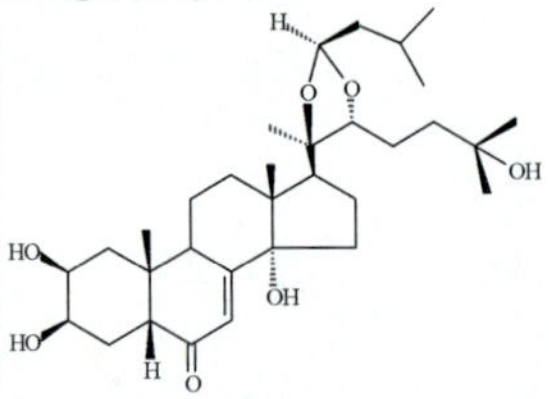	***S. claviformis***	Sadykov et al. (2001)
		S. caucasica	Zibareva (1999)
		S. chamarensis	Akhrem et al. (1973)
		S. chlorantha	Zibareva (1999)
		S. chlorifolia	Zibareva (2000)
		S. ciliata	Zibareva (1999)
		S. coelirosa	Zibareva (1999)
		S. colorata	Zibareva (2000)
		S. colorata	Zibareva (2000)
		S. compacta	Zibareva (2000)
		S. conica	Zibareva (2000)
		S. conoidea	Zibareva (1999)
		S. cretica	Meng et al. (2001)
		S. coronaria	Zibareva (1999)
		S. damboldtiana	Zibareva (2000)
		S. densiflora	Zibareva (1999)
		S. dichotoma	Zibareva (2000)

(continued)

Table 1.1 (continued)

№	Ecdysteroids	Name of plants	References
60	**$C_{30}H_{44}O_9$** **2-Acetate-29-norcyasterone** 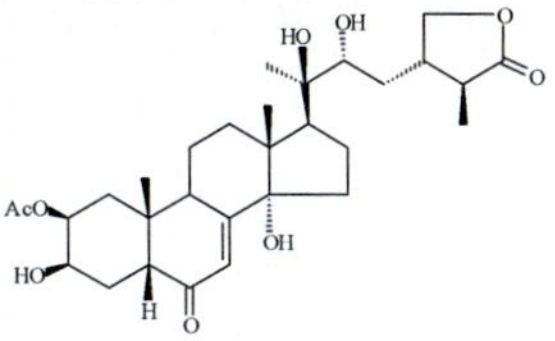	***A. reptans***	Camps and Coll (1993), Calcagno et al. (1995)
61	**$C_{30}H_{44}O_9$** **3-O-Acetate-29-norcyasterone** 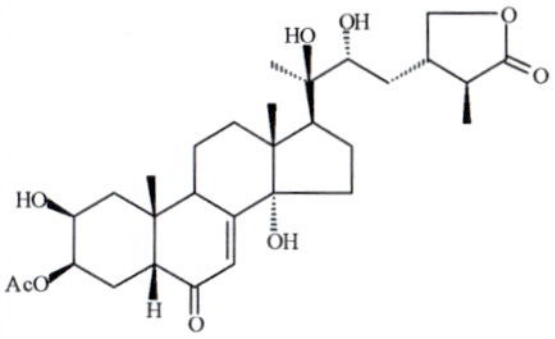	***A. reptans***	Camps and Coll (1993), Calcagno et al. (1995)
62	**$C_{31}H_{46}O_9$** **3-Acetate-cyasterone** 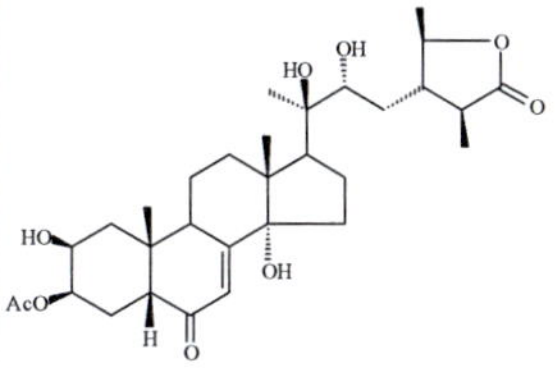	***A. reptans***	Calcagno et al. (1995), Ványolós (2012)
63	**$C_{31}H_{46}O_9$** **22-O-Acetate 3-epi-cyasterone** 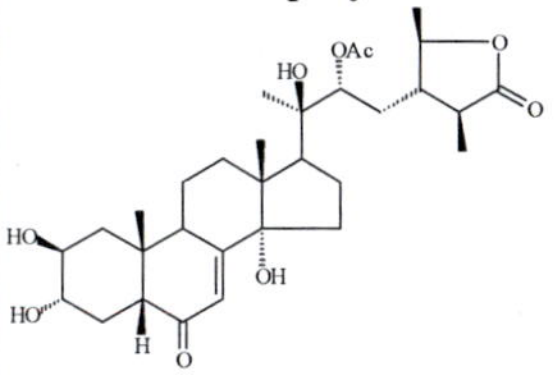	***A. bracteosa***	Kim et al. (2005)
64	**$C_{29}H_{46}O_7$** **Ajugasterone B** 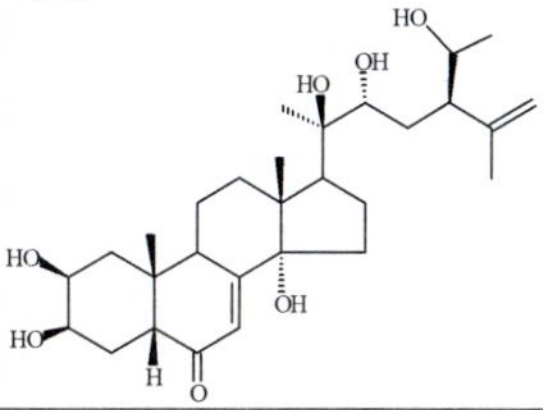	***A. incisa***	Takasaki et al. (2015)
		A. reptans	Alekseeva (1999)
		A. ciliata	Takasaki et al. (2015)
		A. nipponensis	Takasaki et al.; (2015)
		A. bracteosa	Takasaki et al. (2015)

(continued)

Table 1.1 (continued)

№	Ecdysteroids	Name of plants	References
65	**$C_{27}H_{44}O_7$** **Ajugasterone C** 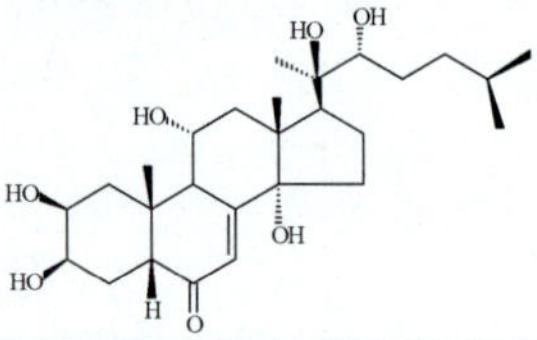	***A. decumbens***	Matsuoka et al. (1969)
		A. remota	Calcagno et al. (1995)
		A. japonica	Imai et al. (1969a, b), Camps and Coll (1993)
		A. bracteosa	Imai et al. (1969a, b)
		A. iva	Imai et al. (1969a, b)
		Serratulla wolffii	Miladera et al. (1992)
66	**$C_{27}H_{42}O_7$** **Ajugasterone D** 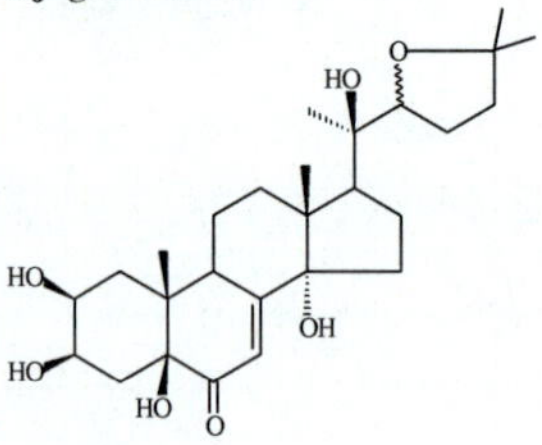	***A. nipponensis***	Keshan and Ray (2000)
67	**$C_{40}H_{58}O_{13}$** **22-Benzoate 25-O-β-D-glucoside-ecdysterone** 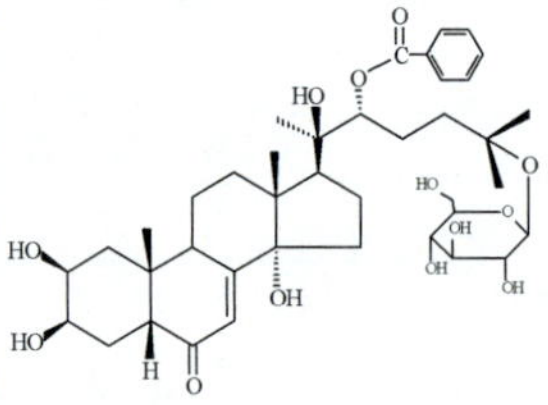	***S. otites***	Girault et al. (1990)
68	**$C_{34}H_{48}O_6$** **22-Benzoate-2-deoxy-α-ecdysone** 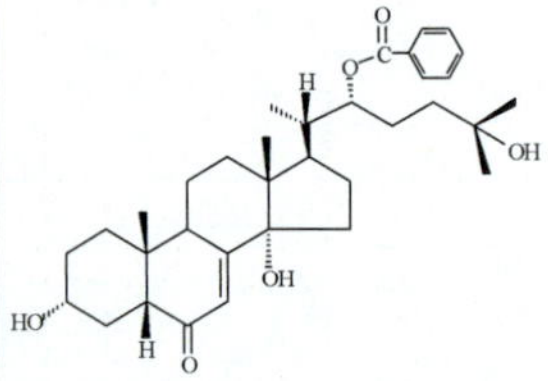	***S. wallichiana***	Saatov et al. (1988)

(continued)

Table 1.1 (continued)

№	Ecdysteroids	Name of plants	References
69	**$C_{34}H_{48}O_7$** **22-O-Benzoate-2-deoxyecdysterone** 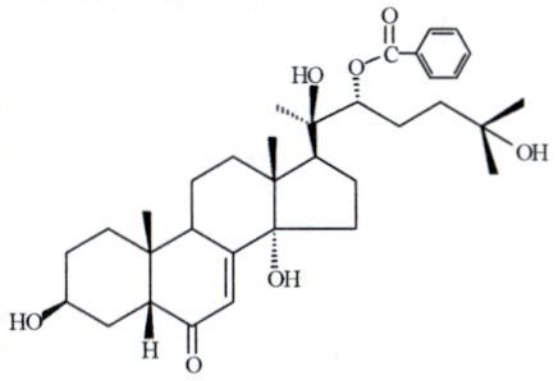	***S. tatarica***	Girault et al. (1990)
70	**$C_{37}H_{52}O_8$** **22-O-Benzoate-2,3-O-monoacetonide-ecdysterone** 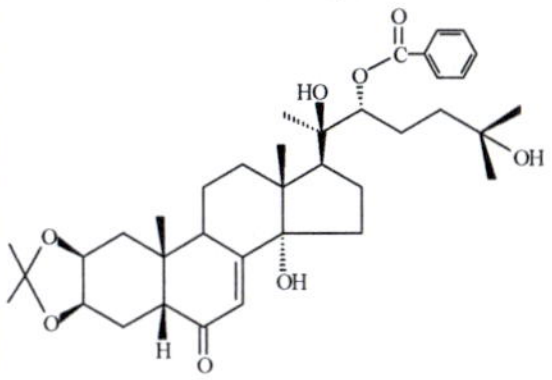	***S. scabrifolia***	Saatov et al. (1987)
71	**$C_{34}H_{48}O_8$** **20-O-Benzoate-ecdysterone** 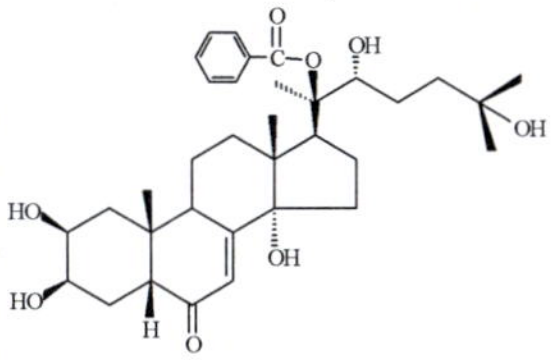	***S. tatarica***	Baltaev and Abubakirov (1987)
72	**$C_{27}H_{44}O_6$** **Brahuisterone** 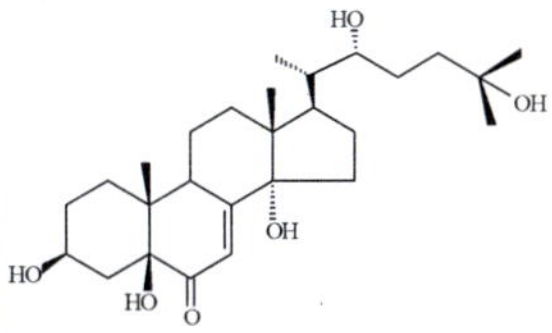	***S. brahuica***	Djukharova et al. (1993)
73	**$C_{27}H_{44}O_6$** **21-hydroxy-2-deoxyecdysone** 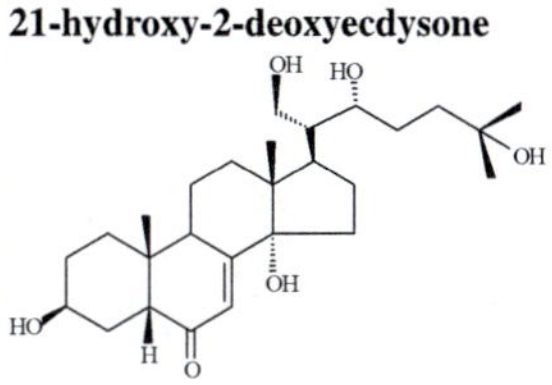	***S. parnassica***	Zibareva (1999, 2000)
		S. potula	Zibareva (1999, 2000)
		S. portensis	Zibareva (1999, 2000)
		S. praemixta	Saatov et al. (1993)
		S. procumbens	Zibareva (2000)
		S. pseudovelutina	Zibareva (2000)
		S. pussila	Zibareva (2000)

(continued)

Table 1.1 (continued)

№	Ecdysteroids	Name of plants	References
		S. pygmaea	Zibareva (2000)
		S. radicosa	Zibareva (2000)
		S. regia	Meng et al. (2001)
		S. teichenbachii	Zibareva (2000)
		S. repens	Zibareva (2000)
		S. reguienii	Zibareva (2000)
		S. roemeri	Zibareva (2000)
		S. rubella	Zibareva (2000)
		S. scabrifolia	Saatov et al. (1993)
		S. sieberi	Zibareva (2000)
		S. schafta	Saatov et al. (1993)
		S. schmuckeri	Zibareva (1999, 2000)
		S. schumacheri	Zibareva (1999, 2000)
		S. schwarzenbergeri	Saatov et al. (1993), Akhrem et al. (1973)
		S. secundiflora	Zibareva (2000)
		S. sedtneri	Zibareva (1999)
		S. sobolevskajae	Saatov et al. (1993)
		S. succulenta	Zibareva (1999)
		S. supine	Akhrem et al. (1973)
74	**$C_{27}H_{44}O_7$** **26-Hydroxy-22-deoxy ecdysterone** 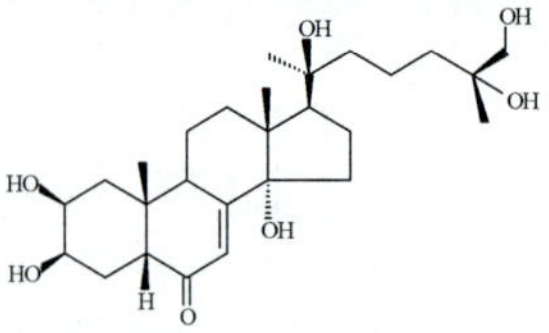	***S. nutans***	Girault et al. (1990)
75	**$C_{27}H_{44}O_9$** **26-hydroxy integristerone A** 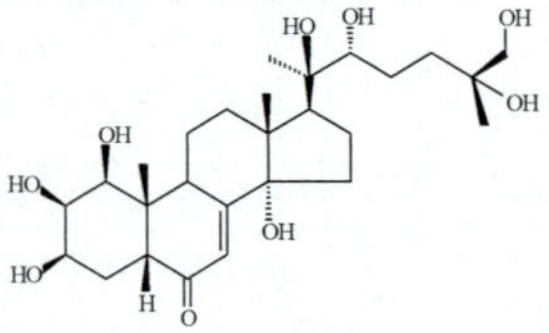	***S. viridiflora***	Zibareva et al. (2009)

(continued)

Table 1.1 (continued)

№	Ecdysteroids	Name of plants	References
76	$C_{27}H_{44}O_9$ **26-hydroxypolypodine B** 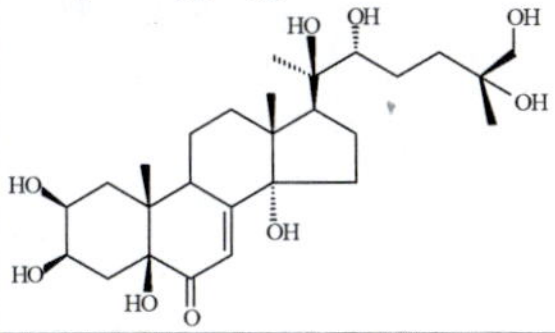	***S. viridiflora***	Mamadalieva et al. (2003, 2004)
		S. vulgaris	Zibareva (2000)
		S. waldsteinii	Zibareva (2000)
77	$C_{33}H_{54}O_9$ **22-β-Glucosil-2-deoxyecdysone** 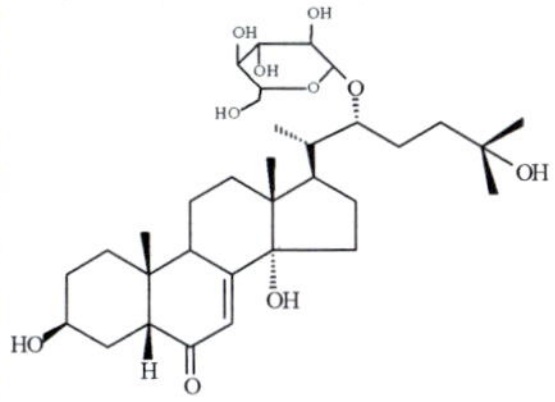	***S. parnassica***	Zibareva (1999, 2000)
		S. potula	Zibareva (1999, 2000)
		S. portensis	Zibareva (1999, 2000)
		S. praemixta	Saatov et al. (1993)
		S. procumbens	Zibareva (2000)
		S. pseudovelutina	Zibareva (2000)
		S. pussila	Zibareva (2000)
		S. pygmaea	Zibareva (2000)
		S. radicosa	Zibareva (2000)
		S. regia	Zibareva (2000)
		S. teichenbachii	Zibareva (2000)
		S. repens	Zibareva (2000)
		S. reguienii	Zibareva (2000)
		S. roemeri	Zibareva (2000)
		S. rubella	Zibareva (2000)
		S. scabrifolia	Saatov et al. (1993)
		S. sieberi	Zibareva (2000)
		S. schafta	Saatov et al. (1993)
		S. schmuckeri	Zibareva (1999, 2000)
		S. schumacheri	Zibareva (1999, 2000)
		S. schwarzenbergeri	Saatov et al. (1993), Akhrem et al. (1973)
		S. secundiflora	Zibareva (2000)
		S. sedtneri	Zibareva (1999)
		S. sobolevskajae	Saatov et al. (1993)
		S. succulenta	Zibareva (1999)
		S. supine	Akhrem et al. (1973)

(continued)

Table 1.1 (continued)

№	Ecdysteroids	Name of plants	References
78	**$C_{33}H_{54}O_{12}$** **3-O-β-D-Glucoside ecdysterone** 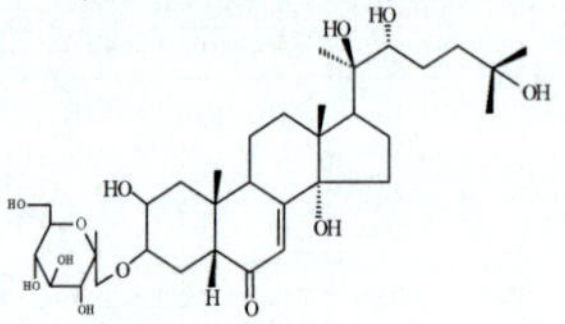	***S. otitis***	Báthori et al. (1997)
79	**$C_{29}H_{44}O_{9}$** **23-hydroxycyasterone** 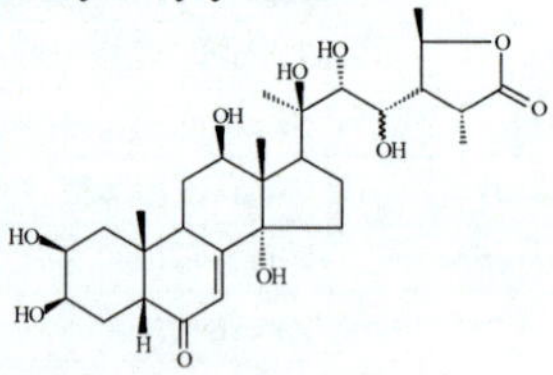	***A. reptans***	Camps et al. (1982)
80	**$C_{29}H_{38}O_{8}$** **2-Dehydroajugalactone** 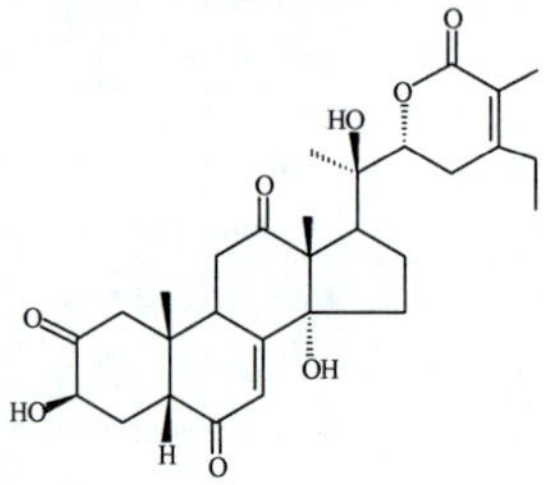	***A. reptans***	Calcagno et al. (1995)
81	**$C_{29}H_{38}O_{8}$** **3-Dehydroajugalactone** 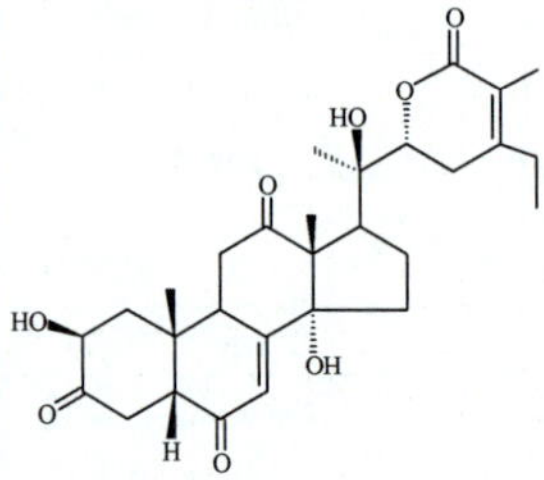	***A. reptans***	Calcagno et al. (1995)

(continued)

Table 1.1 (continued)

№	Ecdysteroids	Name of plants	References
82	**$C_{28}H_{40}O_{10}$** **22-dehydro-12-hydroxy-29-norsengosteron** 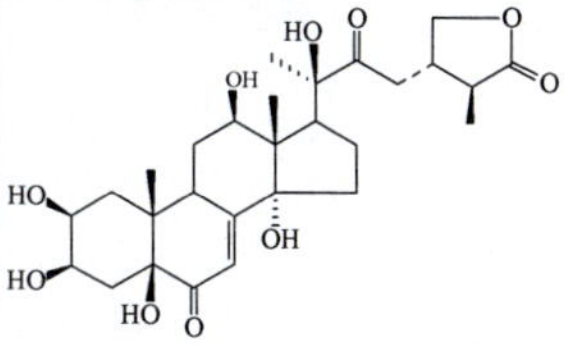	***A. reptans***	Calcagno et al. (1995)
83	**$C_{29}H_{42}O_{10}$** **22-dehydro-12-hydroxy-cyasterone** 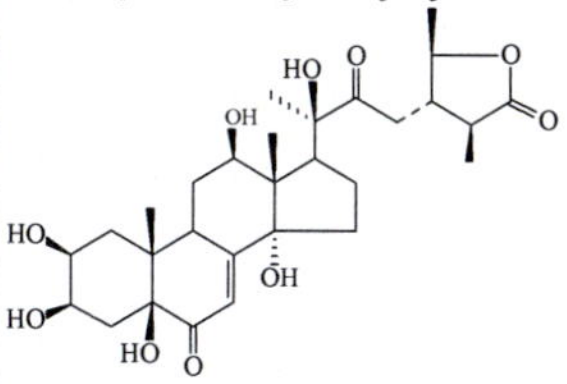	***A. reptans***	Camps et al. (1982)
84	**$C_{33}H_{54}O_{12}$** **2-deoxy-3β-D-glucoside-polypodine B** 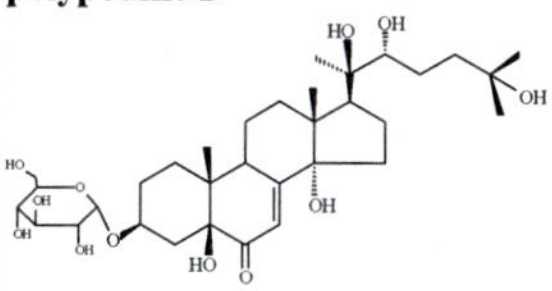	***S. pseudotites***	Meng et al. (2001)
85	**$C_{27}H_{44}O_7$** **2-Deoxy-integristerone A** 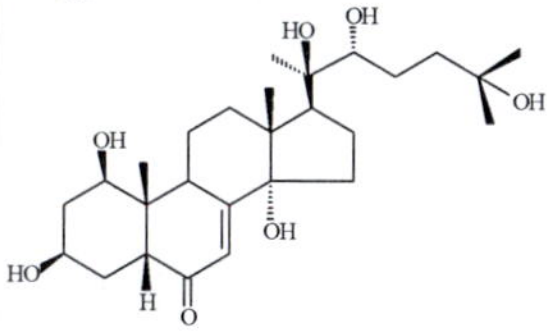	***S. otitis***	Girault et al. (1990)
86	**$C_{27}H_{44}O_7$** **22-Deoxy-integristerone A** 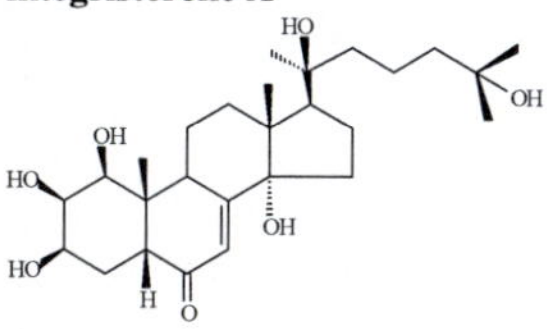	***S. nutans***	Girault et al. (1990)

(continued)

Table 1.1 (continued)

№	Ecdysteroids	Name of plants	References
87	**$C_{27}H_{44}O_7$** **5α-2-Deoxy-integristerone A** 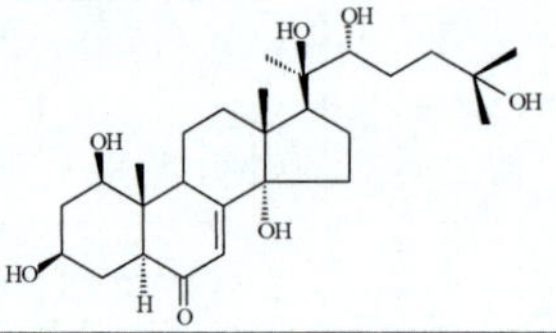	***S. pseudotites***	Meng et al. (2001)
		S. italica ssp	Meng et al. (2001)
88	**$C_{27}H_{44}O_7$** **5α-22-Deoxy-integristerone A** 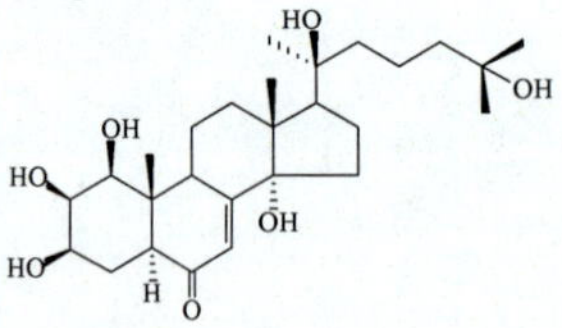	***S. nutans***	Girault et al. (1990)
89	**$C_{29}H_{42}O_8$** **24, 25-Dehydro-precyasterone** 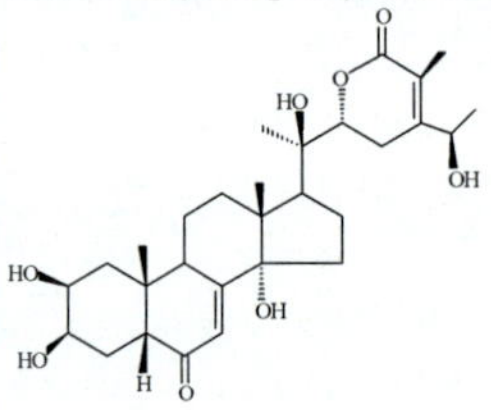	***A. iva***	Bakrim et al. (2014)
90	**$C_{29}H_{42}O_8$** **22-Dehydrocyasterone** 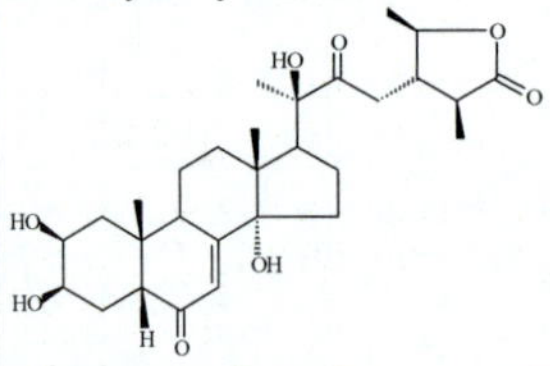	***A. reptans***	Calcagno et al. (1994)
91	**$C_{27}H_{42}O_6$** **3-Dehydroecdysone** 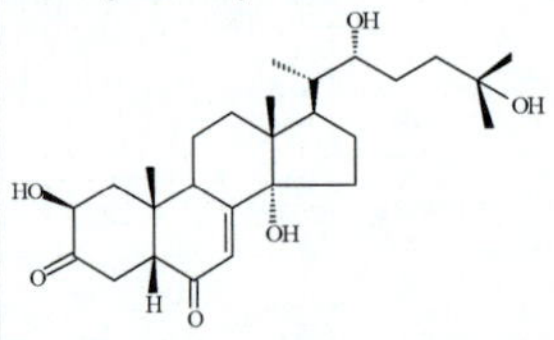	***A. reptans***	Calcagno et al. (1994)

(continued)

Table 1.1 (continued)

№	Ecdysteroids	Name of plants	References
92	**$C_{27}H_{43}O_7$** **3-Dehydroecdysterone** 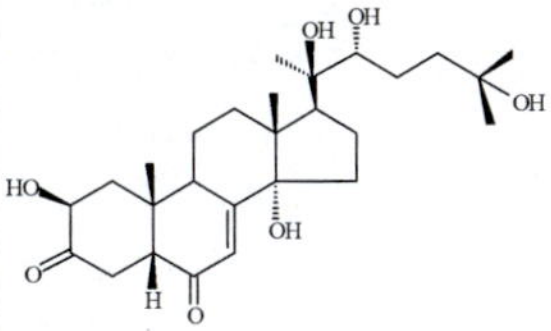	***A. reptans***	Calcagno et al. (1994)
93	**$C_{31}H_{48}O_{10}$** **2,22-Di-O-acetate-26-hydroxyecdysterone** 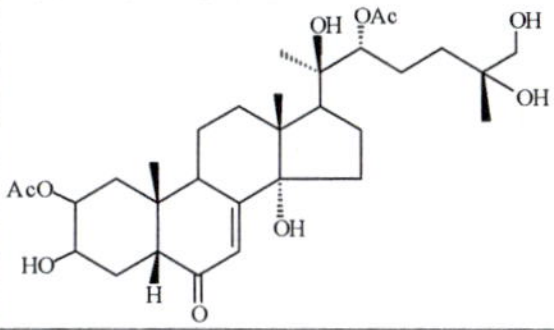	***S. viridiflora***	Mamadalieva et al. (2003, 2004)
		S. vulgaris	Zibareva (2000)
		S. waldsteinii	Zibareva (2000)
94	**$C_{41}H_{52}O_9$** **22,25-Di-O-Benzoate-ecdysterone** 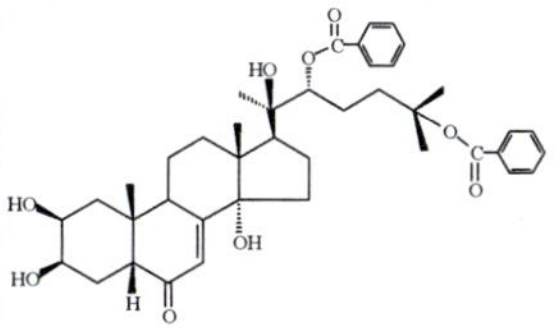	***S. scabrifolia***	Saatov et al. (1990)
95	**$C_{27}H_{44}O_7$** **20,26-dihydro-2-deoxyekdizone** 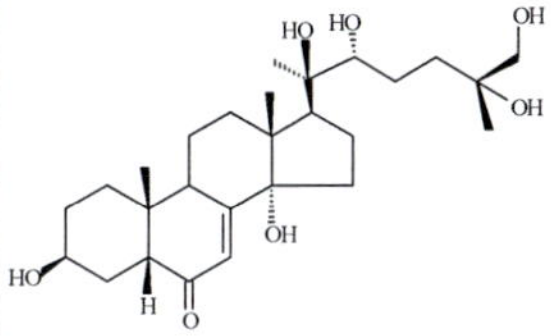	***S. parnassica***	Zibareva (1999, 2000)
		S. potula	Zibareva (1999, 2000)
		S. portensis	Zibareva (1999, 2000)
		S. praemixta	Saatov et al. (1993)
		S. procumbens	Zibareva (2000)
		S. pseudovelutina	Zibareva (2000)
		S. pussila	Zibareva (2000)
		S. pygmaea	Zibareva (2000)
		S. radicosa	Zibareva (2000)
		S. regia	Zibareva (2000)
		S. teichenbachii	Zibareva (2000)
		S. repens	Zibareva (2000)
		S. reguienii	Zibareva (2000)

(continued)

Table 1.1 (continued)

№	Ecdysteroids	Name of plants	References
		S. roemeri	Zibareva (2000)
		S. rubella	Zibareva (2000)
		S. scabrifolia	Zibareva (2000)
		S. sieberi	Zibareva (2000)
		S. schafta	Saatov et al. (1993)
		S. schmuckeri	Zibareva (1999, 2000)
		S. schumacheri	Zibareva (1999, 2000)
		S. schwarzenbergeri	Saatov et al. (1993), Akhrem et al. (1973)
		S. secundiflora	Zibareva (2000)
		S. sedtneri	Zibareva (1999)
		S. sobolevskajae	Saatov et al. (1993)
		S. succulenta	Zibareva (1999)
		S. supine	Akhrem et al. (1973)
96	**$C_{29}H_{44}O_9$** **5,29-Dihydroxy-capitasterone** 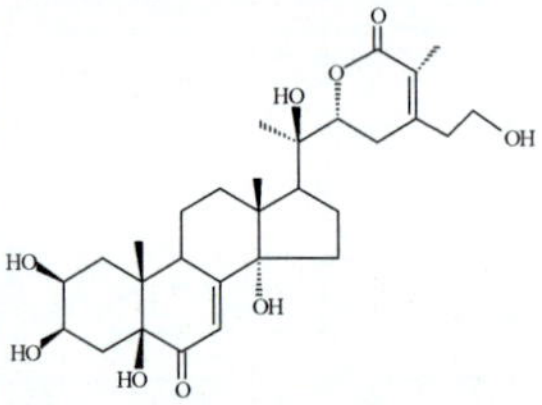	***A. reptans***	Calcagno et al. (1994)
97	**$C_{21}H_{31}O_5$** **Dihydropoststerone** 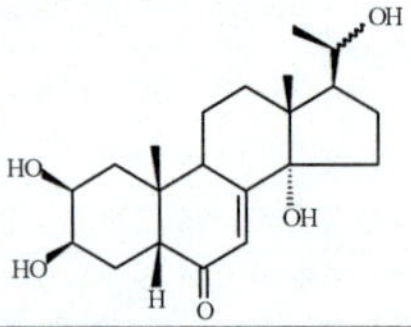	***S. otites***	Girault et al. (1990)
98	**$C_{27}H_{44}O_8$** **20,26-Dihydroxyecdysone** 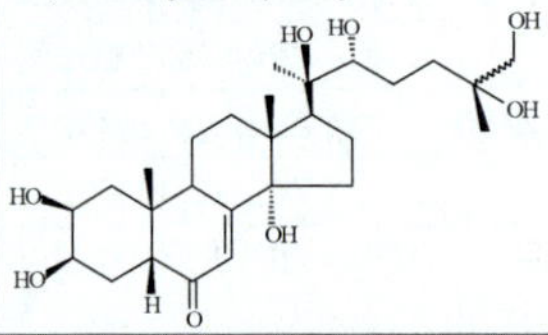	***S. nutans***	Girault et al. (1990)
		S. otites	Girault et al. (1990)
		S. viridiflora	Simon et al. (2009)

(continued)

Table 1.1 (continued)

№	Ecdysteroids	Name of plants	References
99	**$C_{29}H_{44}O_8$** **Carthamosterone** 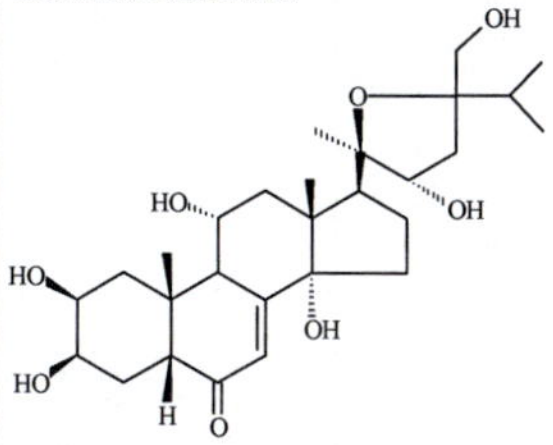	***Rh. carthamoides***	Buděšínsky et al. (2008)
100	**$C_{30}H_{48}O_6$** **20,22-O-Monoacetonide 2-deoxyecdysterone** 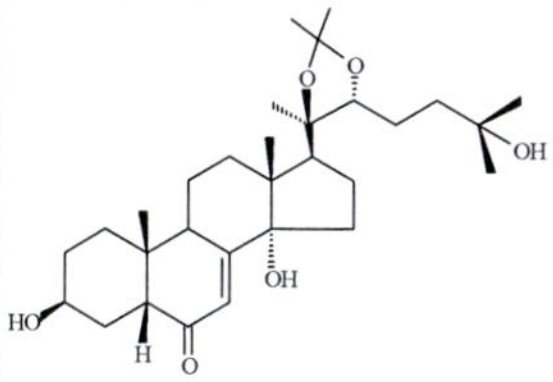	***S. brahuica***	Djukharova et al. (1991)
101	**$C_{30}H_{48}O_7$** **2,3-O-Monoacetonide ecdysterone** 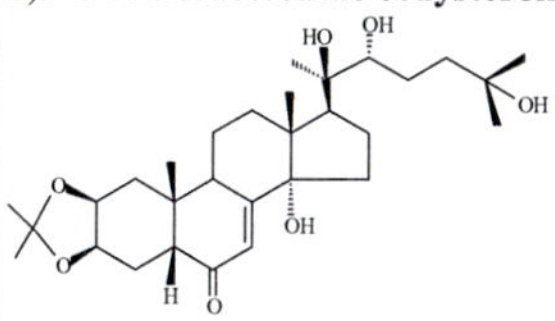	***Rh. carthamoides***	Baltaev and Abubakirov (1987)
102	**$C_{30}H_{48}O_7$** **20,22-O-Monoacetonide ecdysterone** 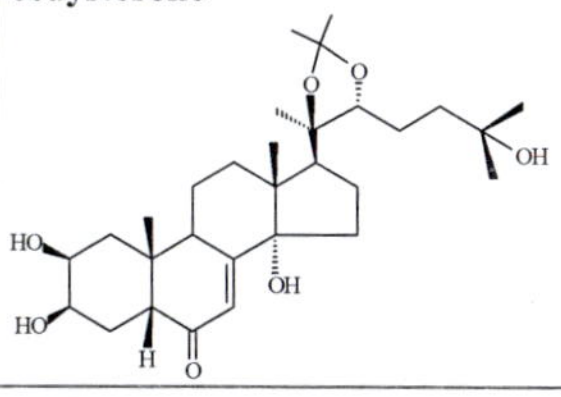	***S. scabrifolia***	Saatov et al. (1987)
		Rh. carthamoides	Baltaev and Abubakirov (1987)
103	**$C_{30}H_{48}O_9$** **20,22-O-Monoacetonide 26-hydroxypolypodine B** 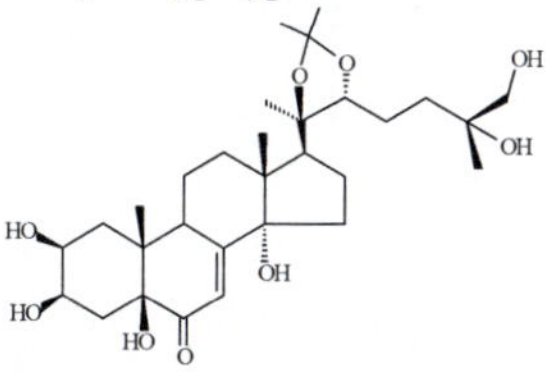	***S. viridiflora***	Simon et al. (2009)

(continued)

Table 1.1 (continued)

№	Ecdysteroids	Name of plants	References
104	**$C_{28}H_{42}O_9$** **29-Norsengosterone** 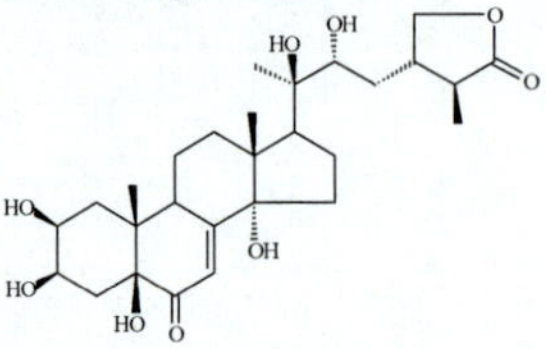	***A. reptans***	Camps et al. (1982)
105	**$C_{28}H_{42}O_8$** **29-Norcyasterone** 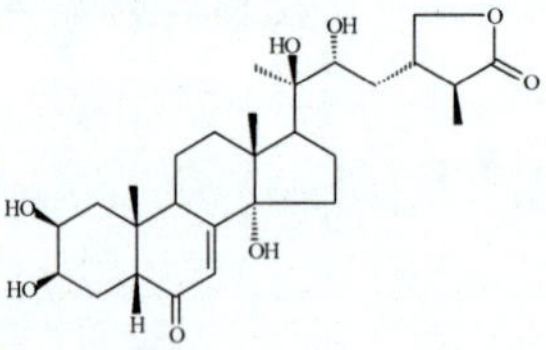	***A. reptans***	Camps et al. (1982)
106	**$C_{27}H_{44}O_9$** **Nusilsterone** 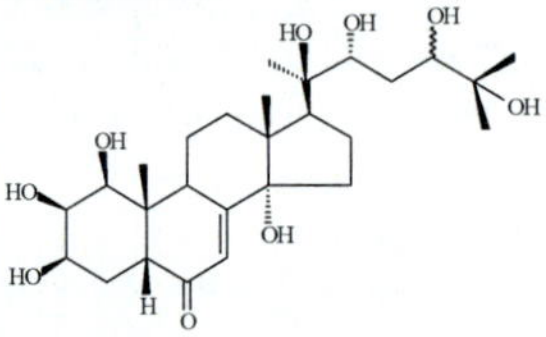	***S. nutans***	Baltaev et al. (1985)
107	**$C_{29}H_{42}O_8$** **22-Oxocysterone** 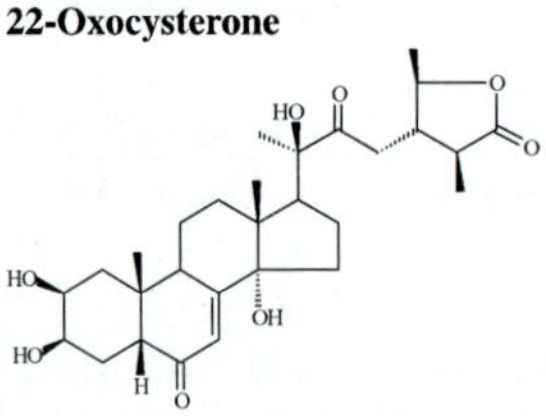	***A. iva***	Bakrim et al. (2014)
108	**$C_{27}H_{44}O_7$** **Paristerone** 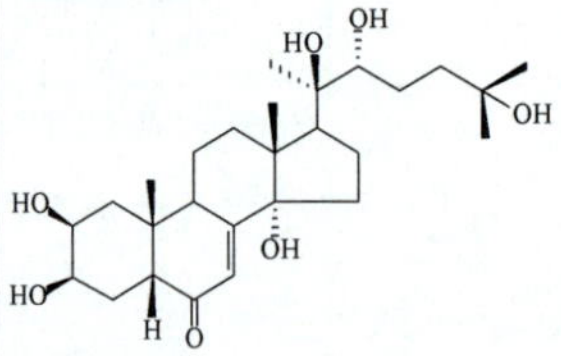	***A. remota***	Sing and Thakur (1982)

(continued)

Table 1.1 (continued)

№	Ecdysteroids	Name of plants	References
109	**$C_{27}H_{44}O_5$** **Praemixsterone** 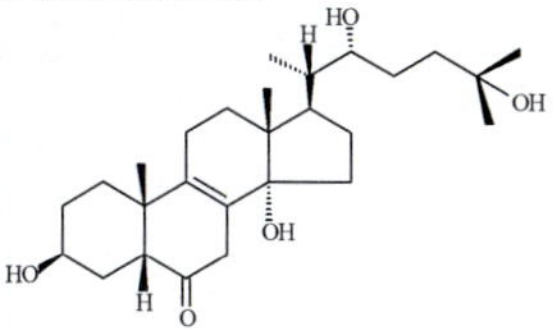	***S. praemixta***	Saatov et al. (1979)
110	**$C_{29}H_{48}O_7$** **Rapisterone** 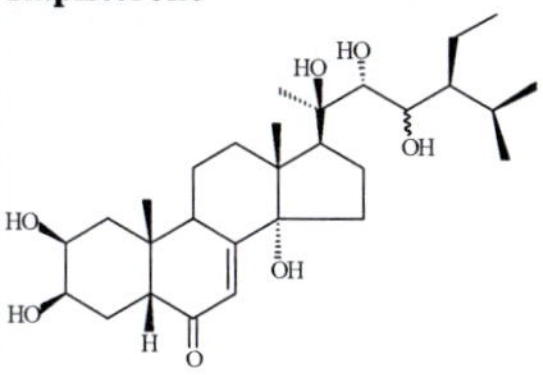	***Rh. carthamoides***	Baltaev and Abubakirov (1987)
111	**$C_{27}H_{44}O_7$** **Rapisterone B** 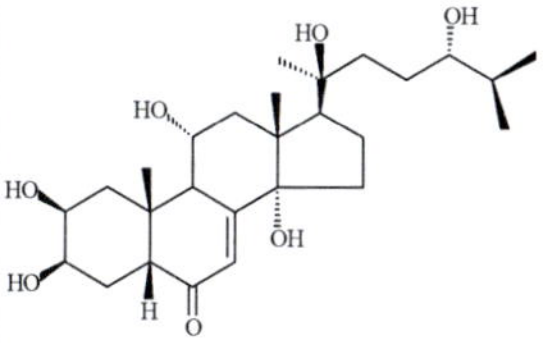	***Rh. carthamoides***	Baltaev (1991)
112	**$C_{29}H_{48}O_7$** **Rapisterone C** 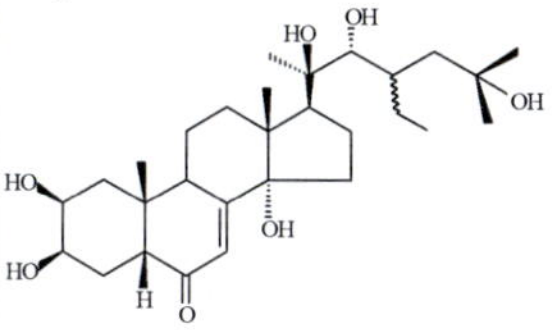	***Rh. carthamoides***	Baltaev (1992)
113	**$C_{27}H_{44}O_8$** **Rapontisterone (Punisterone)** 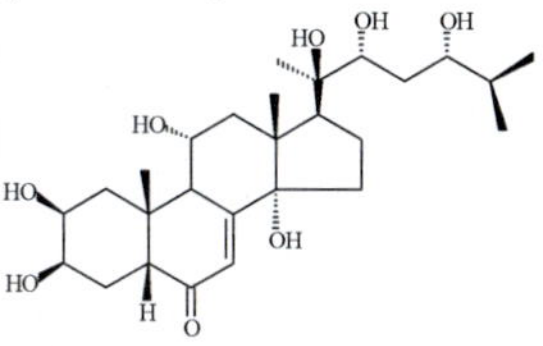	***Rh. uniflorum***	Selivanov et al. (2007)

(continued)

Table 1.1 (continued)

№	Ecdysteroids	Name of plants	References
114	**$C_{29}H_{44}O_8$** **Reptansterone** 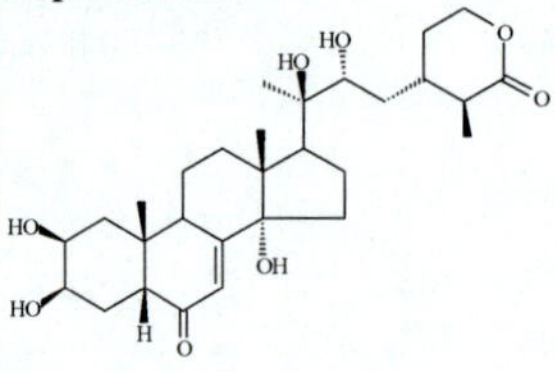	***A. reptans***	Calcagno et al. (1995)
115	**$C_{29}H_{44}O_9$** **Sengosterone** 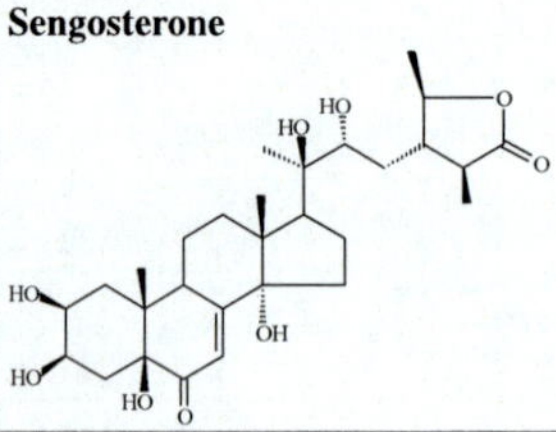	***A. reptans***	Alekseeva et al. (1998), Calcagno et al. (1995)
		A. bracteosa	Kim et al. (2005)
116	**$C_{33}H_{54}O_{10}$** **Sileneoside E** 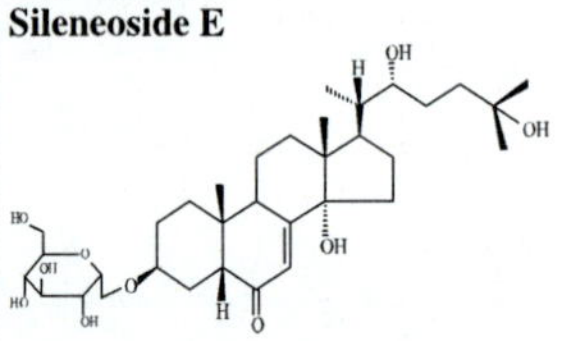	***S. brahuica***	Saatov et al. (1986)
117	**$C_{33}H_{54}O_{10}$** **5α-Sileneoside E** 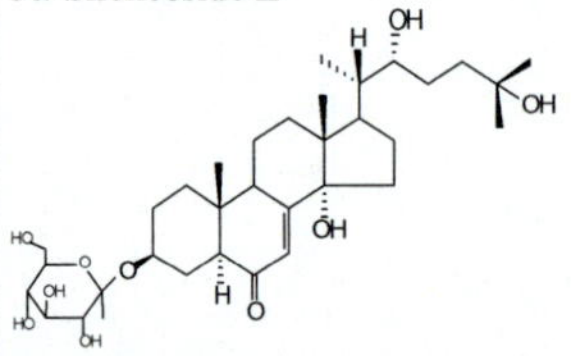	***S. brahuica***	Dzhukharova et al. (1995)
118	**$C_{33}H_{55}O_{11}$** **Sileneoside F** 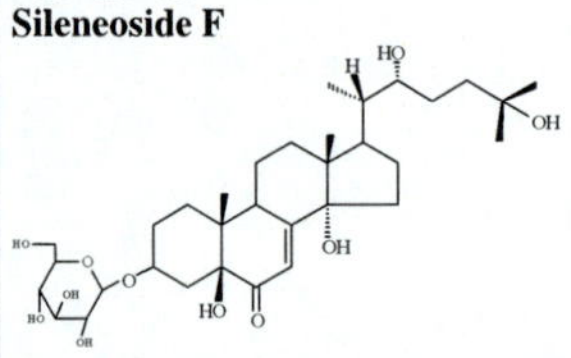	***S. brahuica***	Dzhukharova et al. (1994)

(continued)

Table 1.1 (continued)

№	Ecdysteroids	Name of plants	References
119	**$C_{24}H_{32}O_6$** **Sidisterone** 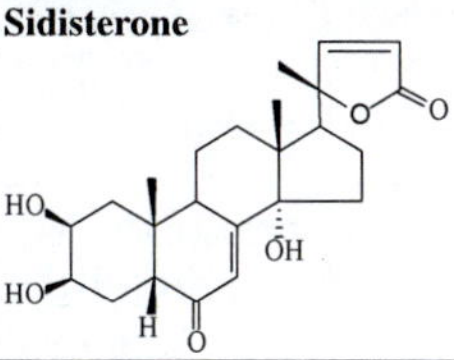	***S. dicica***	Girault et al. (1996)
		S. otites	
120	**$C_{39}H_{64}O_{17}$** **Ecdystereoside** 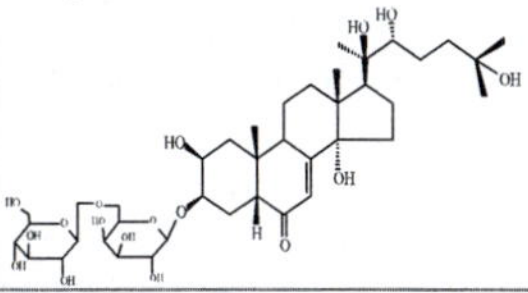	***S. tatarica***	Baltaev (1996)
		S. thessalonica	Zibareva (2000)
121	**$C_{29}H_{44}O_8$** **28-Epi-sengosterone** 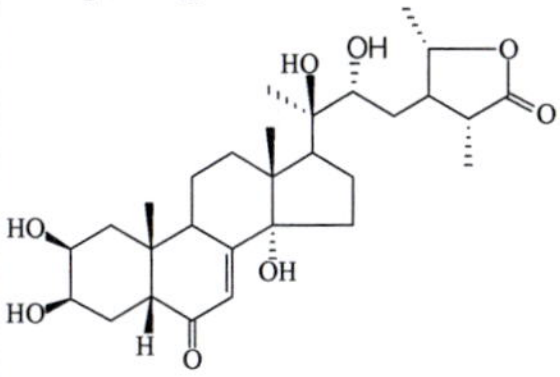	***A. reptans***	Calcagno et al. (1995)
122	**$C_{29}H_{44}O_8$** **3-Epi-cyasterone** 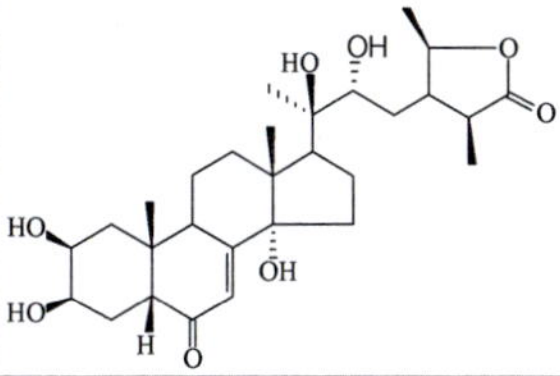	***A. bracteosa***	Camps et al. (1982)
		C. officnalis	Okuzumi et al. (2005)

References

Akhrem AA, Levina IS, Titov YuA (1973) Ecdysones are insect steroid hormones, p 232

Alekseeva LI, Volodin VV, Teteruk LV, Luksha VG, Lafont R (1998) Ecdysteroids of *Ajuga reptans* L. J Plant Physiol 45:372–377 (in Russ)

Alekseeva LI (1999) Ecdysteroids *Serratula coronata* L. and *Ajuga reptans* L. Diss Can Chem Sciences. Syktyvkar, p 114

Bajguz A, Bakala I, Talarek M (2015) Ecdysteroids in plants and their pharmacological effects in vertebrates and humans. Stud Nat Prod Chem 45:121–145. https://doi.org/10.1016/B978-0-444-63473-3.00005-8

Bakrim A, Ngunjiri J, Crouzet S, Guibout L, Balducci Ch, Girault JP, Lafont R (2014) Ecdysteroid Profiles of two *Ajuga* species, *A. iva* and *A. remota*. Nad Prod Commun 9(8):1069–1074. https://doi.org/10.1177/1934578X1400900804

Baltaev UA, Abubakirov NK (1987) Phytoecdysteroids of *Rhaponticum carthamoides*. Chem Nat Compd 23:565–568. https://doi.org/10.1007/BF00598674

Baltaev UA, Rashkes YV, Abubakirov NK (1985) Phytoecdysteroids of *Silene nutans* III. Nusilsterone. Chem Nat Compd 21:489–491. https://doi.org/10.1007/BF00579144

Baltaev UA (1996) Ecdysterioside is a new ecdysteroid from *Silene tatarica* (L) Pers. International meeting on phytoecdysteroids: Abstract, p 78

Baltaev UA (1979) Phytoecdysteroids of plants of the genus *Rhaponticum*: Dis. Cand. Chem. Sciences, p 106

Baltaev UA (1992) Phytoecdysteroids of *Rhaponticum carthamoides* III. Rhapisterone C. Chem Nat Compd 28:198–200. https://doi.org/10.1007/BF00630175

Baltaev UA (1991) Phytoecdysteroids of *Rhaponticum carthamoides* II. Rhapisterone B. Chem Nat Compd 27:712–713. https://doi.org/10.1007/BF00629932

Baltaev UA (2000) Phytoecdysteroids: structure, sources and pathways of biosynthesis in plants. Russ J Bio Chem 26:892–925. https://doi.org/10.1023/a:1026662505403

Barbier M (1978) Introduction to chemical ecology. World, p 232

Báthori M, Girault JP, Kalász H, Máthé I, Lafont R (1997) New minor ecdysteroids from *Silene otites* (L.) Wib. J Pharm Biomed Anal 16(2):327–336. https://doi.org/10.1016/S0731-7085(97)00051-4

Borovikova EB, Shangaraeva GS, Baltaev UA (1999) Rhapisterone D 20-acetate from the seeds of *Leuzea carthamoides*. Chem Nat Compd 35:184–185. https://doi.org/10.1007/BF02234931

Buděšínsky M, Vokáč K, Harmatha J, Cvacka J (2008) Additional minor ecdysteroid components of *Leuzea carthamoides*. Steroids 73:502–514. https://doi.org/10.1016/j.steroids.2007.12.021

Butenandt A, Karlson P (1954) Über die isoilerung eines metamorphose—hormone der insekten in kristallisierter form. Z. Naturforsch 9b(6):389–391

Calcagno MP, Camps F, Coll J, Mele E (1995) Acetylated ecdysteroids from *Ajuga reptans* var. *atropurpurea* (Lamiales: Lamiaceae). Eur J Entomol 92(1):287–293

Calcagno MP, Camps F, Coll J, Mele E (1994) New 22-oxoecdysteroids from *Ajuga reptans*. XI Ecdysone Workshop. Ceske Budejovice, p 41

Camps F, Coll J, Cortel A (1982) 29-norsengosterone and 29-norcyasterone, new c-28 phytoecdysteroids from *Ajuga reptans* (Labiatae). Chem Lett 11(9):1313–1316. https://doi.org/10.1246/cl.1982.1313

Camps F, Coll J, Dargallo O (1984) Phytoecdysteroids from *Ajuga chamaepitys*. Anal De Quim De La Soc Esp De Quim 81:74–75

Camps F, Coll J (1993) Insect allelochemicals from *Ajuga* plants. Phytochemistry 32(6):1361–1370. https://doi.org/10.1016/0031-9422(93)85139-I

Chaubey MK (2018) Role of phytoecdysteroids in insect pest management: a review. J Agron 17(1):1–10. https://doi.org/10.3923/JA.2018.1.10

Cheng DM, Kutzler LW, Boler DD, Drnevich J, Killefer J, Lila MA (2013) Continuous infusion of 20-hydroxyecdysone increased mass of triceps brachii in C57BL/6 mice. Phytother Res 27:107–111. https://doi.org/10.1002/ptr.4679

Das N, Mishra SK, Bishayee A, Ali ES, Bishayee A (2020) The phytochemical, biological, and medicinal attributes of phytoecdysteroids: an updated review. J Acta Pharm Sinica B. https://doi.org/10.1016/j.apsb.2020.10.012

De-fu C, Ming-xue S, Wen-fu X (2002) Pesticidal character of phytoecdysteroids from *Ajuga multiflora* Bunge (Labiatae) on larvae of *Cryptorrhynchus lapathi* L. (Coleoptera: Curculionidae). J For Res 13:177–182. https://doi.org/10.1007/BF02871693

Devarenne TP, Sen-Michael B, Adler H (1995) Biosynthesis of ecdysteroids in Zwa mays. Phytochemistry 40(1):1125–1131. https://doi.org/10.1016/0031-9422(95)00435-A

Dinan L, Harmatha J, Volodin V, Lafont R (2009) Phytoecdysteroids: diversity, biosynthesis and distribution. Ecdysone: structure and functions, p 3–45. https://doi.org/10.1007/978-1-4020-9112-4_1

Dinan L (1998) A strategy towards the elucidation of the contribution made by phytoecdysteroids to the deterrence of invertebrate predators on plants. Rus J Plant Physiol 45(3):347–359

Dinan L (1995) Distrbution and levels of phytoecdysteroids within individual plants of species of the *Chenopodiaceae*. Eur J Entomol 92:295–300

Dinan L (2000) Phytoecdysteroids: biological aspects. Phytochemistry 57:325–339. https://doi.org/10.1016/S0031-9422(01)00078-4

Dzhukharova MK, Saatov Z, Abdullaev ND et al (1994) Phytoecdysteroids of the plants of the genus *Silene* XV. Silenoside F—Brahuisterone 3-O-β-D-glucopyranoside from *Silene brahuica*. Chem Nat Compd 30:680–683. https://doi.org/10.1007/BF00630602

Dzhukharova MK, Saatov Z, Abdullaev ND (1995) Phytoecdysteroids of plants of the genus *Silene* XVI. 5α-sileneoside E from *Silene brahuica*. Chem Nat Compd 31:207–210. https://doi.org/10.1007/BF01170207

Dzhukharova MK, Saatov Z, Gorovits MB et al (1991) Phytoecdysteroids of plants of the genus *Silene*. XVIII. 2-Deoxyecdysterone 20,22-monoacetonide from *Silene brahuica*. Chem Nat Compd 27:207–209. https://doi.org/10.1007/BF00629760

Dzhukharova MK, Tashkhodzhaev B, Saatov Z et al (1993) Phytoecdysteroids of plants of the genus *Silene* XIV. Brahuisterone from *Silene brahuica*. Chem Nat Compd 29:484–489. https://doi.org/10.1007/BF00630575

Erdemoglu N, Sener B, Demirci B et al (2003) The glycosidically bound volatile compounds of *Taxus baccata*. Chem Nat Compd 39:195–198. https://doi.org/10.1023/A:1024870031960

Faux A, Galbraith MN, Horn DHS, Middleton EJ (1970) The structure of two ecdysones analogues, cheilanthones A and B, from the fern *Cheilanthes tenifolia*. J Chem Soc Chem Commun 243–244. https://doi.org/10.1039/C29700000243

Fukuzawa A, Kumagai Y, Masamumo T et al (1981) Acetyl-pinnasterol and pinnasterol, ecdysone-like metabolites from the marine red algae laurencia pinnata yamada. Tetrahedron Lett 22:4085–4086. https://doi.org/10.1016/S0040-4039(01)82071-2

Galbraith MN, Horn DHS (1969) Insect molding hormones: crustecdysone (20-Hydroxyecdysone) from Podocarpus elatus. Aust J Chem 22:1045–1057. https://doi.org/10.1071/CH9691045

Girault JP, Bathori M, Kalasz H et al (1996) Sidisterone, a C-24 ecdysteroid from *Silene dioica* and *Silene otites*. J Nat Prod 59:522–524. https://doi.org/10.1021/np960339s

Girault JP, Bathori M, Varga E, Szendrei K, Lafont R (1990) Isolation and identification of new ecdysteroids from caryophyllaceae. J Nat Prod 53:279–293. https://doi.org/10.1021/np50068a002

Golovatskaya IF (2004) Effect of ecdysterone on morphological and physiological processes in plants. Russ J Plant Physiol 51:407–413. https://doi.org/10.1023/B:RUPP.0000028689.97402.d5

Grebenok R, Adler JH (1991) Ecdysteroid distribution during development of Spinach. Phytochemistry 30(9):2905–2910. https://doi.org/10.1016/S0031-9422(00)98222-0

Grebenok R, Ripa PV, Adler JH (1991) Occurrence and levels of ecdysteroids in Spinach. Lipids 26(8):666–668. https://doi.org/10.1007/BF02536433

Grebenok R, Venkatachari S, Adler JH (1994) Biosynthesis of ecdysone and ecdysone phosphates in Spinach. Phytochemistry 36:1399–1408. https://doi.org/10.1016/S0031-9422(00)89731-9

Grebenok RJ, Galbraith DW, Benveniste I et al (1996) Ecdysone 20-monooxygenase, a cytochrome P450 enzyme from spinach, *Spinacia Oleracea*. Phytochemistry 42(4):927–933. https://doi.org/10.1016/0031-9422(96)00094-5

Hardman R, Benjamin TV (1976) The cooccurrence of ecdysones with bufadienolides and steroidal saponins in the genus *Helleborus*. Phytochemistry 15:1515–1516. https://doi.org/10.1016/S0031-9422(00)88927-X

Hikino H, Nomoto K, Takemoto T (1970) Sengosterone, an insect metamorphosing substance from *Cyathula capitata*: Structure. Tetrahedron 26:887–898

Ikan R, Ravid U (1971) The isolation and identification of cyasterone for *Ajuga chia* (Labiatae). Phytochemistry 10:1659–1661. https://doi.org/10.1016/0031-9422(71)85043-4

Imai S, Hori M, Fujioka S et al (1968) Isolation of four new Phytoecdysones, makisterones A, B, C, D, and the structure of makisterone A, C-28 steroid. Tetrahedron Lett 36:3883–3886

Imai S, Toyasato T, Sakai M et al (1969a) Isolation of cyasterone and ecdysterone from plant materials. Chem Pharm Bull 17:340–342. https://doi.org/10.1248/cpb.17.340

Imai S, Toyosato T, Sakai M et al (1969b) Screening results of plants for phytoecdysones. Chem Pharm Bull 17:335–339. https://doi.org/10.1248/cpb.17.335

Jizba J, Herout V, Storm F (1967) Polypodine B-a novel ecdysterone like substance from plant material. Tetrahedron Lett 18:5139–5143. https://doi.org/10.1016/S0040-4039(01)89630-1

Karlson P, Hoffmeister H, Hoppe W, Huber R (1963) Zur Chemie des Ecdysones. Justus Liebigs Ann Chem 662:1–20. https://doi.org/10.1002/jlac.19636620102

Karlson P, Koolman J (1976) Ecdysone: mode of action, inactivation and elimination. Actualites sur les hormones d'invertebres. Paris: Ed M.M. Durchon, pp 403–412

Karlson P (1976) Chemistry and physiology of insect hormones and insect pheromones. Sem. Etude Theme Prod Nature et prot Plant 41:17–28. https://doi.org/10.1351/pac196714010075

Karlson P. On the use of ecdysteroid nomenclature. XI Ecdysone Workshop: Program and Abstracts. Ćeske Budèjovice; 1994. p.7–8

Keshan B, Ray AK (2000) Estradiol-17β in Bombyx mori: possible significance and its effect on silk production. J Insect Physiol 46(6):1061–1068. https://doi.org/10.1016/S0022-1910(99)00218-8

Kim OT, Manickavasagm M, Kim YJ et al (2005) Genetic transformation of *Ajuga multiflora* Bunge with Agrobacterium rhizogenes and 20-hydroxyecdysone production in hairy roots. J Plant Biol 48(2):258–262. https://doi.org/10.1007/BF03030416

Koreeda M, Nakanishi K, Goto M (1970) Ajugalactone, an insect molding inhibitor as tested by the chile dipping method. J Am Chem Soc 92(25):7512–7513. https://doi.org/10.1021/ja00728a068

Kreis W, Muller-Uri F (2018) Biochemistry of sterols, cardiac glycosides, brassinosteroids, phytoecdysteroids and steroid saponins. In Annual plant reviews. Biochemistry of plant secondary metabolism. Wiley-Blackwell, vol 40, pp 304–363. http://dx.doi.org/https://doi.org/10.1002/9781119312994.apr0428

Kubo I, Klocke J, Asano S (1981) Insect ecdysis inhibitors from the east African medicinal plant *Ajuga remota* (Labiatae). Agric Biol Chem 45:1925–1927. https://doi.org/10.1271/bbb1961.45.1925

Kubo I, Klocke JA, Ganjian I, Ichikawa N, Matsumoto T (1983) Efficient isolation of phytoecdysones from *Ajuga* plants by high-performance liquid chromatography and droplet counter-current chromatography. J Chromatogr A 257:157–161. https://doi.org/10.1016/S0021-9673(01)88168-1

Lafont R, Bouthier A, Wilson ID (1991) Phytoecdysteroids: structures, occurrence, biosynthesis and possible ecological significance. Conf Insect Chem Ecol. Program and Abstracts, pp 197–214

Lafont R, Ho R, Raharivelomanana P, Dinan L (2011) Ecdysteroids in ferns: distribution, diversity, biosynthesis, and functions. In: Kumar A, Fernández H, Revilla M (eds) Working with ferns. Springer, New York. https://doi.org/10.1007/978-1-4419-7162-3_22

Lafont R, Horn DHS (1989) Phytoecdysteroids: structures and occurrence. Ecdysone. From chemistry to mode of action. Koolman J (ed). Stuttgart, New York, pp 39–64

Lafont R, Wilson ID (1996) The ecdysone handbook. Nottingham, p 14

Lafont R (1997) Ecdysteroids and related molecules in animals and plants. Arch Insect Biochem Physiol 35(1–2):3–20. https://doi.org/10.1002/(SICI)1520-6327(1997)35:1/2

Lafont R (1998) Phytoecdysteroids and world flora: diversity, distribution, and evolution. Plant Physiol 45:326–346

Lloyd-Jones JG, Rees HH, Goodwin TW (1973) Biosynthesis of ecdysterone from cholesterol in *Taxus baccata*. Phytochemistry 12:569–572. https://doi.org/10.1016/S0031-9422(00)84445-3

Lukyantseva IA, Rakhimova EB, Zarembo EN et al (2001) Phytoecdysteroids and plant taxa *Rhaponticum carthamoides* Willd. (Iljin), *Chenopodium album* L., *Stemmacantha uniflora* L., *Matteuccia strupthiopteris* L. Tod. and Taxus cuspidate (Seib and Zucc.). Abstract. doc. celebrated. scientific conf. young scientists of Bashkir Gos Ufa 1:196–197

Mamadalieva NZ, Ramazanov NS, Dinan LN et al (2000) Phytoecdysteroids of plants of the *Silene* Genus. 2-Dehydroxyecdysterone-3-O-benzoate from *Silene wallichiana*. Chem Nat Compd 36:513–515. https://doi.org/10.1023/A:1002847725472

Mamadalieva NZ, Zibareva LN, Saatov Z et al (2003) Phytoecdysteroids of *Silene viridiflora*. Chem Nat Compd 39:199–203. https://doi.org/10.1023/A:1024822116031

Mamadalieva NZ, Zibareva LN, Saatov Z (2002) Phytoecdysteroids of *Silene linicola*. Chem Nat Compd 38:268–271. https://doi.org/10.1023/A:1020436128797

Mamadalieva NZ, Zibareva LN, Todeschi NE et al (2004) New minor ecdysteroids from *Silene viridiflora*. Cool Czechosl Chem Comm. 69:1675–1680. https://doi.org/10.1002/chin.200505176

Matsuoka T, Imai S, Sakai M, Kamada M (1969) Studies on phytoecdysones: a review of our works. Ann Rep Takeda Res Lab 28:221–271

Meng Y, Whiting P, Zibareva L et al (2001) Identification and quantitative analysis of the phytoecdysteroids in *Silene* species (*Caryophyllaceae*) by hidh-performance liquid chromatography. Novel ecdysteroids from *S. pseudotites*. J Chromatogr 935:309–319. https://doi.org/10.1016/s0021-9673(01)00893-7

Miladera K, Saatov Z, Kholodova YD et al (1992) Phytoecdysteroids of plants of the genus Serratula. Ajugasterone C 20,22-monoacetonide from Serratura wolffii. Chem Nat Compd 28:59–63. https://doi.org/10.1007/BF00629794

Murtazina EP, Sidorova YS, Zhuravlev BV et al (2014) Study of behavioral parameters during learning of rats in an operant feeding task and evaluation of biochemical indexes after dietary consumption of the phytoecdysteroid extract. Bull Exp Biol Med 158:4–8. https://doi.org/10.1007/s10517-014-2678-7

Nachtigall M, Bulow L, Schubert J, Frese L (2016) Development of SSR markers for the genus *Patellifolia (Chenopodiaceae)*. Appl Plant Sci 48(8). https://doi.org/10.3732/apps.1600040

Nakanishi K, Koreeda M, Sasaki S, Chang ML, Hsu HY (1966) Insect hormones. The structure of ponasterone A, insect-moulting hormone from the leaves of *Podocarpus nakaii* Hay. Chem Commun 24:915–917. https://doi.org/10.1039/C19660000915

Ohsawa T, Yukawa M, Takao C et al (1992) Studies on constituents of fruit body of polyporus umbellatus and their cytotoxic activity. Chem Pharm Bull 40:143–147. https://doi.org/10.1248/cpb.40.143

Okamoto N, Viswanatha R, Bittar R, Li Z, Haga-Yamanaka S, Perrimon N, Yamanaka N (2018) A membrane transporter is required for steroid Hormone Uptake in *Drosophila*. Dev Cell 47(3):294–305. https://doi.org/10.1016/j.devcel.2018.09.012

Okuzumi K, Hara N, Uekusa H, Fujimoto Y (2005) Structure elucidation of cyasterone stereisomers isolated form *Cyathula officinalis*. J Org Biomol Chem 3(7):1227–1232. https://doi.org/10.1039/b416868b

Otaka T, Okui S, Uchiyama M (1969) Stimulation of protein synthesis in mouse liver by ecdysterone. Chem Pharm Bull 17:75–81. https://doi.org/10.1248/cpb.17.75

Qi Y, Liu X, Lu Y, Zhao L (2011) Research on teaching reform on environmental biochemistry based on teaching materials. In: Lin S, Huang X (eds) Advances in computer science, environment, ecoinformatics, and education. CSEE 2011. Communications in computer and information science. Springer, Berlin, Heidelberg, p 218. https://doi.org/10.1007/978-3-642-23357-9_5

Ramazanov NS, Maksimov ES, Saatov Z et al (1995) Phytoecdysteroids of plants of the genus *Silene*. XVII. Tomentesterone from *Silene tomentella*. Chem Nat Compd 31:600–603. https://doi.org/10.1007/BF01164888

Ramazanov NS, Maksimov ES, Saatov Z et al (1997) Phytoecdysteroids of plants of the genus *Rhaponticum* I. Carthamosterone A from *Rh. carthamoides*. Chem Nat Compd 33:301–302. https://doi.org/10.1007/BF02234880

Ramazanov NS (2005) Phytoecdysteroids and other biologically active compounds from plants of the genus *Ajuga*. Chem Nat Compd 41:361–369. https://doi.org/10.1007/s10600-005-0153-4

Ramazonov NS, Mamadalieva NZ, Bobaev ID (2007) Phytoecdysteroids from five species of the genus *Silene*. Chem Nat Compd 43:117–118. https://doi.org/10.1007/s10600-007-0049-6

Rao R (1967) Two new species of *Patellaria* from India. Mycopathol Mycol Appl 31:29–32. https://doi.org/10.1007/BF02050281

Rashkes YV, Gorovits MB, Makarichev GK et al (1971) Some questions of the mass spectrometry of the cardenolides. Chem Nat Compd 7:724–732. https://doi.org/10.1007/BF00567925

Rees HH (1995) Ecdysteroid biosynthesis and inactivation in relation to function. Eur J Entomol 92:9–39

Reggi ML, Hirn DMH, Delaage MA (1975) Radioimmunoassay of ecdysone an application to Drosophila larvae and pupae. Biochem Biophys Res Commun 66(4):1307–1315. https://doi.org/10.1016/0006-291X(75)90502-1

Russell GB, Fenemore PG (1970) Insect moulting activity in some New Zealand gymnosperms. New Zealand J Sci 13(1):61–68

Russell GB, Fenemore PG (1971) Insect moulting activity in some New Zealand ferns. New Zealand J Sci 24(1):31–35

Saatov Z, Abdullaev ND, Gorovits MB et al (1984) Phytoecdysteroids of plants of the genus *Silene*. VII. Sileneoside D—ecdysterone 3-O-α-D-galactopyranoside from *Silene brahuica*. Chem Nat Compd 20:700–703. https://doi.org/10.1007/BF00580027

Saatov Z, Abdullaev ND, Gorovits MB et al (1990) Phytoecdysteroids of plants of the genus *Silene*. XVII. Ecdysterone 22,25-di-O-benzoate from *Silene scabrifolia*. Chem Nat Compd 26:301–303. https://doi.org/10.1007/BF00597854

Saatov Z, Abdullaev ND, Gorovits MB et al. (1986) Phytoecdysteroids of plants of the genus *Silene* X. Sileneoside E—2-Deoxy-α-ecdysone 3-O-β-D-glucopyranoside—From *Silene brahuica*. Chem Nat Compd 22:297–300. https://doi.org/10.1007/BF00598300

Saatov Z, Agzamkhodzhaeva DA, Syrov VN (1999) Distribution of phytoecdysteroids in plants of Uzbekistan and the possibility of using drugs based on them in neurological practice. Chem Nat Compd 35:186–191. https://doi.org/10.1007/BF02234932

Saatov Z, Gorovits MB, Abdullaev ND et al (1981) Phytoecdysteroids of plants of the genus *Silene*. III. Sileneoside A—A new glycosidic ecdysteroid of *Silene brachuica*. Chem Nat Compd 17:534–539. https://doi.org/10.1007/BF00574372

Saatov Z, Gorovits MB, Abdullaev ND et al. (1982) Phytoecdysteroids of plants of the genus *Silene*. IV. Sileneoside C—A galactoside of intergristerone A from *Silene brahuica*. Chem Nat Compd 18:193–196. https://doi.org/10.1007/BF00577192

Saatov Z, Gorovits MB, Abdullaev ND et al (1987) Phytoecdysteroids of plants of the genus *Silene* XII. 5α-Ecdysterone 22-O-benzoate from *Silene scabrifolia*. Chem Nat Compd 23:563–565. https://doi.org/10.1007/BF00598673

Saatov Z, Gorovits MB, Abdullaev ND et al (1985) Phytoecdysteroids of plants of the genus *Silene*. VIII. 2-Deoxyecdysterone 3-acetate from *Silene praemixta*. Chem Nat Compd 21:56–58. https://doi.org/10.1007/BF00574249

Saatov Z, Gorovits MB, Abubakirov NK (1993) Phytoecdysteroids of plants of the genus *Silene*. Chem Nat Compd 29:551–557. https://doi.org/10.1007/BF00630195

Saatov Z, Gorovits MB, Abubakirov NK (1988) Phytoecdysteroids of plants of the genus *Silene*. XVI. Viticosterone E 22-O-benzoate from *Silene wallichiana*. Chem Nat Compd 24:463–465. https://doi.org/10.1007/BF00598532

Saatov Z, Usmanov BZ, Abubakirov NK (1977) Phytoecdysones of *Ajuga turkestanica*. IV. Chem Nat Compd 13:359. https://doi.org/10.1007/BF00573570

Saatov Z, Usmanov BZ, Abubakirov NK (1979) Phytoecdysones of *Silene praemixta* I. Silenosterone. Chem Nat Compd 15:700–703. https://doi.org/10.1007/BF00565565

Sadikov ZT, Saatov Z (1999) Phytoecdysteroids of plant of the genus Silene brahuica. Chem Nat Compd 35:440- 441. https://doi.org/10.1007/BF02282513

Sadykov Z, Saatov Z, Garcia M et al (2001) Ecdysteroids from *Silene claviformis*. Chem Nat Compd 37:262–265. https://doi.org/10.1023/A:1012526325597

Sadykov Z, Saatov Z, Girault JP, Lafont R (2000) Silenoside H, a New phytoecdysteroid from *Silene brachuica*. J Nat Prod 63:987–988. https://doi.org/10.1021/np990609h

Sadykov ZT, Ramazanov NS, Saatov Z (1997) Phytoecdysteroids of plants of the genus *Rhaponticum* polypodine B 22-O-benzoate from *Rhaponticum carthamoides*. Chem Nat Compd 33:665–666. https://doi.org/10.1007/BF02249636

Sadykov ZT, Saatov Z (1998) Phytoecdysteroids of plants of the *Silene* genus. XIX. The structure of sileneoside G. Chem Nat Compd 34:602–604. https://doi.org/10.1007/BF02319284

Sauer H, Bennet R, Heftman E (1968) Ecdysterone biosynthesis in Pdocarpus elata. Phytochemistry 7:2027–2030. https://doi.org/10.1016/S0031-9422(00)90763-5

Selivanov SI, Solov'ev AY, Morozkina SN et al (2007) An NMR study of the conformational mobility of steroid estrogen 7α-methyl-8α analogues. Russ J Bioorg Chem 33:302–309. https://doi.org/10.1134/S1068162007030053

Simon A, Toth N, Toth G, Groska J, Bathori M (2009) Ecdysteroids from *Silene viridiflora*. Helv Chim Acta 92:753–761. https://doi.org/10.1002/hlca.200800362

Sing SB, Thakur RS (1982) Structure and stereochemistry of paristerone, a novel phytoecdysone from the tubers of *Paris polyphylla*. Tetrahedron 38:2189–2194. https://doi.org/10.1016/0040-4020(82)85166-1

Spindler KD, Hönl C, Tremmel C et al (2009) Ecdysteroid hormone action. Cell Mol Life Sci 66:3837–3850. https://doi.org/10.1007/s00018-009-0112-5

Suksamrarn A et al (2002) Ecdysteroids of *Silene bupleroides*. J Nat Prod 65:1194–1197

Takasaki M et al (2015) The ecdysone handbook. Nottingham, p 32

Takemoto T, Ogawa S, Nishimoto N (1967) Isolation of the moulting hormones of insects from Achyranthes radix. Yakugaku Zasshi 87(12):1469–1473. https://doi.org/10.1248/yakushi1947.87.12_1469

Takhtadzhyan AL (1987) Magnoliophyte system. Science, p 438

Thiem B, Kikowska M, Maliński MP et al (2017) Ecdysteroids: production in plant in vitro cultures. Phytochem Rev 16:603–622. https://doi.org/10.1007/s11101-016-9483-z

Tomas J, Camps F, Claveria E, Coll J, Mele E, Messeguer J (1992) Composition and location of phytoecdysteroids in *Ajuga reptans* in vivo and in vitro cultures. Phytochemistry 31(5):1585–1591. https://doi.org/10.1016/0031-9422(92)83112-C

Troczka B, Zimmer ChT, Elias J et al (2012) Resistance to diamide insecticides in diamondback moth, Plutella xylostella (Lepidoptera: Plutellidae) is associated with a mutation in the membrane-spanning domain of the ryanodine receptor. Insect Biochem Mol Biol 42(11):873–880. https://doi.org/10.1016/j.ibmb.2012.09.001

Usmanov BZ, Rashkes YV, Abubakirov NK (1978) Phytoecdysones of *Ajuga turkestanica* VI. 22-Acetylcyasterone. Chem Nat Compd 14:175–178. https://doi.org/10.1007/BF01134622

Usmanov BZ, Saatov Z, Abubakirov NK (1977) Phytoecdysones of *Ajuga turkestanica*. V. Chem Nat Compd 13:595. https://doi.org/10.1007/BF00569612

Usmanov VZ, Gorovits MB, Abubakirov NK (1971) Phytoecdysones of *Ajuga turkestanica*. Chem Nat Compd 7:520. https://doi.org/10.1007/BF00564772

Ványolós A (2012) Investigation of plants containing unusual ecdysteroids with ring in their side-chain. PhD thesis, Szeged, p 54. https://doi.org/10.14232/phd.1514

Vokáč K, Buděšínsky M, Harmatha J (2002) Minor ecdysteroid components of *Leuzea carthamoides*. J Collect Czechasl Chem Commun 67(1):124–139. https://doi.org/10.1135/CCCC20020124

Volodin V, Chadin I, Whiting P, Dinan L (2002) Screening plants of European North-East Russia for ecdysteroids. Biochem Syst Ecol 30:525–578. https://doi.org/10.1016/S0305-1978(01)00128-4

Yusupova UYu, Usmanov DA, Ramazonov NSh (2019) Phytoecdysteroids from the Plant *Dianthus helenae*. Chem Nat Compd 55(2):393–394. https://doi.org/10.1007/s10600-019-02701-y

Yusupova U, Usmanov D, Azamatov A, Ramazonov N, Rejepov J (2020a) Phytochemical constituents and biological activities of *Dianthus helenae* Vved., growing in Uzbekistan, Nat Prod Res. https://doi.org/10.1080/14786419.2020.1862834

Yusupova UYu, Usmanov DA, Ramazonov NSh (2020b) Phytoecdysteroids from the aerial part of *Silene popovii*. Chem Nat Compd 56(3):562–563. https://doi.org/10.1007/s10600-020-03092-1

Zhang T, Xing Y, Xu L et al (2019) Comparative analysis of the complete chloroplast genome sequences of six species of *Pulsatilla* Miller *Ranunculaceae*. Chin Med 14:53. https://doi.org/10.1186/s13020-019-0274-5

Zhu TT, Liang H, Zhao YY, Wang B (2004) Isolation and structure identification of C25 epimers of inokosterone from *Achyranthes bidentata* Blume. Yao Xue Xue Bao 39:913–916

Zibareva L, Yeriomina VI, Munkhjargal N, Girault JR, Dinan L, Lafont R (2009) The phytoecdesteroid profiles of 7 species of *Silene* (*Caryophyllaceae*). Arch Insect Biochem Physiol 72(4):234–248. https://doi.org/10.1002/arch.20331

Zibareva L (2000) Distrbution and levels of phytoecdysteroids in plants of the genus silene during development. Arch Insect Biochem Phisiol 43(1):1–8. https://doi.org/10.1002/(SICI)1520-6327(200001)43:1

Zibareva LN (1991) Phytoecditeroids and other chemical components of plants of the genus *Lychnis* L. Abstract. Dis. Cand. Biol. Sciences. Novosibirsk, p 24

Zibareva L (1999) Occurrence of phytoecdysteroids in Silene L. genus and dynamics of their content. *Rastit Resus*35(1):79–87

Chapter 2
Spectral Methods for Studying Phytoecdysteroids

Abstract Modern physicochemical spectroscopic methods highly accelerated the pace at which isolated compounds can be identified nowadays. Structural elucidation of the phytoecdysteroids has been done using spectroscopic methods such as ^{1}H, ^{13}C-NMR, mass spectroscopy, etc. Analysis of the fragmentation ions of the phytoecdysteroids side chain makes it possible to distinguish C-27-ecdysteroids from C-28 and C-29, since the transition to the next row causes a shift of 14 mass units, as well as establishing some other structural details. The structure of ecdysteroids, the NMR spectra are of great help, the analysis of which allows one to determine the presence and stereochemistry of structural elements present in the steroid skeleton and side chain. The resonance frequencies of the protons of methyl groups depend both on the nature and the orientation of the substituents in the skeleton and side chain. However, ^{1}H NMR spectroscopy, ^{13}C NMR spectroscopy is very widely used to elucidate various structural problems of ecdysteroids. it is possible to identify the atoms not related to protons, as well as to distinguish methyl, methylene, methane, and quaternary carbon atoms by DEPT methods.

Keywords Spectral methods · Mass spectra · ^{1}H spectroscopy · ^{13}C spectroscopy · DEPT method

2.1 Mass Spectra of Ecdysteroids

An important role in structural studies of ecdysteroids belongs to mass spectrometry. The main information obtained by studying the mass spectra of ecdysteroids is the molecular weight, number, and position of oxy groups in the steroid skeleton, and since the most significant difference between ecdysteroids is the structure of their side chains, the peculiarities of its structure (Fig. 2.1).

The nature of the decay of the ecdysone molecule under the influence of electron impact is similar to the decay of other types of steroid compounds (Johnson and Carlson 2015), but there are also distinctive features.

There are a number of methods for ionizing molecules of organic compounds: electron impact, chemical ionization, field desorption, and bombardment by fast atoms.

U. Y. Yusupova et al., *Phytoecdysteroids*,
https://doi.org/10.1007/978-981-16-6711-4_2

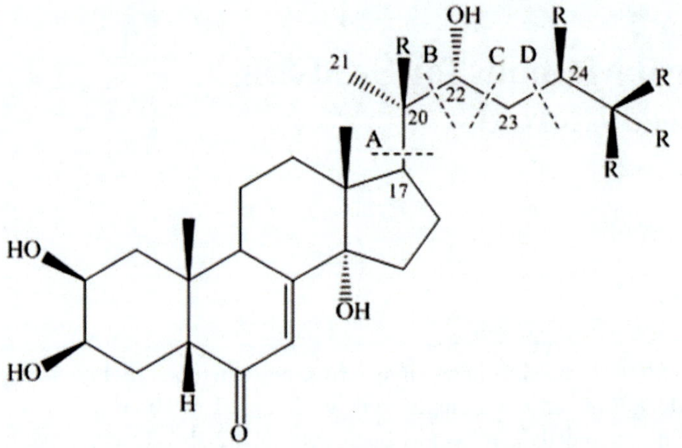

Fig. 2.1 The preferred directions in which the fragmentation of the ecdysteroid molecule occurs

Recently, new methods of "soft" ionization, i.e., secondary emission mass spectrometry, which includes accelerated atom bombardment (FAB) (Martin et al. 1982), and secondary ion mass spectrometry (SIMS) (Mil'grom et al. 1991).

These methods significantly expanded the possibilities of mass spectrometry in the analysis of unmodified polar and thermally labile natural compounds. They are widely used to determine the molecular weights of various natural compounds (Mil'grom et al. 1991).

In the mass spectra of free ecdysteroids, peaks are observed that correspond to:

(1) molecular ion (M^+);
(2) the ions of the steroid nucleus of the molecule;
(3) ions, side chain.

A peak of a molecular ion having a very low intensity is often absent altogether due to the ease of dehydration of ecdysteroid molecules by electron impact. Each main fragment can produce a series of peaks consisting of ions formed by sequential cleavage of several water molecules (for example, $[M\text{–}n\ H_2O]^+$ where n = 1, 2, 3, 4 …). The mass spectra of ecdysteroids are also characterized by ions corresponding to the cleavage of one of the quaternary methyl groups $(M\text{–}CH_3\text{–}n\ H_2O)^+$.

Cleavage according to type (A) leads to the breaking of the C_{17}–C_{20} bonds, according to the type (B)–C-20–C-22 bonds, according to the type (B)–C_{22}–C_{23} bonds, according to type (D)–C_{23}–C_{24}—communications (Fig. 2.2).

The most often decay occurs according to type (B), less often according to type (A) and (C) (Nakanishi 1971). If there is a substituent at position 24, then there is an additional gap of type (G) such decays are more common in C_{28} and C_{29} ecdysteroids.

The main directions of fragmentation of ecdysteroids can be seen in the mass spectrometric decay of ecdysterone (2) (Fig. 2.3) (Warren and Gilbert 1998; Galbraith et al. 1973; Kaplanis et al. 1966).

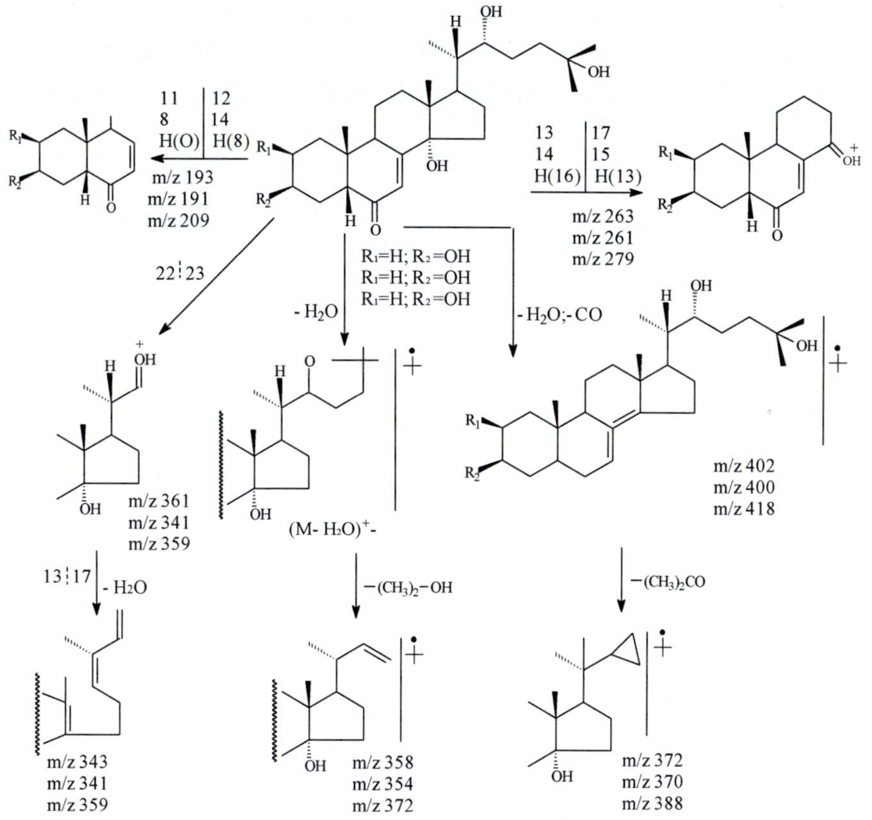

Fig. 2.2 The main ways of mass spectrometric decay of ecdysteroids

There is no peak of a molecular ion with m/z 480 in the spectrum, but there are peaks corresponding to the sequential cleavage of four water molecules (c m/z 462, 444, 426, 408). The appearance of fragments with m/z 411 and 393 can be explained by the cleavage of the methyl group and three and four water molecules ($M–CH_3–nH_2O$)+ from the ecdysterone molecule (2).

Intense peaks with m/z 363 $(M\text{-}117)^+$, 345 ($M\text{-}117\text{-}H_2O$), and 327 $(M\text{-}117\text{-}2H_2O)^+$ correspond to ions formed as a result of cleavage of the C20–C22 bond with the subsequent loss of 1 and 2 molecules of water. The corresponding side-chain fragments are represented by peaks with m/z 99 ($117\text{-}H_2O$) and 81 ($117\text{-}2H_2O$). Peaks with m/z 363 and 3345 correspond to ions formed due to the complete loss of the side chain, water molecule, and hydrogen atoms. The side-chain fragments in this decay are represented by peaks with m/z 161,143,125, and 107 (Fig. 2.3).

Easy removal of water from the tetracyclic core occurs due to the 14-α-hydroxy group, which is confirmed by the study of mass spectra of model compounds. It is known that the 2,3-diol group under conditions of mass spectrometry does not undergo easy dehydration (Faux et al. 1970).

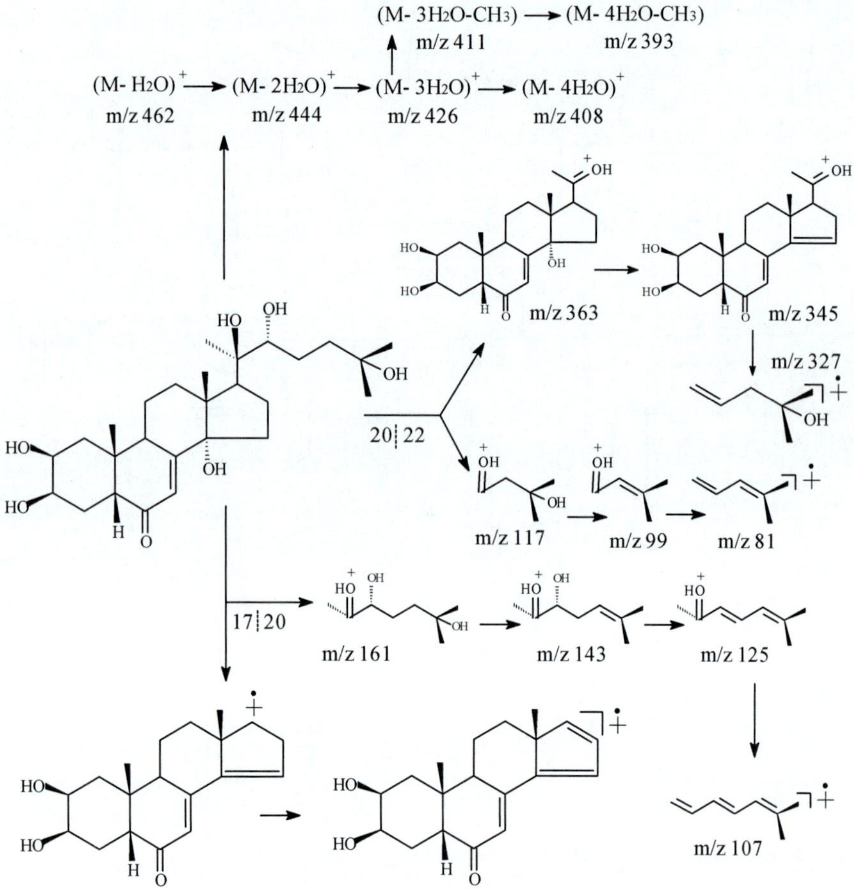

Fig. 2.3 The main ways of mass spectrometric decay of ecdysterone (2)

Analysis of the fragmentation ions of the side chain makes it possible to distinguish C27-ecdysteroids from C28 and C29, since the transition to the next row causes a shift of 14 mass units, as well as establishing some other structural details.

An example of the direct application of mass spectrometry to determine the structure of the side chain are makisterone A–D (12, 17, 18, 19) and 24,28-dihydromakisterone A (20) (Imai et al. 1968; Danili et al. 1974; Burns and Gildan 1977) (Table 2.1).

The transition from the makisterones A (12) and B (17)–C28–ecdysteroids to the makisterones C (18) and D (19) C29–ecdysteroids is accompanied by a shift of the peaks of all side chain fragments to high masses by 14 units. Makisterones having a substituent at position 24 undergo a gap of type G. This makes it possible to distinguish ecdysteroids that are isomeric in the position of hydroxy groups in the side chain (Fig. 2.4).

Table 2.1 The main pathways of makisterone A–D side-chain fragmentation (12, 17, 18, and 19)

Compound	Decay A	m/z	Decay B	m/z	Decay G	m/z
72	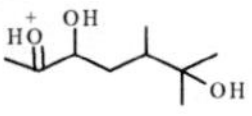	175 157 139 121	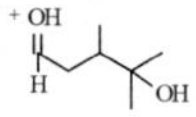	131(24) 113(59) 95(30)	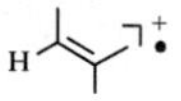	131 (100) 131(24)
74	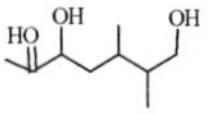	175 157 139 121	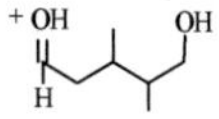	131(19) 113(94) 95(39)		69
75	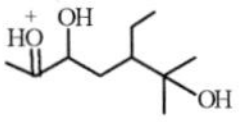	189 171 153 135	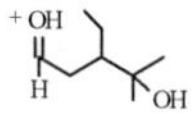	145(13) 127(100) 109	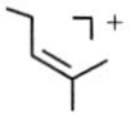	84 (84)
76	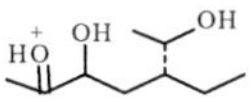	189 171 153 135	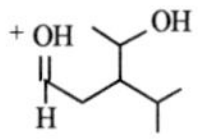	145(22) 127(100) 109	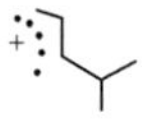	83

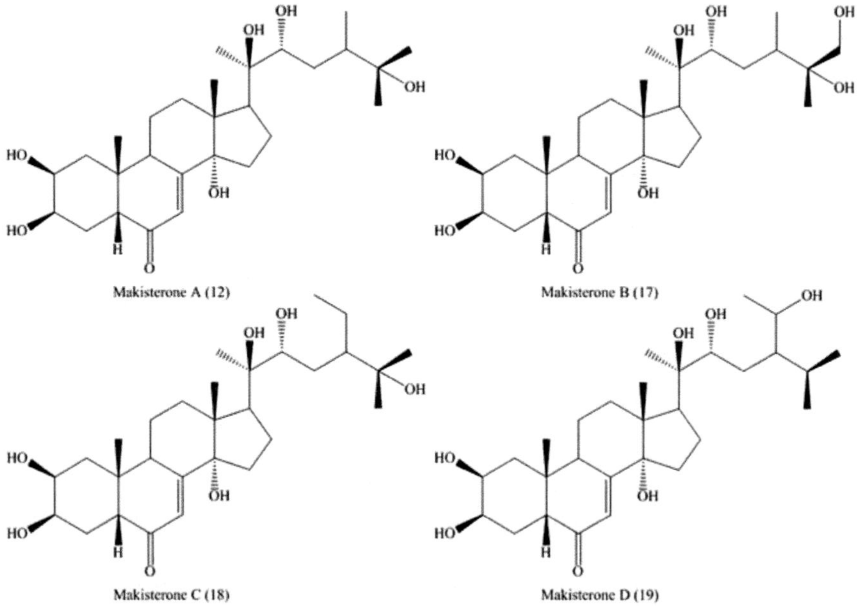

Fig. 2.4 The structure of makisterone A, B, C, and D

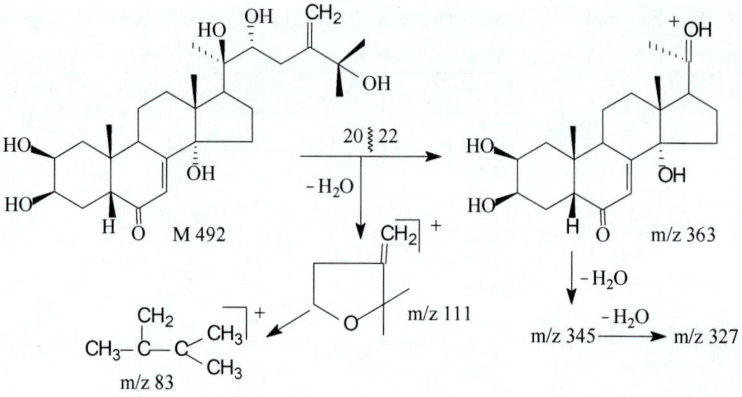

Fig. 2.5 The main ways of fragmentation of 24 (28)-dehydromakisterone A (20) under electron impact

In the mass spectrum of 24(28)-dehydromakisterone A (20) (Fig. 2.5), an ion with m/z 99 characteristics of the side chain of ecdysterone is shifted by 12 m.y. and has m/z 111. The ejection of a C=O molecule from this fragment leads to a peak of an ion with m/z 83 (Burns and Gildan 1977; Spiteller 1963).

The formation of ions with m/z 111 and 83 indicates the presence of a hydroxyl group at C-25 and an additional carbon atom compared to ecdysterone.

With a decrease in the number of OH-groups, the role of skeletal ruptures increases, which are reflected in the appearance of a number of ions in the mass spectra of 2- and 20-deoxyecdysteroids. In the mass spectrum of 2-deoxy-α-ecdysone, the peak of the ion with m/z 263 has the composition $C_{16}H_{23}O_3$. The result of the decay of the D ring at the C_{13}–C_{17} and C_{14}–C_{15} bonds is subject to the appearance of similar fragments with m/z 261 and 279 in the spectra of silenosterone and α-ecdysone (Fig. 2.2) (Rashkes et al. 1971; Ohnishi et al. 1977).

An ion with m/z 193 of the composition $C_{12}H_{17}O_2$ of 2-deoxy-α-ecdysone (22) and its analogues in the spectra of silenosterone (21) and α-ecdysone (1) can be the result of the decay process indicated in (Fig. 2.1) (Rashkes et al. 1971; Ohnishi et al. 1977).

In some works (Rashkes et al. 1971; Ohnishi et al. 1977), it was noted that ions with m/z 361 in the mass spectrum of 2-deoxy-α-ecdysone arise as a result of the separation of $C_5H_{11}O$ fragment (87 amu) from M^+ with a C_{22} gap–C_{23}, in other works it was indicated that this shred is peculiar only to 20-hydroxyecdysteroids (Koreeda et al. 1969).

The difficulty in detecting this type of ions lies in the fact that $(M\text{-}87)^+$ ions can also appear as a result of the removal of $(4H_2O+CH_3)$, which is characteristic of any polyoxysteroids. Measurements of the elemental composition of ions in this spectral region indicated that the cleavage of the C_{22}–C_{23} bond is a fairly common phenomenon during the decay of ecdysteroids (Rashkes and Abubakirov 1980).

The peaks of ions $(M\text{-}87)^+$ in the spectra of 2-deoxy-α-ecdysone (22), silenosterone (21), and α-ecdysone (1) are completely due to the elimination of the $C_5H_{11}O$ fragment (Fig. 2.2). Ions $(M\text{-}105)^+$ appearing in the spectra of these compounds are formed along the same path, but with the loss of an additional water molecule. The distinct signs of decomposition at the C_{22}–C_{23} bond are exhibited by the compounds 22-deoxy-α-ecdysone, 3.22-O-diacetate-2-deoxy-α-ecdysone, 22-O-monoacetate of silenosterone, and 3.22-Premixisterone O-diacetate, but with simultaneous removal of the ketene molecule due to the 22-acetoxy group (Fig. 2.6).

In diacetate of 2-deoxyecdysterone (23), ecdysterone triacetate (Fig. 2.7), and integristerone pentaacetate A, the C_{22}–C_{23} bond cleavage follows the removal of acetic acid molecules—substituents at C_{20} and C_{22} (Rashkes and Abubakirov 1980).

In ecdysteroids—2-deoxy-α-ecdysone (22), silenosterone (21), and α-ecdysone (1)—peaks of even-mass and low-intensity ions are detected that occur by removing one water molecule and a molecule of tert-butyl alcohol.

In contrast to the spectra of free ecdysones or their acetates, the spectra of acetonides are characterized, which is very important, by the presence of a peak in the molecular ion. In addition, a number of intense peaks above m/z 300 and a large number of metastable peaks facilitate the interpretation of the spectrum. The

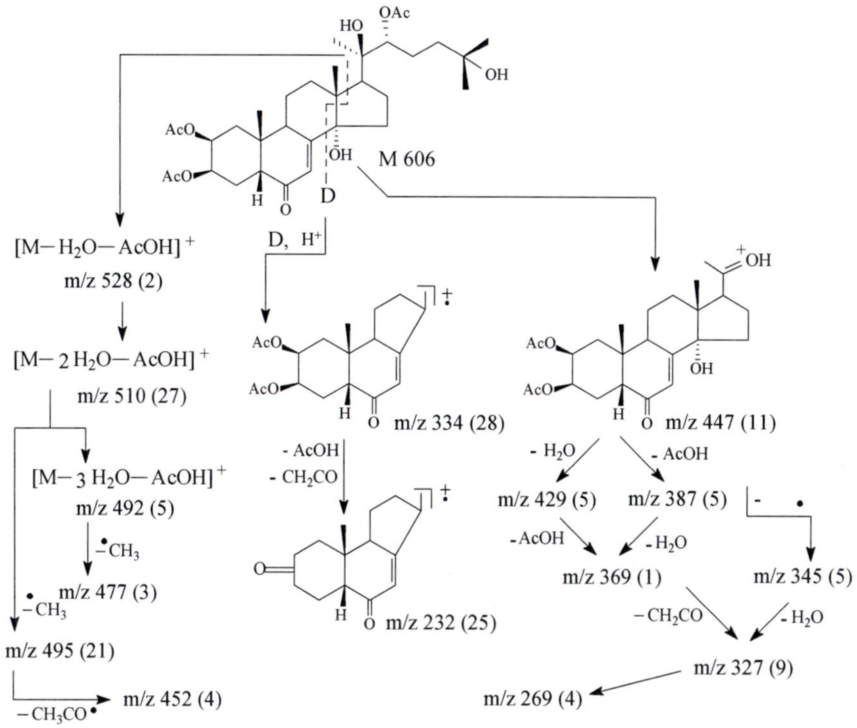

Fig. 2.6 Fragmentation of ecdysteroid acetates (for example, ecdysterone triacetate)

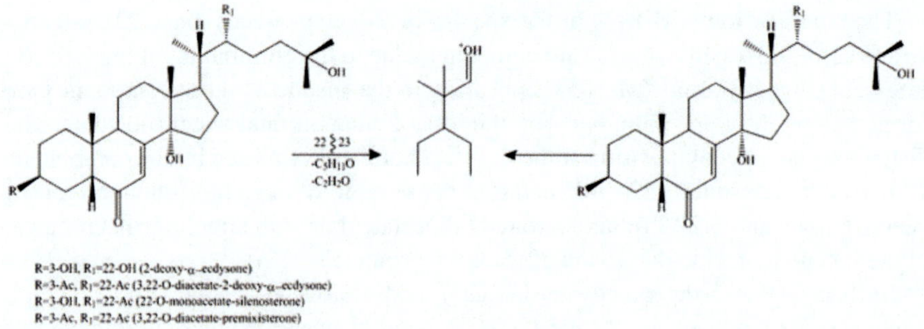

Fig. 2.7 The decay of ecdysteroids in the bonds C_{22} and C_{23}

main directions of acetonide fragmentation are seen from (Fig. 2.8), which shows the main types of ruptures of ponasterone A diacetonide (Koreeda et al. 1969).

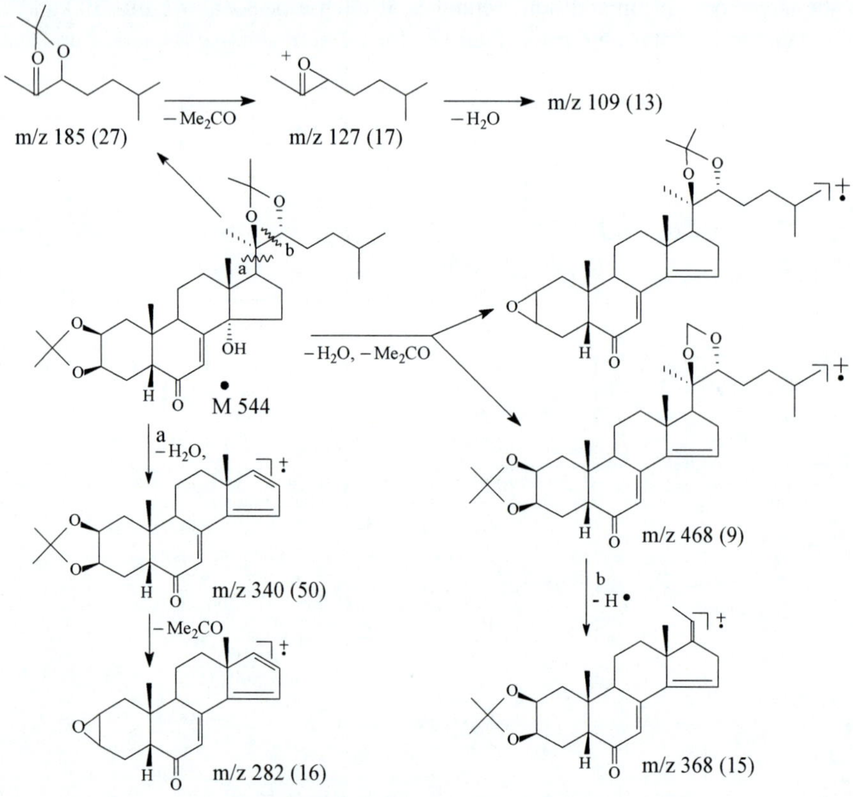

Fig. 2.8 Schematic diagram of fragmentation 2,3,20,22-O-diacetonide-ponasterone A

Fig. 2.9 The structure of ecdysteroids silenosides A, B, C, and D

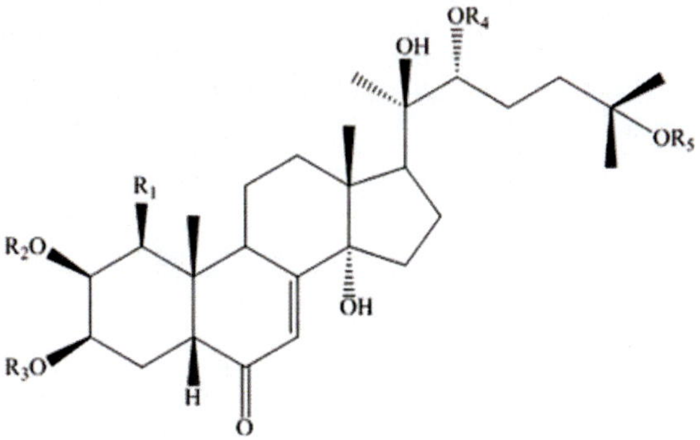

R_1=R_2=R_3=R_5=H; R_4=α-D-Gal (Silenoside A) (24)
R_1=R_2=R_5=H; R_3=R_4=α-D-Gal (Silenoside B) (25)
R_1=OH; R_2=R_3=R_5=H; R_4=α-D-Gal (Silenoside C) (26)
R_1=R_2=R_4=R_5=H; R_3=α-D-Gal (Silenoside D) (27)

It should be noted that the most intense ions, in this case, are formed due to the cleavage of the C_{17}–C_{20} bond.

At present, the spectra of glycoecdysteroids obtained using mild ionization methods are well studied. In the spectra of 3-β-D-glucopyranoside 2,20-dideoxy-ecdysterone, α- and 5β-hydroxyecdysteroids obtained by plasma desorption with California 252, peaks of ions $(M^+ Na)^+$, $(M^+ H^+Na)$ were found + $(M^+ 2Na)^+$, $(M^+ H^+ 2Na)^+$ (Rashkes and Abubakirov 1980). In the desorption chemical ionization spectra of 3-β-D-glucopyranoside-2,20-dideoxyecdysterone and its peracetate, there are peaks of ions $(M^+ H)^+$, or $(M^+ H–H_2O)^+$, as well as ions $(M–Glc)^+$ (Kibmer and Wichtl 1987; Suksamravn et al. 1986).

The LSI–MS spectra of selenosides A–D (24, 25, 26, 27) (Fig. 2.9) were obtained using both a pure glycerol matrix and glycerol supplemented with NaCI. Mass numbers of ion peaks in the spectra of silenosides A–D (98, 99, 100, 101) are given in Table 2.2. The spectra obtained using the glycerol matrix are of the same type, with the exception of some features due to structural differences. They contain peaks of ions $(M^+ H)^+$ and products of sequential cleavage of two water molecules, peaks of ions $(M–GalO)^+$, as well as products of their dehydration. The spectra of silenosides B (25) and D (27), whose molecules contain a carbohydrate unit at C-3, are also characterized by the presence of ions $(M–Gal^+ 2H)^+$. In the spectrum of sileneoside B (25), cleavage of $(Gal^+ 2H)$, (GalO), and $(GalO^+ H_2O)$ is accompanied by the removal of the second carbohydrate unit in the form of a GalOH molecule (Table 2.2).

The LSI–MS spectra of sileneosides in the glycerol matrix with the addition of NaCl differ in the presence of peaks of ions $(M^+ Na)^+$ and $(M^+ 2Na–H)^+$, as well as the dehydration products of the first of them. All spectra contain doublets with a height ratio corresponding to the ^{35}Cl and ^{37}Cl isotopes, and according to mass numbers, ions of the composition $(M^+ Na^+ NaCl)^+$ (Koreeda et al. 1969).

Additional valuable information can be obtained by studying the mass spectra of ecdysone derivatives—acetates and acetonides (Rashkes and Abubakirov 1980).

Table 2.2 Mass spectra of characteristic ion peaks of sileneoside A (24), B (25), C (26), and D (27)

Compounds	Matrix-glycerin							Matrix-glycerin + NaCl			
	(M +H$^+$)	(M+H–H_2O)$^+$	(M+2H–H_2O)$^+$	(M–Gal+2H$^+$)	(M–GalO)$^+$	(M–+GalO+2H_2O)$^+$	(M–GalO–2H_2O)+	(M+Na)$^+$	(M+Na–H_2O)$^+$	(M+2Na–H)$^+$	(M+Na+Cl)$^+$
98	643	625	607	–	463	445	427	665	647	687	723/725
99	805	787	769	643	625	607	589	827	809	849	885/887
100	659	641	623	–	479	461	443	681	663	703	739/741
101	643	625	607	481	463	445	427	665	647	687	723/725

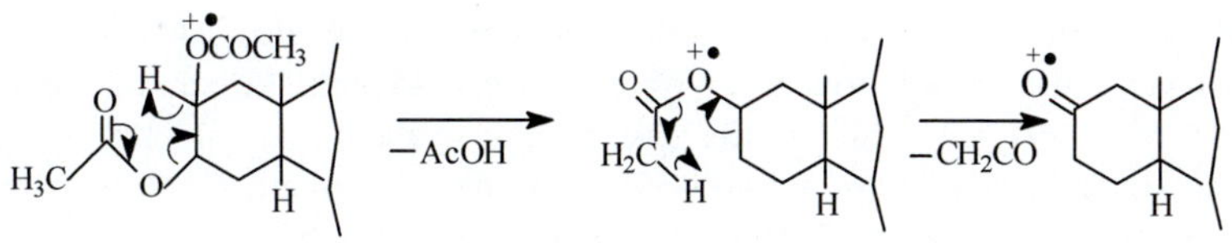

Fig. 2.10 An alternative process involves the formation of keto group at C-3

The high volatility of acetates during mass spectra recording compared with free ecdysones makes them more suitable for mass spectrometric studies. Although the molecular ion (M^+) is also usually absent, there is a characteristic series of weak peaks corresponding to the elimination of water, acetic acid, methyl radical, and ketene (Fig. 2.10). The largest mass number has the fragment (M–H_2O–AcOH) + (M-78)+. Probably, the initial release of acetic acid and ketene proceeds according to the following Fig. 2.10 (Rashkes and Abubakirov 1980).

Ecdysone acetates (Fig. 2.10) are characterized by decomposition along the C-17–C-20 bonds; C-20–C-22 and, if there is a substitute at C-24, then on the connection C-23–C-24*. The decay of ecdysone acetate molecules is accompanied by the splitting of the rings of the tetracyclic system. Intense ion peaks with m/z 334 and 232 are formed upon cleavage of the rings C and D. It is interesting to note that for acetates of ajugasterone C-20 (Mil'grom et al 1991), and Turkesterone C-17 (Kibmer and Wichtl 1987), having 11-α-hydroxy group, the corresponding ion peaks are shifted by two units. They appear at m/z 332 and 230, respectively. This is obviously explained by the easy cleavage of the 11-acetyl group in the form of acetic acid.

2.2 Proton Magnetic Resonance Spectroscopy

When determining the structure of ecdysteroids, the NMR spectra are of great help, the analysis of which allows one to determine the presence and stereochemistry of structural elements present in the steroid skeleton and side chain. When studying the ^{1}H NMR spectrum of an unknown ecdysteroid, a conclusion about its structure is made primarily on the basis of signals from angular methyl groups, as well as methyl groups in the side chain. The resonance frequencies of the protons of methyl groups depend both on the nature and on the orientation of the substituents in the skeleton and side chain. It worth to note that there is characteristic differences in this by two ancestors of the ecdysteroid class: α-ecdysone (1) and ecdysterone (2).

In the ^{1}H NMR spectrum of α-ecdysone (1) in a strong field at δ 0.74 and 1.07 ppm in deuterpyridine, C-18 and C-19 signals of angular methyl groups appear. The CH_3 group present in the molecule at C-21 appears as a doublet of 1.22 ppm The next single signal at 1.34 ppm (6H) is due to two identical tertiary methyl groups at C-26 and C-27 (Kholodova 1979).

The ^{1}H NMR spectrum of ecdysterone (2) differs from the spectrum of α-ecdysone (1) in practical terms only by the position of the signals of C-18 and C-21 methyl groups. The hydroxyl group present in the ecdysterone molecule (2) at C-20 causes a shift of C-18 of the methyl group into a weak field by 0.45 ppm The paramagnetic shift also undergoes CH_3-21, which in this case manifests itself as a singlet at 1.59 ppm, while the signals of methyl groups at C-19, C-26, and C-27 practically do not undergo bias (Warren and Gilbert 1998; Galbraith et al. 1973).

It should be noted that along with free ecdysteroids, the NMR spectra of their derivatives—acetates and acetonides were studied, which provide significant additional information on the stereochemistry of substituents (Yusupova et al. 2020; Volodin 2003).

The Δ7-6-keto grouping characteristic of these compounds appears as a broadened singlet from the vinyl proton at C-7, which for all studied ecdysteroids is in the range of 6.1–6.3 ppm Evidence of this is the fact that in the ^1H NMR spectra of the heilantons A (28) and B (29) (Faux et al. 1970), which are 7.8-dihydroecdysteroids, there are no vinyl proton signals in this region. Signal CH_3-18 is shifted to a weak field by 0.09–0.11 ppm in comparison with those for α-ecdysone. In the ^1H-NMR spectrum of acetylpinnasterol, which does not have a 14α-hydroxy group, the vinyl proton signal at C-7 (6.03 ppm) appears as a triplet due to allylic interaction with methine protons at C-9 and C-14 (Fukuzawa et al. 1981).

One of the characteristic structural elements of ecdysteroids is the 2,3-diol system.

In the ^1H NMR spectrum, carbinol protons of the glycol moiety appear as multiplet signals at 4.0–4.2 ppm. In the analysis of free alcohols, the resonance lines under consideration are in most cases superimposed on each other and on the signals of other carbinol protons. It is carried out by analysis of the ^1H NMR spectra of their acetates and acetonides (Akhrem and Kovganko 1989; Yusupova et al. 2020).

The use of the intramolecular nuclear effect of Overhauser (Akhrem et al. 1973) provides very useful information for elucidating stereochemistry at C-2, C-3, and C-5. Additional irradiation of the signal from 9α-H ecdysteroids in the ^1H NMR spectrum of 2,3-diacetate-20,22-acetonide of ponasterone A (Akhrem and Kovganko 1989), causes a 10% increase in the integral signal intensity from 2α-H (Fig. 2.11).

The signal intensity from 3α-H does not change.

This effect is observed due to the close spatial arrangement of protons at C-2 and C-9 and is possible only in the case of cis-jointing of the A/B rings and 2β-, 3β-orientation of the glycol group. The Overhauser effect has been used to elucidate

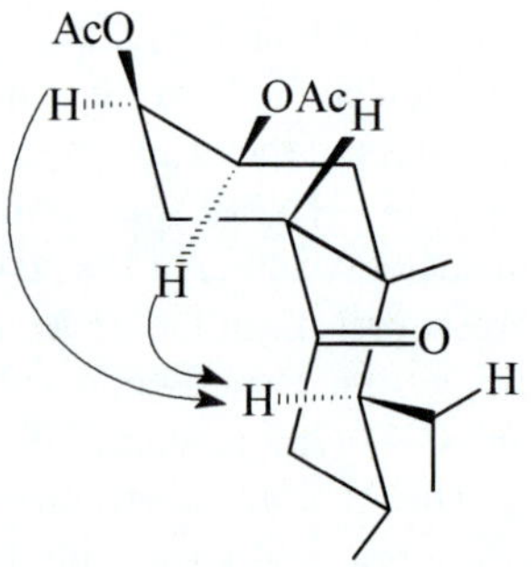

Fig. 2.11 2,3-diol system of ecdysteroids

the stereochemistry of a number of other ecdysteroids (Koreeda et al. 1970; Koreeda et al. 1969; Cherbes et al. 1982).

A change in the stereochemistry (α- or β-configuration) of the 2,3-glycol group and the joint of the A/B rings is reflected in the chemical shift of the protons of the angular C-19 methyl group.

In ponasterone B (4) containing a 2α, 3α-glycol group, the C-19 signal of the methyl group is shifted to a weaker field by 0.08 ppm compared with that for ponasterone A (3) (Nakanishi et al. 1966). trans-articulation of A/B rings in 28-epicycasterone (6) (Okuzumi et al. 2005), 5α-ecdysterone (Ogawa et al. 1971), and 22-O-benzoate-5α-ecdysterone (30) (Saatov et al. 1987), causes a noticeable paramagnetic shift (0.34–0.35 ppm) of the C-19 signal of the methyl group.

Ecdysteroids having a 5β-OH group detect a shift of the C-19 signal of the methyl group to a weak field by 0.05–0.07 ppm (Akhrem et al. 1973).

The presence of a hydroxyl group at C-11 in 22-O-benzoate-viticosterone E (31) (Saatov et al. 1988), and turkesterone (32) (Usmanov et al. 1978), noticeably affects the signal shift of C- protons 19 methyl group, de-screening it at 0.20–023 ppm.

The lateral chains of ecdysteroids with respect to structural elements are more diverse than elements of the steroid skeleton. They differ in the presence of alkyl substituents, double bonds, lactone rings, the number and different arrangement of hydroxyl groups, and some other structural units. These features characteristically manifest themselves in the ^{1}H NMR spectra and are described in detail in monographs (Akhrem et al. 1973; Akhrem and Kovganko 1989; Lozano et al. 1988).

2.3 Carbon Magnetic Resonance Spectroscopy

At present, along with ^{1}H NMR spectroscopy, ^{13}C NMR spectroscopy is very widely used to elucidate various structural problems of ecdysteroids (Wells et al. 2018).

Recently, increasing attention has been paid to the use of two-dimensional (2D) ^{1}H NMR correlation methods. In the most complex cases, ^{13}C NMR studies are required. This method was first applied in 1972 for α-ecdysone (1). While most protons in the spectra are in the range of 3 ppm, the carbon atoms in the ^{13}C NMR spectra are located in the region of up to 100 ppm, with the carbonyl and olefin carbon atoms being even further in the low-field region (up to 220 m.d.). It is obvious that ^{13}C NMR spectra have better resolution and are easier to interpret, when dealing with a structural core of the molecule. In addition, it is possible to identify atoms not related to protons, as well as to distinguish methyl, methylene, methine, and quaternary carbon atoms by DEPT methods.

The advantage of ^{13}C NMR spectroscopy is that this method allows direct observation of C atoms of the core/skeleton, carbon-containing functional groups, and proton-free carbon atoms.

In many cases, the minimum spectral data of high-quality one-dimensional NMR, including signals of both methyl groups and a number of characteristic protons, is sufficient to establish the structure of the ecdysteroid, provided that these data are correlated with other independent information. However, two-dimensional ^{1}H NMR correlation methods are of great importance for those purposes when the number of samples is significantly limited, and the ^{13}C study is not feasible. For the reliable establishment of the structure, it is necessary to have a network of neighboring protons identified by the COESY method.

^{13}C NMR spectroscopy has greater potential for the study of ecdysteroids with a complex structure than ^{1}H NMR spectroscopy.

Stereochemical problems of the side chain are more complex and require additional information for their resolution. This information can be obtained using the Overhauser nuclear effect (NOE) technique, in experiments with the Overhauser difference nuclear effect, NOE in two-dimensional spectroscopy (NOESY), as well as indirect methods.

To remove the ^{13}C NMR spectrum of substances of plant origin, the required amount of substance (4–20 mg) is available. The task of obtaining a sufficient amount of material from insects is much more difficult, although there are examples of ^{13}C NMR of insect ecdysteroids. More recently, improved ^{13}C NMR methods have been developed in which information on C nuclei is obtained from signals of more sensitive H nuclei in inversion correlation experiments (Galyautdinov 2000).

Modern ^{13}C NMR spectroscopy makes it possible to successfully investigate all the complex problems of the chemistry of ecdysteroids and their glycosides due to the appearance of numerous varieties of 2D methods.

In the molecule of 25-O-acetate-22-O-α-D-galactopyranosyl integristerone A (sileneoside H) (33), the location of the sugar residue was established on the basis of the ^{13}C NMR spectrum. The carbon signals of the steroid skeleton 25-O-acetate-22-O-α-D-galactopyranosyl integristerone A (33) were identical to those of integristerone A, except for the side chain signals. The location of the acetyl group at C-25 was confirmed by a shift in the signals of the methyl groups at C-26 and C-27 (by +0.2 ppm for protons and +1.8–2.6 ppm for carbons) compared to integristerone A (Table 2.3).

The presence of a hexose residue was proven by ^{13}C NMR spectra, where six additional oxygenated carbons were observed in the region of δ 61.8 δ 102.9 ppm.

The C-22 signal (δ 90.6) was shifted to a weak field (by +12.3 ppm compared with integristerone A) and this suggested the location of the hexose residue at C-22, which was finally confirmed by ^{1}H-^{13}C correlations H-1′ → C-22 and H-22 → C-1′ observed in the PFG-HMBC spectrum of 25-acetate-22-O-α-D-galactopyranosyl integristerone A (33).

The identity of the sugar residue with α-D-galactopyranoside was determined from the characteristic signal of the anomeric proton at δ 5.10 (d, J = 3.3 Hz) (Jones et al. 1993), and from the chemical shifts C-2′, C-3′, C-4′ and C-5′ in the ^{13}C NMR spectrum (Schneider et al. 2003).

A phase-sensitive NOESY experiment performed comparatively for integristerone A and 25-O-acetate-22-O-α-D-galactopyranosyl-integristerone A (33) confirmed the 5α configuration of the hydrogen atom due to the presence of strong NOE between H-5 and Me-19 observed for both compounds and between H-2 and H-9 for integristone A. The conclusion about the 1β-OH configuration was concluded from the average

Table 2.3 ^{13}C data spectra of 25-O-acetate-22-O-α-D-galactopyranosyl integristerone A (104) (D_2O)

No	Carbon	δ_C, p.pm	No	Carbon	δ_C, p.pm
1	CH	75.6	21	CH_3	21.8
2	CH	68.2	22	CH	90.6
3	CH	70.0	23	CH_2	26.4
4	CH_2	32.7	24	CH_2	38.5
5	CH	46.5	25	C	85.9
6	C	–	26	CH_3	26.5
7	CH	122.1	27	CH_3	26.7
8	C	–		AcO–CH_3	23.1
9	CH	35.4		CO	174.9
10	C	43.2		1′	102.9
11	CH_2	21.3		2′	70.9
12	CH_2	31.9		3′	70.1
13	C	48.4		4′	70.3
14	C	86.0		5′	72.2
15	CH_2	31.5		6′	61.8
16	CH_2	21.2			
17	CH	50.1			
18	CH_3	18.2			
19	CH_3	20.0			
20	C	80.6			

NOE observed between H-1 and H-9 and the strong NOE between H-1 and H-11 (1.80 ppm) for both compounds.

For 25-O-acetate-22-O-α-D-galactopyranosyl integristerone A (33), the position of hexose at C-22 was determined from the average NOE between H-1′ and H-22′ and the location of the acetyl group at C-25 is confirmed from NOE between MeCO and Me-27 and Me-26 (Bakrim et al. 2014).

In the ^{13}C NMR spectrum, sileneoside G (34), the signal of the carbon atom C-22 is at 90, 96 ppm and compared with a similar signal in the spectrum of ecdysterone at 77.5 ppm shifted to a weak field by 13.46 ppm (Table 2.4).

In the literature (Saatov et al. 1984), it was shown that in Sileneoside D (27), the galactose residue is attached to the hydroxyl group at C-3 ecdysterone by the α-glycosidic bond. Ecdysterone at C-2 77.5 ppm shifted to a weak field by 13.46 ppm.

The signal of the anomeric proton of D-glucose in the ^1H NMR spectrum resonates in the form of a doublet with coupling constants of $^3J = 4.03$ Hz at 5.69 ppm, which indicates the $^4C_1$—configuration of the glycosidic bond and the $^4C_1$ conformation of the monosaccharide in pyranose form.

Table 2.4 The values of chemical shifts of ^{13}C NMR signals of sileneoside G (34) and ecdysterone (2) (C_5D_5N, δ, ppm TMS)

Atom C	Compounds		Atom C	Compounds	
	34	**2**		**34**	**2**
1	39.36	38.8	21	22.28	21.5
2	68.05	68.0	22	90.96	77.5
3	79.12	68.0	23	27.11	27.2
4	31.61	32.3	24	41.75	42.4
5	51.41	51.1	25	69.72	69.7
6	203.01	203.2	26	29.73	29.9
7	121.46	121.6	27	29.73	29.9
8	166.54	165.7		3-O-α-D-galactose	
9	34.22	34.5	1	103.68	
10	38.50	35.5	2	70.93	
11	21.01	21.0	3	71.75	
12	31.97	31.6	4	71.15	
13	48.05	48.1	5	72.44	
14	84.15	84.2	6	62.59	
15	31.72	32.0		22-O-α-D-glucosa	
16	21.42	21.6	1	104.04	
17	49.85	50.1	2	73.44	
18	18.05	17.6	3	79.10	
19	24.24	24.3	4	71.19	
20	77.81	76.9	5	78.30	
			6	62.74	

Thus, Silenoside G (34) is 3,22-di-O-(α-D-galactopyranoside)-ecdysterone (Sadykov and Saatov 1998).

References

Akhrem AA, Kovganko NV (1989) Ecdysteroids: chemistry and biological activity [in Russian]. Nauka i Tekhnika, Minsk, p 325

Akhrem AA, Levina IS, Titov YuA (1973) Ecdysones are insect steroid hormones, p 232

Bakrim A, Ngunjiri J, Crouzet S, Guibout L, Balducci Ch, Girault JP, Lafont R (2014) Ecdysteroid profiles of two *Ajuga* species, *A. iva* and *A. remota.* Nad Prod Commun 9(8):1069–1074. https://doi.org/10.1177/1934578X1400900804

Burns BG, Gildan MW (1977) The isolation and identification of makisterone A from the yew *Taxus cuspidate.* Can J Chem 55:1129–1134. https://doi.org/10.1139/v77-159

Cherbes P, Trainor DA, Stanard RJ, Nakanishi K (1982) 14-Deoxymuristerone, a compound exhibiting exceptional molding hormonal activity. J Chem Soc Chem Commun 1307–1308. https://doi.org/10.1039/C39820001307

Danili B, Ferrari G, Krepinsky J et al (1974) The determination of absolute configuration at C-24 of the phytoecdysone makisterone A. J Chem Soc Chem Commun 745–746. https://doi.org/10.1039/C39740000745

Faux A, Galbraith MN, Horn DHS, Middleton EJ (1970) The structure of two ecdysones analogues, cheilanthones A and B, from the fern *Cheilanthes tenifolia*. J Chem Soc Chem Commun 243–244. https://doi.org/10.1039/C29700000243

Fukuzawa A, Kumagai Y, Masamumo T et al (1981) Acetyl-pinnasterol and pinnasterol, ecdysone-like metabolites from the marine red algae laurencia pinnata yamada. Tetrahedron Lett 22:4085–4086. https://doi.org/10.1016/S0040-4039(01)82071-2

Galbraith MN, Horn DHS, Middleton EJ et al (1973) Structure of podecdysone C, a steroid with moulting hormone activity from the bark of *Podocarpus elatus* R. Br Experientia 29:782. https://doi.org/10.1007/BF01946280

Galyautdinov IV (2000). Phytoecdysteroids *Serratuala coronata* and the transformation of 20-hydroxyecdysone and its acetonides under the action of trifluoroacetic anhydride. Dis Cand Chem Sciences. Russia, p 81

Imai S, Fujioka S, Murotata E et al (1968) The strustures of three additional phytoecdysons from *Podocarpus macrophyllus*, makisterone B, C and D. Tetrahedron Lett 3887–3890. https://doi.org/10.1016/S0040-4039(00)72357-4

Johnson AR, Carlson EE (2015) Collision-induced dissociation mass spectrometry: a powerful tool for natural product structure elucidation. Anal Chem 87(21):10668–10678. https://doi.org/10.1021/acs.analchem.5b01543

Jones C, Mulloy B, Thomas AH (1993) Methods in molecular biology. Spectrosc Methods Anal (Humana Press, Totowa) 17(7):366

Kaplanis JN, Thompson NJ, Yamamoto RT et al (1966) Ecdysterones from the pupa of the Tobacco hornworm Manduca Sexta (Jomannson). Steroids 8:605–623. https://doi.org/10.1016/0039-128X(66)90002-X

Kholodova IuD (1979) Fitoekdizony-biologicheski aktivnye poligidroksilirovannye steriny [Phytoecdysones--biologically active polyhydroxylated sterols]. Ukr Biokhim Zh 51(5):560–75. Russian. PMID: 390803

Kibmer B, Wichtl M (1987) Ecdysone aus Wurzeln und Samen von Helleborns-Arten. Ach Pharm. 320:541–546

Koreeda M, Nakanishi K, Goto M (1970) Ajugalactone, an insect molding inhibitor as tested by the Chile dipping method. J Am Chem Soc 92(25):7512–7513. https://doi.org/10.1021/ja00728a068

Koreeda M, Narada N, Nakanishi K (1969) Interaction between non-adjacent benzoate groups: an extension of the Dibenzoate cherally rule. J Chem Soc Chem Commun 548–549. https://doi.org/10.1039/C29690000548

Lozano R, Thompson MJ, Lusby WR, Svoboda JA (1988) Metabolism of Cholesterol in Manduca sexta pupae: Isolation and identification of sterol sulfates, free ecdysteroids, and ecdysteroid acids. Insect Biochem Physiol 7(4):249–266. https://doi.org/10.1002/arch.940070404

Martin SA, Costello CE, Biamann K (1982) Optimization of experimental procedures of fast atom bombardment mass spectrometry. Anal Chem 54(13):2363–2368. https://doi.org/10.1021/ac00250a050

Mil'grom YM, Sadovskaya VL, Rashkes YV et al (1991) Secondary-ion mass spectrometry of lycotetraosides of the spirostan and furostan series. Chem Nat Compd 27:45–49. https://doi.org/10.1007/BF00629829

Nakanishi K, Koreeda M, Sasaki S, Chang ML, Hsu HY (1966) Insect hormones. The structure of ponasterone A, insect-moulting hormone from the leaves of *Podocarpus nakaii* Hay. Chem Commun 24:915–917. https://doi.org/10.1039/C19660000915

Nakanishi K (1971) The Ecdysones. Pure and Appl Chem 25:256–312. https://doi.org/10.1351/pac197125010167

Ogawa S, Nishimato N, Okamoto N et al (1971) Studies on the constituents of Achyranthis radix. VIII. The insectmoulting substances in *Achyranthes* genus. J Pharm Soc Jpn 91(Suppl 2):916–920. https://doi.org/10.1248/yakushi1947.88.10_1293

Ohnishi E, Mizuno T, Chatani F et al (1977) 2-Deoxy-α-ecdysone from ovaries and egges of the silkworm. Bombyx Mori. Scince 17:66–67. https://doi.org/10.1126/science.867050

Okuzumi K, Hara N, Uekusa H, Fujimoto Y (2005) Structure elucidation of cyasterone stereisomers isolated form *Cyathula officinalis*. J Org Biomol Chem 3(7):1227–1232. https://doi.org/10.1039/b416868b

Rashkes YV, Gorovits MB, Makarichev GK et al (1971) Some questions of the mass spectrometry of the cardenolides. Chem Nat Compd 7:724–732. https://doi.org/10.1007/BF00567925

Rashkes YV, Abubakirov NK (1980) Characteristics of the mass spectra of ecdysteroids with different numbers of -OR groups. Chem Nat Compd 16:372–381. https://doi.org/10.1007/BF00568371

Saatov Z, Abdullaev ND, Gorovits MB et al (1984) Phytoecdysteroids of plants of the genus *Silene*. VII. Sileneoside D—ecdysterone 3-O-α-D-galactopyranoside from *Silene brahuica*. Chem Nat Compd 20:700–703. https://doi.org/10.1007/BF00580027

Saatov Z, Gorovits MB, Abdullaev ND et al (1987) Phytoecdysteroids of plants of the genus *Silene* XII. 5α-Ecdysterone 22-O-benzoate from *Silene scabrifolia*. Chem Nat Compd 23:563–565. https://doi.org/10.1007/BF00598673

Saatov Z, Gorovits MB, Abubakirov NK (1988) Phytoecdysteroids of plants of the genus *Silene*. XVI. Viticosterone E 22-O-benzoate from *Silene wallichiana*. Chem Nat Compd 24:463–465. https://doi.org/10.1007/BF00598532

Sadykov ZT, Saatov Z (1998) Phytoecdysteroids of plants of the *Silene* genus. XIX. The structure of sileneoside G. Chem Nat Compd 34:602–604. https://doi.org/10.1007/BF02319284

Schneider B, Gershenzon J, Graser G et al (2003) One-dimensional ^{13}C NMR and HPLC-^{1}H NMR techniques for observing carbon-13 and deuterium labelling in biosynthetic studies. Phytochem Rev 2:31–43. https://doi.org/10.1023/B:PHYT.0000004196.73829.4e

Spiteller G (1963) Strukturuntersuchung organischer Verbindungen mit Hilfe der Massenspektrometrie. Z Anal Chem 197:1–15. https://doi.org/10.1007/BF00468721

Suksamravn A, Wilkie JS, Horn DHS (1986) Blechnosides A and B: ecdysteroid Glycosides from Blechnum minus. Phytochemistry 25:1301–1304. https://doi.org/10.1016/S0031-9422(00)81277-7

Usmanov BZ, Rashkes YV, Abubakirov NK (1978) Phytoecdysones of *Ajuga turkestanica* VI. 22-Acetylcyasterone. Chem Nat Compd 14:175–178. https://doi.org/10.1007/BF01134622

Volodin VV (2003) Phytoecdysteroids. Nauka, Moscow, p 293

Warren JT, Gilbert LI (1998) Radioimmunoassay: ecdysteroids. In: Gilbert LI, Miller TA (eds) Immunological techniques in insect biology. Springer Series in experimental entomology. Springer, New York. https://doi.org/10.1007/978-1-4612-3798-3_6

Wells RD, Bond JS, Klinman J, Masters BS (2018) Nuclear magnetic resonance. Molecular life sciences. Springer, New York. https://doi.org/10.1007/978-1-4614-1531-2_100109

Yusupova U, Usmanov D, Azamatov A, Ramazonov N, Rejepov J (2020) Phytochemical constituents and biological activities of *Dianthus helenae* Vved., growing in Uzbekistan, Nat Prod Res. https://doi.org/10.1080/14786419.2020.1862834

Chapter 3
Chemical Transformations of Phytoecdysteroids

Abstract Currently, there is great interest in chemical modifications of the derivatives of 20-hydroxyecdysone, the most common phytoecdysteroid, which is a convenient synthon for producing new biologically active compounds for the needs of medicine and the national economy. The study of the biological activity of ecdysteroids and the possibility of their practical use depends on their availability. Since the natural sources of ecdysteroids are very limited, the task of their chemical synthesis from available raw materials seems quite urgent. Given the uniqueness of the chemical structure of ecdysteroids, its successful solution is impossible without the creation of special synthetic methods that allow you to enter the desired functional groups into the molecules of the starting compounds. To this end, special studies are constantly undertaken to find new methods for the synthesis of ecdysteroids and their structural analogs.

Keywords Phytoecdysteroids · Ring A and B · Ring C and D · Chemical transformation

20-Hydroxyecdysone (2) is a polyhydroxy compound containing six hydroxyl groups: three more reactive secondary at C_2, C_{22}, and C_3 and three less reactive at C_{25}, C_{20}, and C_{14} (Akhrem and Kovganko 1989; Kovganko et al. 2003).

A study of the reactions of the chemical modification of ecdysterone made it possible to find out the reactivity of the hydroxy groups to conjugate and place them in the following series by decreasing reactivity: $C_2 > C_{22} > C_3 > C_{25} > C_{20} > C_{14}$ (Shirshova et al. 2006; Dinan 1988). Therefore, with the direct esterification of ecdysterone with a large excess of an acid chloride or anhydride of an organic acid, tri- and tetra-ethers are rather easily obtained (Bobaev et al. 2015).

In natural objects, among free ecdysteroids of both plant and animal origin, their conjugates were found, the most common of which are the monoesters of α-ecdysone (1) and 20-hydroxyecdysone (2), namely: 2-, 3-, and 22, 25-aromatic carboxylic acid acetates or 22 esters and monosaccharides (Kim et al. 2005; Calcagno et al. 1995; Ramazanov et al. 1997; Dinan 2000).

In recent years, the development of new methods of chemical transformation of the most common phytoecdysteroids, in particular 20-hydroxyecdysone (2). The

U. Y. Yusupova et al., *Phytoecdysteroids*,
https://doi.org/10.1007/978-981-16-6711-4_3

synthesis of derivatives of ecdysteroids consists in the absence of the need to introduce into the starting molecules a large number of functional groups that are already present in the starting compounds. This can significantly reduce the total number of stages of synthesis. However, due to the presence of several hydroxy groups in ecdysteroid molecules, which usually need to be protected, and the labile 14α-hydroxy-Δ7-6-keto group, only some organic reactions can be used. With insignificant chemical transformations of ecdysteroids, different reactivity of hydroxy groups is usually used, due to their unequal position and conformations (Akhrem and Kovganko 1989).

The steroid backbone of ecdysteroids includes cis-linked rings A, B and trans-linked rings C and D. In the ring A of 20-hydroxyecdysone (2), β-OH groups are present at the C-2 and C-3 atoms. Many works are devoted to transformations involving secondary hydroxyls.

3.1 Transformations in Ring A

The most reactive group in ecdysteroid molecules is the 2β-hydroxy group, through which it is possible to selectively carry out the reaction. For example, in Harmatha et al. (2002) describes the selective conversion of 20-hydroxyecdysone (2) to 2-dansiloxy-20-hydroxycdysone.

Studying the effect of the configuration of—OH groups in ring A and the articulation of rings A and B on 20-hydroxyecdysone activity, the epimers of 2α, 3α-20-hydroxyecdysone and 2α-, 3α-, 5α-20-hydroxyecdysone based on 20-hydroxyecdysone were synthesized (Fig. 3.1) (Homvisasevongsa et al. 2004).

20-Hydroxyecdysone (2) was converted to 20,22-O-acetonide-20-hydroxyecdysone, mesylation of which led to a mixture of 2-O-methyl-20,22-acetonide-20-hydroxyecdysone and 2,3-O-dimethyl 20,22-O-monoacetonide-20-hydroxyecdysone derivatives, separated by column chromatography. The reduction of 2,3-O-dimethyl-20,22-O-monoacetonide-20-hydroxyecdysone with zinc in the presence of NaI in DMF at 80 °C yielded the olefin—(20,22-O-monoacetonide) -14α, 22,25-trihydroxy -5β-cholest-2,7-dien-6-one with a yield of 75% (Fig. 3.2).

Its dihydroxylation (OsO_4–Py) followed by treatment with aqueous $NaHSO_3$ gave two compounds, one of which is 20,22-monoacetonide-20-hydroxyecdysone (58%),

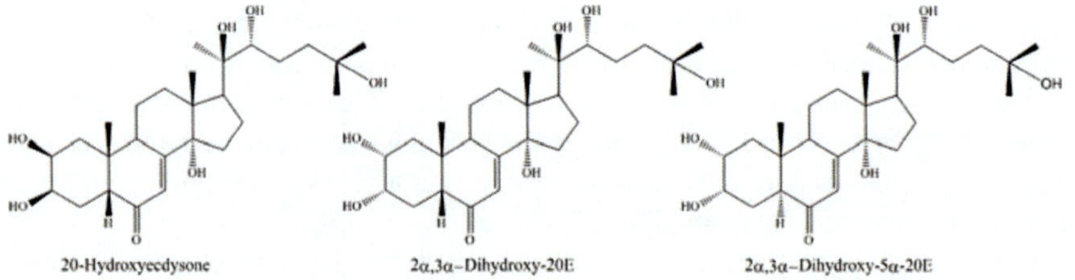

Fig. 3.1 Synthesis of 2α, 3α-20-hydroxyecdysone epimers and 2α-, 3α-, 5α-20-hydroxyecdysone

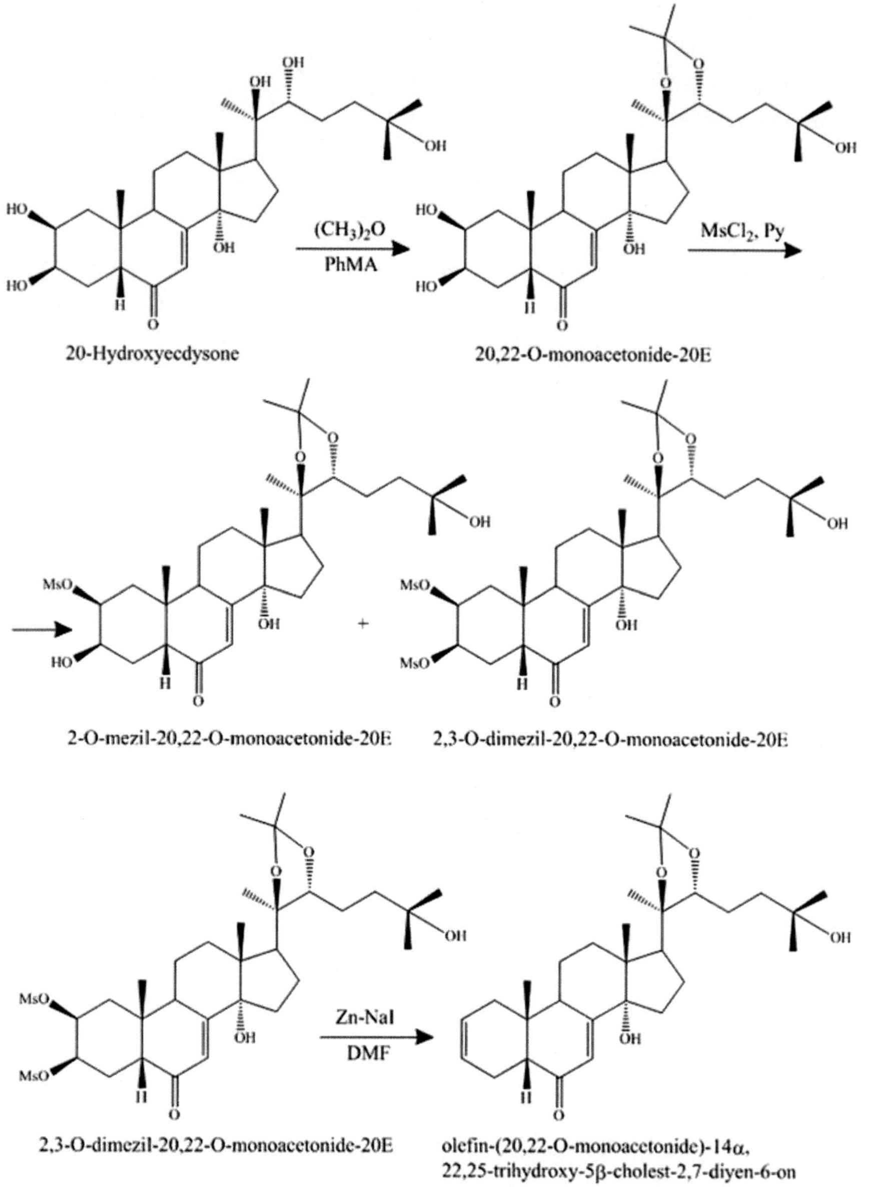

Fig. 3.2 Obtaining olefin—(20,22-O-monoacetonide) -14α, 22,25-trihydroxy-5β-cholest-2,7-dien-6-one

and the other is the 2α-, 3α-epimer of 20,22-monoacetonide-20-hydroxyecdysone (32%) (Fig. 3.3).

The maximum yield of the epimer 2α-, 3α-epimer 20,22-O-mono-acetonide-20-hydroxyecdysone (68%) is achieved by carrying out the dihydroxylation reaction in a solvent mixture of tert-butyl alcohol—tetrahydrofuran—water in the presence of a three-fold excess of chiral ligand—diester 1,4-phthalazinediyl dihydroquinidine. The advantage of the latter was obvious in comparison with other ligands—dihydroquinidine and dihydroquinine. The 2α-, 3α-epimer of 20,22-O-monoacetonide-20-hydroxyecdysone was converted to 2,3,25-O-triacetate of 2α-, 3α-epimer of 20,22-O-monoacetonide-20-hydroxy-ecdysone, in which was established 5β-H configuration based on the NOE experiment, which showed the correlation of protons $C^{19}H_3$ (δ 0.94 ppm) and H-5 (δ 2.20 ppm), as well as protons H-9 (δ 3.75 ppm) and methyl protons of the 2α-$OCOCH_3$ group (δ 1.98 ppm). Under the action of 2% Na_2CO_3 in methanol, 2α-, 3α-, 5β epimer of 20,22-monoacetonide-20-hydroxyecdysone turns into 2α-, 3α-, 5α-epimer of 20-hydroxyecdysone, in the NOE spectrum of which there is a correlation peak between proton signals $C^{19}H_3$ (δ 0.95 ppm) and H-5 (δ

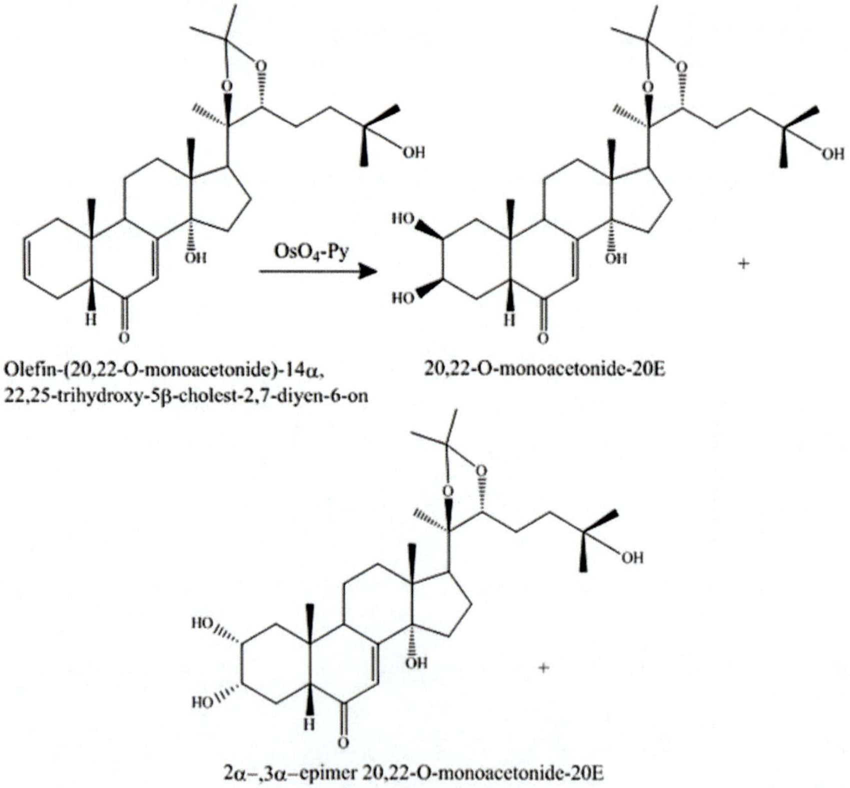

Fig. 3.3 Synthesis of 20,22-monoacetonide-20-hydroxyecdysone, -2α-, 3α-epimer of 20,22-monoacetonide-20-hydroxyecdysone

3.02 ppm) are not detected, while the correlation between the protons H-5 and H-9 (δ 3.23 ppm) is observed.

Hydrolysis of the 2α-, 3α-epimer of 20,22-O-monoacetonide-20-hydroxyecdysone and 2α-, 3α-, 5α-epimer of 20-hydroxyecdysone in 70% acetic acid in the presence of benzyltrimethylammonium chloride led to 2α-, 3α-, 5β to 20-hydroxyecdysone and 2α-, 3α-, 5α-epimer to 20-hydroxyecdysone, respectively. The total yield of the epimer—2α-, 3α-, 5α-epimer of 20-hydroxyecdysone, based on the starting compound 20-hydroxyecdysone, is 18%.

A test for Musca domestica showed that the epimer of 2α-, 3α-, 5β-20-hydroxyecdysone is 30 times less active than the starting compound 20-hydroxyecdysone, and the epimer 2α-, 3α-, 5α-20-hydroxyecdysone was 42 times smaller than the compound of 20-hydroxyecdysone, and only 1.5 times smaller than the epimer of 2α-, 3α-, 5β-20-hydroxyecdysone, although it was expected to be inactive in this test.

In a series of ecdysteroids, acetylation using chemicals (acetic anhydride, pyridine, 40 °C, reaction time from 30 min to 20 h) was studied in detail more than 40 years ago (Galbraith and Horn 1969). Mixtures of mono-, di-, and polyacetates are obtained, separated by column chromatography. A study of the relative reactivity of hydroxyl groups in the acetylation reaction showed that the 2β-hydroxy group is most rapidly acetylated, the 22-OH group is less reactive, and the 3β-OH group is acetylated more slowly (Charoensuk et al. 2000). When protecting 2-, 3-, 20-, 22-hydroxy groups, monoacetates can be obtained. Under these conditions, 20-hydroxyecdysone diacetonide turns into a 25-O-acetate derivative—viticosterone E (Zatsny et al. 1973), 22-O-acetate-20-hydroxyecdysone obtained using phenylborate and acetone protection of diol groups (Volodin 2003) (Fig. 3.4).

An increase in the excess of the acylating reagent makes it possible to increase the yield of acylation products (Akhrem and Kovganko 1989). The introduction of acyl groups into ecdysteroid molecules by enzymatic methods has not been studied. A study of the hydrolytic stability of 2,3,22,25-O-tetraacetate-20-hydroxyecdysone in the presence of the Helix complex preparation showed that all ester bonds undergo hydrolysis within 4 days. The hydrolysis of 2,3,22-O-tripalmitoyl-20-hydroxyecdysone was carried out in the presence of pancreatic lipase (on the fifth day, the amount of free 20-hydroxyecdysone was 68% of the theoretically possible).

3.2 Transformations in Ring B

A ring of the conjugated enon is present in ring B of the steroid core, the chemical properties of which have been studied relatively little until recently (Akhrem and Kovganko 1989; Silva et al. 2005).

Hydride reduction. In the work (Odinokov et al. 2005), the stereochemistry of hydride reduction of the 6-oxo group was studied. When interacting with sodium borohydride in methanol in the presence of $CeCl_3$ at −5 °C, the 6-oxo group of 2,3,20,22-O-diacetonide-20-hydroxyecdysone is stereospecifically reduced to form

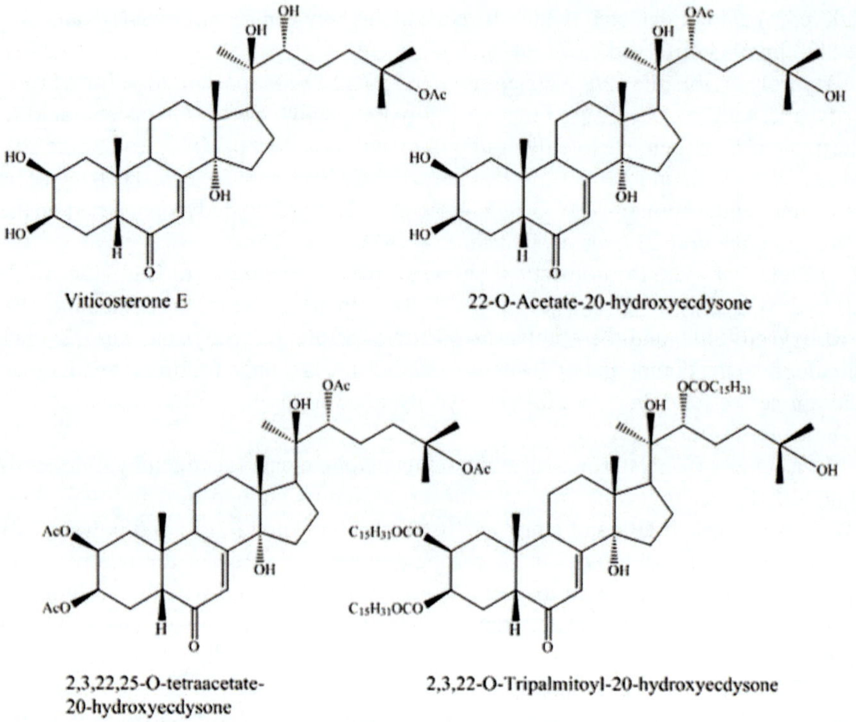

Fig. 3.4 Conversion to 25-O-acetate derivative-viticosterone E

the corresponding 6α-alcohol. In this case, dehydration occurs according to the 14α-OH group, and the reaction product is 14,15-anhydro-6α-hydroxy-2,3,20,22-O-diacetonide-20-hydroxyecdysone (75%). The results are probably due to the formation on the α-side of the intermediate intramolecular complex of cerium with the participation of Δ14 and Δ7 bonds and the 25-OH group. The hydride attack on the carbonyl group in this case proceeds from the β-side with the formation of 6α-alcohol (Fig. 3.5).

In the absence of $CeCl_3$, the reaction proceeds non-stereospecifically, and a hydride attack from the spatially less difficult α-side leads to a mixture of 6α and 6β-epimeric alcohols 2,3,20,22-O-diacetonide-20-hydroxyecdysone in a 1:2 ratio, the yield of which after separation by column chromatography, the obtained product is 46%.

The need for the presence of two double bonds for stereo-specific reduction with the reagent $NaBH_4$-$CeCl_3$ is confirmed by the reaction with diacetonides Δ24.25/Δ25.26-anhydro derivatives of 2,3,20,22-O-diacetonide-20-hydroxy-ecdysone. In this case, dehydration does not occur with cleavage of the

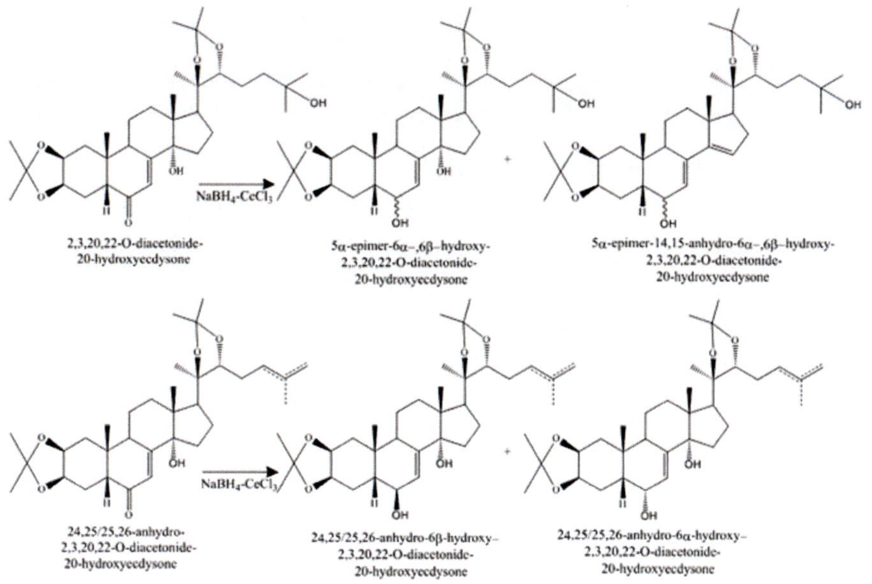

Fig. 3.5 Hydride attack on a carbonyl group

14α-OH group, and the reaction product is 6α-alcohol 2,3,20,22-O-diacetonide-20-hydroxyecdysone. In the absence of $CeCl_3$, a mixture of 6α- and 6β-alcohols of 2,3,20,22-O-diacetonide-20-hydroxyecdysone is formed with a predominance of 6β-epimer (~1:2). For 14α-trimethylsilyl-2,3,20,22-O-diacetonide-20-hydroxyecdysone with a protected group 14α-OH, the reagent $NaBH_4$–$CeCl_3$ acts non-stereospecifically and an equimolar mixture of 6α- and 6β-alcohols 14α-trimethylsilyl-2,3, 20,22-O-diacetonide-20-hydroxyecdysone (Fig. 3.6).

Based on the data of 1D and 2D ^{1}H and ^{13}C NMR spectra, it was established that the obtained hydride reduction products belong to compounds of the 5α series with trans-coupling of rings A and B. The interactions detected between the H-5 protons and the hydroxyl group at C-6 are 5α, 6α-the epimer of 2,3,20,22-O-diacetonide-20-hydroxyecdysone and between the protons H-5 and H-6 in the 5α, 6β-epimer of 2,3,20,22-O-diacetonide-20-hydroxyecdysone, and there is no correlation between H-5 and H-19 in both isomers that indicate epimerization at the C-5 atom during hydride reduction. The position of the H-9 signal in the ^{1}H NMR spectra of hydride reduction products in the region of δ 2.3–2.5 ppm also testifies in favor of their assignment to the 5α series; for compounds of the 5β series, this signal is in a noticeably stronger field—δ 2.8–3.2 ppm (Suksamrarn and Yingyongnarongkul 1995).

In the protection of hydroxyl groups at C-14 and C-25 atoms, the reduction with the reagent $NaBH_4$–$CeCl_3$ proceeds non-stereospecifically. Upon the reduction of

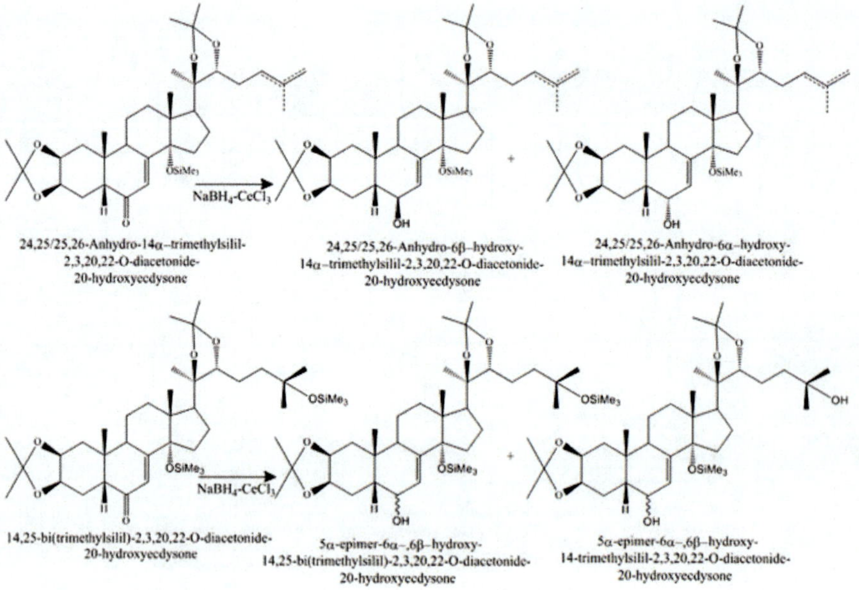

Fig. 3.6 Formation of an equimolar mixture of 6α- and 6β-alcohols 14α-trimethylsilyl-2,3,20,22-O-diacetonide-20-hydroxyecdysone

2,3,20,22-O-diacetonide-20-hydroxyecdysone and trimethylsilyl ether 2,3,20,22-O-diacetonide-20-hydroxyecdysone, a mixture of 6α and 6β was obtained -hydroxy derivatives of diacetonide-20-hydroxyecdysone (1:1).

In the reduction of the 6β-hydroxy derivatives of the 14.25-bis (trimethylsilyl) ether 2,3,20,22-O-diacetonide-20-hydroxyecdysone, the hydroxyl group is partially released at the C-25 to form 6β-hydroxy-14-trimethylsilyl ether 2,3,20,22-O-diacetonide-20-hydroxy-ecdysone. The yield of 6α/6β-hydroxy-14,25-bis (trimethyl-silyl) ester mixture of 2,3,20,22-O-diacetonide-20-hydroxyecdysone increases to 85% upon reduction of the 6β-hydroxy derivatives of 14,25 bis (trimethylsilyl) ether 2,3,20,22-O-diacetonide-20-hydroxyecdysone $LiAlH_4$ in ether, the ratio of stereoisomers being reversed (6α/6β 1:2).

As in the case of ecdysteroids (Akhrem and Kovganko 1989), the Δ7 bond in their 6β-hydroxy-14-trimethylsilyl ether 2,3,20,22-diacetonide-20-hydroxyecdysone could not be restored under the conditions of catalytic hydrogenation. During the reduction of ecdysteroids with lithium aluminum hydride in ether, along with the reduction of the keto group (1,2-reduction), the Δ7 bond (1,4-reduction) is restored (Afonkina et al. 2006; Shafikov et al. 2007).

In allyl 5α-epi-6α-hydroxy-7-ene-2,3,20,22-O-diacetonide-20-hydroxyecdysone and 5α-epi-6β-hydroxy-7-ene 2,3,20,22-, instead of reducing the Δ7 bond of O-diacetonide-20-hydroxyecdysone, the hydrogenolysis of the 14α-hydroxyl group occurs with a shift of the double bond to position 8, 14 and the formation of homoallyl alcohols -5α-epi-6α-hydroxy-8-en-14-deoxy- 2,3,20,22-O-diacetonide-20-hydroxyecdysone and 5α-epi-6β-hydroxy-8-en-14-deoxy-2,3,20,22-O-diacetonide-20-hydroxyecdysone, respectively (Odinokov et al. 2008). According to the ROESY spectrum, in 6α-alcohol 5α-epi-6α-hydroxy-8-ene-14-deoxy-2,3,20,22-O-diacetonide-20-hydroxyecdysone, the proton H-6 is axial and has a β configuration, as evidenced by the nuclear Overhauser effect (NEA) between H-6 and the H_3C^{19} group. The hydroxyl group at the C-6 atom of 5α-epi-6α-hydroxy-8-en-14-deoxy-2,3,20,22-O-diacetonide-20-hydroxyecdysone has the α configuration and is equatorial The large value of the coupling constant ($^3J = 5.6, 11$ Hz) in the spectrum of 5α-epi-6α-hydroxy-8-en-14-deoxy-2,3,20,22-O-diacetonide-20-hydroxyecdysone indicates axial-axial trans the location of the H^5 and H^6 protons in the ring B and the α-orientation of the H^5 proton, that is, the trans-junction of the rings A and B.

During the hydrogenolysis of 6β-allyl alcohol 5α-epi-6β-hydroxy-7-en-2,3,20,22-O-diacetonide-20-hydroxyecdysone, epimerization occurs at the atom, as evidenced by the presence of correlation peaks of the H^5 and H_3C^{19} protons in the spectrum ROESY 5α-epi-6β-hydroxy-8-en-14-deoxy-2,3,20,22-O-diacetononide-20-hydroxyecdysone. The absence of correlation between the H^6 proton and the H_3C^{19} protons confirms the α-configuration of H^6, and the hydroxyl group at the C^6 atom in 5α-epi-6β-hydroxy-7-en-2,3,20,22-O-diacetonide-20-hydroxyecdysone has β-orientation (Fig. 3.7).

Oximation of ecdysteroids and their 7.8-dihydro analogs. Steroid oximes attract attention as potential antitumor agents (Krstić et al. 2007; Levina et al. 2008). In this regard, the oximation of ecdysteroids was studied (Galyautdinov et al. 2006), and their 7.8α-dihydroanalogs (Shafikov et al. 2009).

The reaction of 20-hydroxyecdysone with $NH_2OH \cdot HCl$ in a water–ethanol solution in the presence of NaOH yielded the 20-hydroxyecdysone E-oxime. For 2,3,20,22-O-diacetonide 20-hydroxyecdysone, poorly soluble in an aqueous-alcoholic medium, the oximation was carried out in pyridine and the E-oxime 2,3,20,22-O-diacetonide-20-hydroxyecdysone and 14 were obtained, 15-anhydro-Z-oxime-diacetononide-20-hydroxyecdysone. 20-Hydroxyecdysone in pyridine is oxidized to form a mixture of 14.15-anhydro-E/Z-oximes of 20-hydroxyecdysone (Galyautdinov et al. 2006) (Fig. 3.8).

7,8-Dihydro-2,3,20,22-O-diacetonide-20-hydroxyecdysone, when oximated in pyridine, is converted exclusively to the corresponding E-oxime 7,8-dihydro-2,3,20,22-O-diacetonide- 20-hydroxyecdysone. Processing the reaction mixture after oximation of 7,8-dihydro-2,3,20,22-O-diacetonide-20-hydroxy-ecdysone in pyridine with a solution of KOH in ethanol leads to the cleavage of the 14-hydroxy group to form Δ14 (15)-E-oxime 7,8-dihydro-20-hydroxyecdysone diacetonide. The catalytic hydrogenation of the latter over Raney nickel led to diacetononide 7,8α-dihydrostachysterone B, i.e., only the oxime group undergoes the transformation (Shafikov et al. 2009).

Reliable spectral criteria have been identified for assigning oxides of ecdysteroids and 7,8-dihydroecdysteroids to the Z- and E-series. In the (Z) -oxides of ecdysteroids,

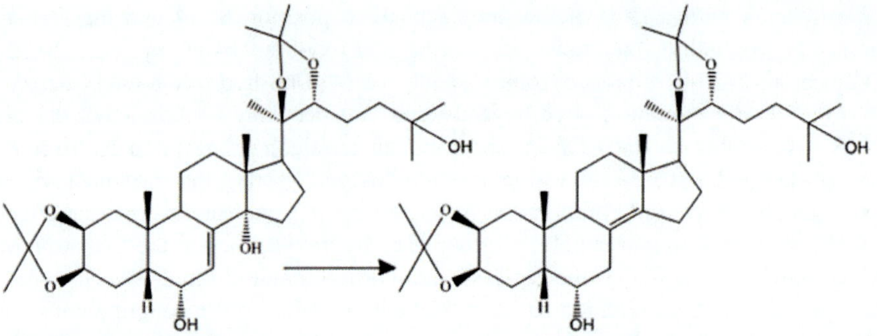

5α-Epi-6α-hydroxy-7-en-2,3,20,22-O-diacetonide-20-hydroxyecdysone

5α-Epi-6α-hydroxy-8-en-14-deoxy-2,3,20,22-O-diacetonide-20-hydroxyecdysone

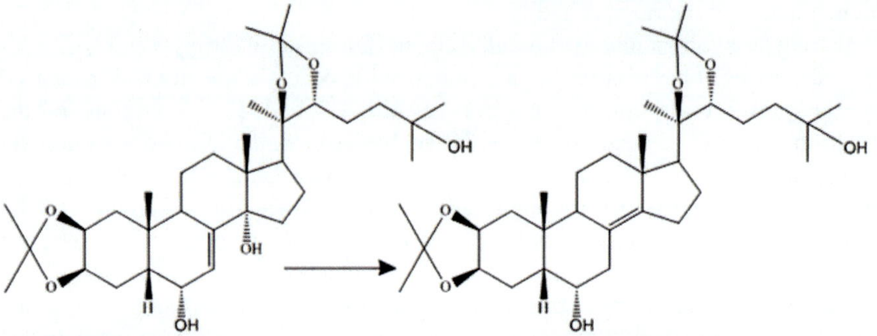

5α-Epi-6β-hydroxy-7-en-2,3,20,22-O-diacetonide-20-hydroxyecdysone

5α-Epi-6β-hydroxy-8-en-14-deoxy-2,3,20,22-O-diacetonide-20-hydroxyecdysone

Fig. 3.7 Origin epimerization at the atom

the HC_5 α-methine group is in the syn position to the OH oxime group, and its signal in the ^{13}C NMR spectra is in the region of ~36 ppm, while the signal of the HC_5 methine group, which is in the anti-position with respect to the hydroxyl of the oxime group in the (E) oximes, is found in a significantly weaker field (δ ~ 42 ppm) (Galyautdinov et al. 22006. In the ^{13}C (E)-oxime NMR spectra of 7,8-dihydro-ecdysteroids, the HC_5 methine signal is detected in the region of ~42 ppm (anti-hydroxyl arrangement of the oxime group), and the carbon atom of the methylene group H_2C_7 resonates in a stronger field (δ ~ 26 ppm) (syn-orientation to the hydroxyl group of oximes) (Shafikov et al. 2009) (Fig. 3.9).

The resulting E-oximes of 7.8-dihydro-2,3,20,22-O-diacetonide-20-hydroxyecdysone and Δ14 (15)-2,3,20,22-O-diacetonide 7,8-dihydro-20-hydroxyecdysone were involved in Beckman's rearrangement in order to convert ring B into a seven-membered lactam cycle, a structural fragment of azabrassinosteroids. It turned out that the Beckman reaction in the presence

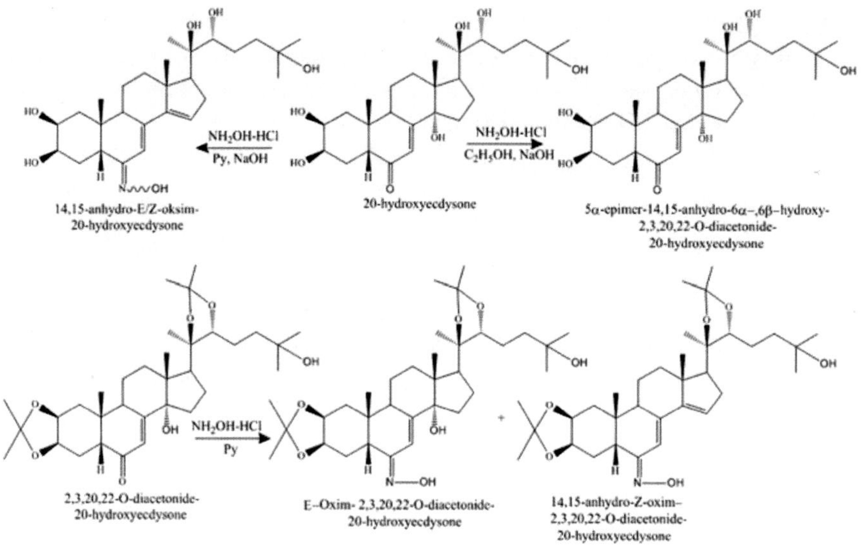

Fig. 3.8 Oximation of ecdysteroids and their 7.8-dihydro analogs

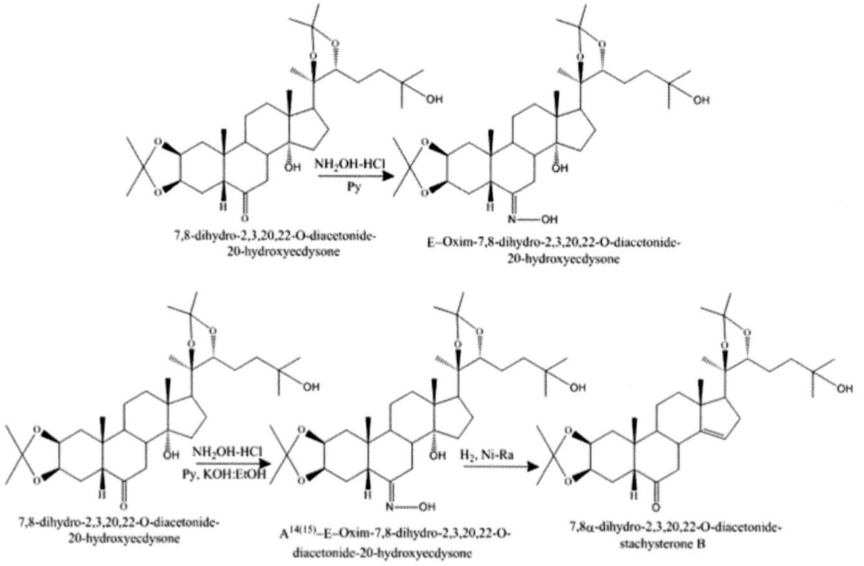

Fig. 3.9 The resulting E-oximes of 7.8-dihydro-2,3,20,22-O-diacetonide-20-hydroxyecdysone and Δ14 (15)-2,3,20,22-O-diacetonide 7,8-dihydro-20-hydroxyecdysone

of $BF_3 \cdot Et_2O$ is of little use for ecdysteroid oximes. E-oxime 2,3,20,22-O-diacetononide-20-hydroxyecdysone undergoes the release of 2,3-hydroxy groups, 14,15-dehydration, and E/Z-isomerization with the formation of the Z-oxime 14,15-anhydro-20,22-O-acetonide-20-hydroxyecdysone.

The action of P_2O_5, PCl_5, $SOCl_2$, TsCl on E-oxime diacetonide 20-hydroxyecdysone and E-oxime diacetonide 7,8-dihydro-20-hydroxyecdysone led to the formation of a mixture of compounds.

The rearrangement of oximes 2,3,20,22-O-diacetonide-20-hydroxyecdysone and 2,3,20,22-O-diacetonide-7,8-dihydro-20-hydroxyecdysone with the transformation of ring B into a seven-membered lactam cycle has performed the effects of tosyl chloride in acetone in the presence of sodium carbonate (Fig. 3.10).

The structures of the obtained lactams of 2,3,20,22-O-diacetonide-20-hydroxyecdysone and 7,8-dihydro-2,3,20,22-O-diacetonide-20-hydroxyecdysone were established based on the analysis of ^1H and ^{13}C NMR spectra (Shafikov et al. 2009).

In contrast to E-oximes, the diene Z-oxime Δ14 (15), the E-oxime 2,3,20,22-O-diacetonide-7,8-dihydro-20-hydroxyecdysone does not undergo Beckmann rearrangement, and the final product the reaction under the same conditions is (Z)-O-tosyloxime Δ14 (15)-E-oxime 20,22-monoacetonide 20-hydroxyecdysone with unlocked 2,3-hydroxy groups (Shafikov et al. 2009) (Fig. 3.11).

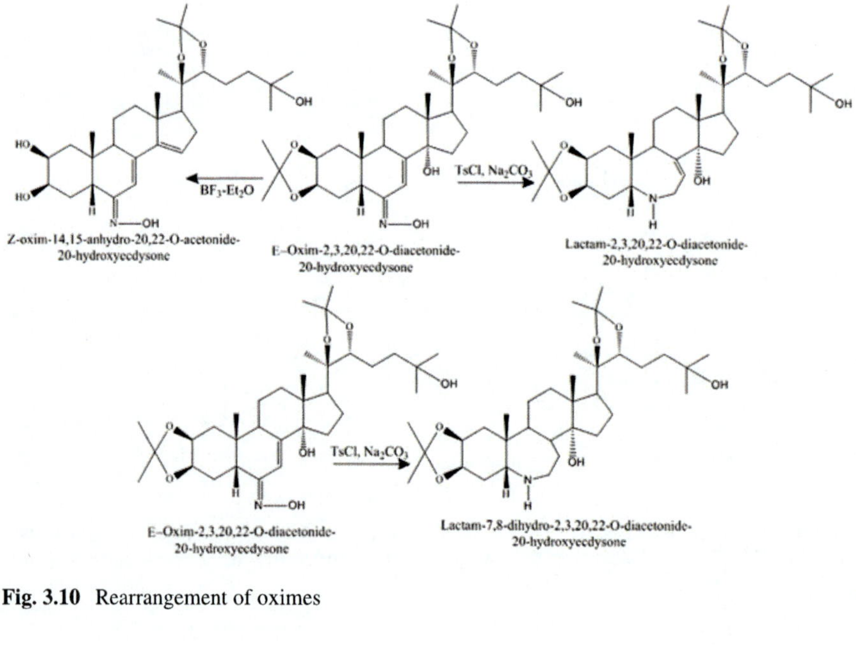

Fig. 3.10 Rearrangement of oximes

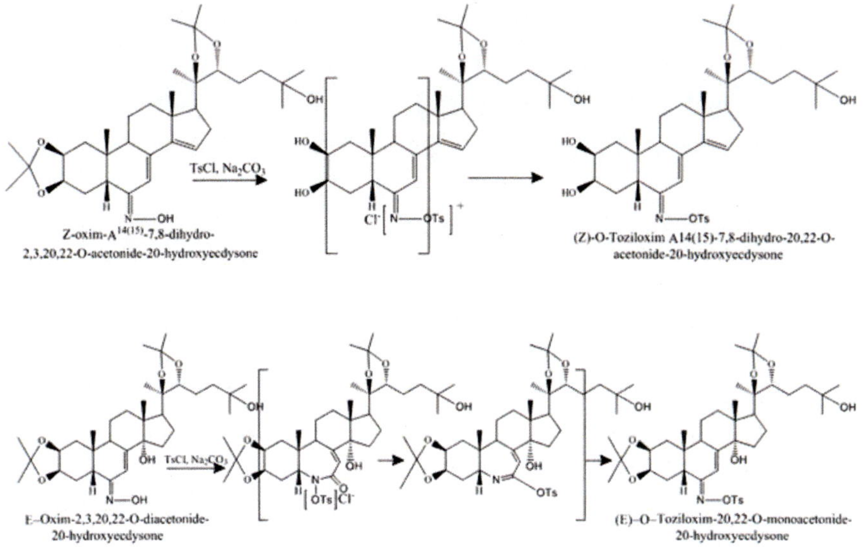

Fig. 3.11 The final reaction product

3.3 Transformations in Rings C and D

The reduction of the double bond in α, β-unsaturated ketones with alkali metals in liquid ammonia (Birch reaction) is widely used in steroid chemistry (Salvador et al. 2011). However, in a series of ecdysteroids, using this reaction, it was not possible to perform selective restoration of the Δ7 bond and to obtain the corresponding 7.8-dihydro analogs. When 20-hydroxyecdysone diacetonide is reacted with lithium in liquid ammonia, instead of the characteristic reduction of the conjugated enon double bond to the corresponding 7.8-dihydro14α-hydroperoxide 2,3,20,22-O-diacetonide-20-hydroxyecdysone is formed due to the auto-oxidation of the intermediate 6.8-dienolate-2,3,20,22-O-diacetonide-20-hydroxyecdysone (Suksamrarn et al. 1994) (Fig. 3.12).

3.4 Side-Chain Transformations

Transformations in side chains and ecdysteroids in the synthesis of inaccessible ecdysteroids in monographs and reviews (Volodin 2003; Akhrem and Kovganko 1989; Ves'kina and Odinokov 2012; Kovganko et al. 2003). Later, new data were obtained on the synthesis of viticosterone E (Galyautdinov et al. 2004). The problem of the synthesis of this ecdysteroid is the allocation of acetylation of a 25-hydroxy group in 20-hydroxy-ecdysone water. 25-Hydroxy groups kill the defense with the help of vicinal 2,3- and 20,22-OH groups. The total yield of viticosterone E when

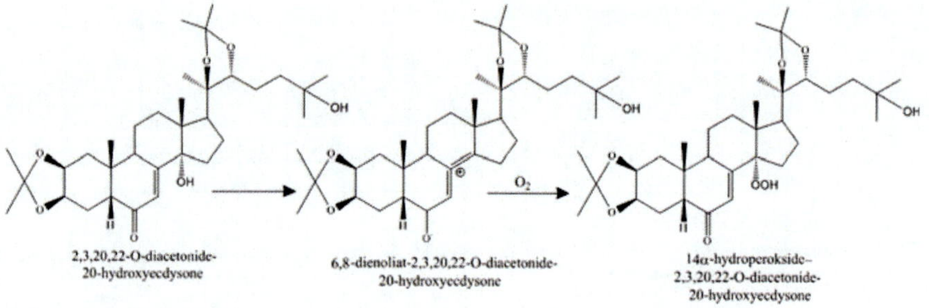

Fig. 3.12 Formations of 14α-hydroperoxide 2,3,20,22-O-diacetonide-20-hydroxyecdysone

using acetonide and phenylborate protection does not exceed 24% (Zatsny et al. 1973; Politova et al. 1997). In (Galyautdinov et al. 2004), protective groups formed during the interaction of 20-hydroxyecdysone with various ketones and aldehydes were used. The most promising is the synthesis of viticosterone E through 2,3,20,22-bis(ethylidenedioxy)-20-hydroxyecdysone. Its conversion to 25-acetate-2,3,20,22-bis(ethylidenedioxy)-20-hydroxyecdysone followed by acid hydrolysis led to the target ecdysteroid, Viticosterone E. The total yield of viticosterone E in this three-stage synthesis was 84% (Fig. 3.13).

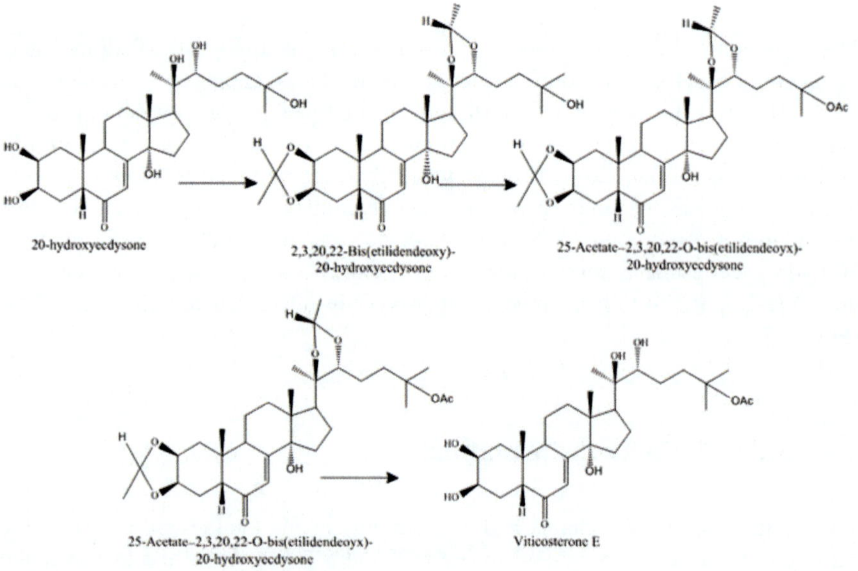

Fig. 3.13 The output of viticosterone E

Partial hydrolysis of bisacetal 2,3,20,22-bis(ethylidenedioxy)-20-hydroxyecdysone yielded 20,22-ethylenedeoxy-20-hydroxyecdysone, a phytoecdysteroid isolated from the plant *Serratula coronata* (Ves'kina and Odinokov 2012).

Thus, studies of the chemistry of ecdysteroids in recent years have significantly enriched the arsenal of methods that made it possible to efficiently synthesize a number of minor natural ecdysteroids and previously unknown analogues by transforming available ecdysteroids.

The result obtained paves the way to new possibilities in the synthesis of natural ecdysteroids and their analogues.

References

Afonkina CP, Shafikov PB, Savchenko RG, Galyautdinov IV, Odinokov V (2006) Obtaining 7,8-dihydro analogs in the interaction of derivatives of 20-hydroxyecdysone with lithium aluminum hydride. Russ J Org Chem 42(8):1252–1253

Akhrem AA, Kovganko NV (1989) Ecdysteroids: chemistry and biological activity [in Russian]. Nauka i Tekhnika, Minsk, p 325

Bobaev ID, Bobakulov KM, Ramazanov NS et al (2015) Synthesis of new 20-hydroxyecdysone o-chlorobenzoyl derivatives. Chem Nat Compd 51:706–710. https://doi.org/10.1007/s10600-015-1389-2

Calcagno MP, Camps F, Coll J, Mele E (1995) Acetylated ecdysteroids from *Ajuga reptans* var. *atropurpurea* (Lamiales: Lamiaceae). Eur J Entomol 92(1):287–293

Charoensuk S, Yingyongnarongkul B, Suksamrarn A (2000) Synthesis of 2-Dehydro-3-epi-20-hydroxyecdysone. Tetrahedron 56(7):9313–9317. https://doi.org/10.1016/S0040-4020(00)00845-0

Dinan L (2000) Phytoecdysteroids: biological aspects. Phytochemistry 57:325–339. https://doi.org/10.1016/S0031-9422(01)00078-4

Dinan L (1988) The chemical synthesis of ecdysone 22-long-chain fatty acyl esters in high yield. J Steroid Biochem 31(2):237–245

Galbraith MN, Horn DHS (1969) Insect molding hormones: crustecdysone (20-Hydroxyecdysone) from Podocarpus elatus. Aust J Chem 22:1045–1057. https://doi.org/10.1071/CH9691045

Galyautdinov IV, Nazmeeva SR, Savchenko RG et al (2004) New derivatives of 20-hydroxyecdyzone. Viticosterone E synthesis. Russ J Org Chem 40:675–684. https://doi.org/10.1023/B:RUJO.0000043713.14760.02

Galyautdinov IV, Ves'kina NA, Afon'kina SR et al (2006) Synthesis of 20-hydroxyecdysone oxime, its diacetonide, and their 14,15-anhydro derivatives. Russ J Org Chem 42:1333–1339. https://doi.org/10.1134/S1070428006090132

Harmatha J, Budesinsky M, Vokac K (2002) Photochemical transformation of 20-hydroxyecdysone: production of monomeric and dimeric ecdysteroid analogues. Steroids 67(2):127–135. https://doi.org/10.1016/S0039-128X(01)00140-4

Homvisasevongsa S, Chuaynugul A, Chimnoi N, Suksamrarn A (2004) Stereoselective synthesis and molding activity of 2,3-diepi-20-hydroxyecdysone and 2,3-diepi-5α-20-hydroxyecdysone. Tetrahedron 60:3433–3438. https://doi.org/10.1016/j.tet.2004.02.042

Kim OT, Manickavasagm M, Kim YJ et al (2005) Genetic transformation of *Ajuga multiflora* Bunge with Agrobacterium rhizogenes and 20-hydroxyecdysone production in hairy roots. J Plant Biol 48(2):258–262. https://doi.org/10.1007/BF03030416

Kovganko NV, Kashkan ZN, Chernov YG et al (2003) Synthesis of ecdysteroids and related compounds. Chem Nat Compd 39:411–437. https://doi.org/10.1023/B:CONC.0000011114.96205.98

Krstić NM, Bjelaković MS, Pavlović VD, Žižak Z, Juranić ZD, Pavlović MD (2007) Synthesis of some steroidal oximes, lactams, thiolactams and their antitumor activities. Steroids 72:406–411. https://doi.org/10.1016/j.steroids.2007.02.005

Levina TS, Kulikova LE, Kamernitzky AV, Kachala VV, Pokrovskaya EV, Smirnov AN (2008) 3-and 19-Oximes of 16α,17α-cyclohexanoprogesterone derivatives: synthesis and interactions with progesterone receptor and other proteins. Steroids 73:815–827. https://doi.org/10.1016/j.steroids.2008.03.003

Odinokov VN, Savchenko RG, Shafikov RV et al (2005) Stereochemistry of hydride reduction of 20-hydroxyecdysone derivatives. Russ J Org Chem 41:1296–1305. https://doi.org/10.1007/s11178-005-0338-0

Odinokov VN, Shafikov RV, Savchenko RG et al (2008) Analogs of ecdysteroids with a tetra-substituted $\Delta^{8,14}$-bond. Russ J Org Chem 44:671–674. https://doi.org/10.1134/S1070428008050059

Politova NK, Punegov VV, Volodin VV et al (1997) Synthesis of 20-hydroxyecdysone 25-acetate. Chem Nat Compd 33:57–60. https://doi.org/10.1007/BF02273925

Ramazanov NS, Maksimov ES, Saatov Z et al (1997) Phytoecdysteroids of plants of the genus *Rhaponticum* I. Carthamosterone A from *Rh. carthamoides*. Chem Nat Compd 33:301–302. https://doi.org/10.1007/BF02234880

Salvador JAR, Silvestre SM, Pinto RMA, Santos RC, LeRoux C (2011) New applications for Bismuth(III) salts in organic synthesis: from bulk chemicals to steroid and terpene chemistry. In: Ollevier T (eds) Bismuth-mediated organic reactions. Topics in current chemistry, vol 311. Springer, Berlin, Heidelberg. https://doi.org/10.1007/128_2011_170

Shafikov RV, Afonkina SR, Savchenko RG, Galyautdinov IV, Odinokov VN (2007) Stereoselectivity of the hydrogenolysis of the 14α-hydroxyl group in the diacetonides of 6-dihydro-5α-20-hydroxyecdysones. J Organic Chem 43(2):310–311

Shafikov RV, Urazaeva YR, Afonkina SR, Savchenko RG et al (2009) Oximes of 20-hydroxyecdysone derivatives and their re-arrangement into lactams. J Organic Chem 45:1473–1476

Shirshova TI, Politova NK, Burtseva SA et al (2006) Antimicrobial activity of natural ecdysteroids from *Serratula coronate* L. and their acyl derivatives. Pharm Chem J 40:268–271. https://doi.org/10.1007/s11094-006-0106-7

Silva MC, Riva S, Melo ML (2005) Regioselective enzymatic acylation of vicinal diols of steroids. Tetrahedron 61:3065–3073. https://doi.org/10.1016/j.tet.2005.01.104

Suksamrarn A, Ganpinyo P, Sommechai C (1994) Base-catalyzed autooxidation of 20-hydroxyecdysone. Synthesis of calonysterone and 9,20-dihydroecdysone. Tetrahedron Lett 35(25):4445–4448. https://doi.org/10.1016/S0040-4039(00)73380-6

Suksamrarn A, Yingyongnarongkul B (1995) Selective acetalation of 20-hydroxyecdysone. Partial synthesis of some minor ecdysteroids and analogues. Tetrahedron 51(38):10633–10650. https://doi.org/10.1016/0040-4020(95)00635-L

Ves'kina NA, Odinokov VN (2012) Transformations of ecdysteroids in the synthesis of their structural analogs and minor phytoecdysteroids. Russ J Org Chem 48:1137–1161. https://doi.org/10.1134/S1070428012090011

Volodin VV (2003) Phytoecdysteroids. Nauka, Moscow, p 293

Zatsny IL, Gorovits MB, Abubakirov NK (1973) Phytoecdysones of *Serratula* II. Viticosterone E from *Serratula sogdiana* and its partial synthesis. Chem Nat Compd 9:170–173. https://doi.org/10.1007/BF00563337

Chapter 4
Some Aspects of Isolation and Chemical Transformation of Phytoecdysteroids

Abstract One of the most promising sources of phytoecdysteroids is the plant's *Ajuga, Rhapoticum,* and *Silene*, which are traditionally used in medicine of many peoples as general strengthening and wound-healing agents. Ten species of plants of the genus *Ajuga* grow in the common of independent states (CIS), two of them in Central Asia. To date, ecdysteroids have been found in 17 species of the genus *Ajuga. Rhaponticum* (fam. *Asteraceae*) - rhizome herbaceous plants with large spherical flower baskets. The flora of the CIS includes 17 species of *Rhaponticum*, of which 9 grow in Central Asia. From the point of view of the search for ecdysteroid-containing plants among the representatives of the domestic flora, plants of the genus *Silene* (Smolevka, family *Caryophyllaceae*) seem promising. This genus, with about 400 representatives in the world flora, is represented by 153 in the CIS and 84 species in Central Asia. In this part of the monograph, we describe ecdysteroids isolated from various species of *Silene, Rhaponticum,* and *Ajuga.* Given methods isolation of phytoecdysteroids, structures of compounds which isolated from these genera and their chemical transformations.

Keywords Rhaponticum carthamoides · Ajuga turkestanica · Silene brahuica · Silene tomentella · Silene wallichiana

4.1 General Information

The quantitative content of ecdysteroids in plants is usually small and is calculated in tenths or thousandths of a percent relative to the weight of the raw material, although in some cases their share can reach 2.5–2.9% or more (Dinan et al. 2009).

Knowledge of the composition and dynamics of the content of ecdysteroids in plants of the studied species is of practical importance from the point of view of assessing the possibility of their use as plant materials to obtain valuable biologically active compounds. Ecdysteroids exhibit adaptogenic, tonic, cardiotropic, immunomodulatory, analgesic, and wound-healing effects (Dragana et al. 2015; Pongrácz et al. 2003).

U. Y. Yusupova et al., *Phytoecdysteroids*,
https://doi.org/10.1007/978-981-16-6711-4_4

For the first time, the presence of ecdysteroids in plants of the genus *Ajuga* was indicated by Japanese chemists. From the aerial organs of *Ajuga decumbens*, they isolated ecdysterone and cyasterone (Imai et al. 1969).

Ten species of plants of the genus *Ajuga* grow in the CIS, two of them in Central Asia. To date, ecdysterods have been found in 17 species of the genus *Ajuga* (Ramazanov 2005).

For the first time, phytoecdysteroids in the plant *Ajuga turkestanica* (family *Labiatae*) were discovered in the roots by Usmanov et al. (1971), where they described turkesterone isolated from these organs with a yield of 0.052% (Usmanov et al. 1973; Politova et al. 1997).

Rhaponticum (fam. *Asteraceae*)—rhizome herbaceous plants with large spherical flower baskets. The flora of the CIS includes 17 species of *Rhaponticum*, of which 9 grow in Central Asia (Borisov 1963).

Rhaponticum carthamoides (Willd) Iljin, *Leuzea* safflower perennial herbaceous plant 50–180 cm high. Grows in subalpine and alpine meadows and forest outskirts of Western and Eastern Siberia and Central Asia.

Rhaponticum integrifolium C. Winkl, which is a tall plant with large flower baskets, grows in Central Asia (Vasilchenko 1936).

From the point of view of the search for ecdysteroid-containing plants among the representatives of the domestic flora, plants of the genus *Silene* (Smolevka, family *Caryophyllaceae*) seem promising. This genus, with about 400 representatives in the world flora, is represented by 153 in the CIS and 84 species in Central Asia (Ramazonov et al. 2020).

In this part of the monograph, we describe ecdysteroids isolated from various *Silene* species. For the first time, Abubakirov N.K. pointed out the presence of ecdysteroids in plants of this genus using *S. praemixta* as an example in 1979 (Saatov et al. 1979). The plant turned out to be very rich in ecdysteroid compounds (Saatov et al. 1982), the most common ecdysteroid in the animal and plant worlds. The results of the analysis (Yakubova et al. 1978) showed that 20-hydroxyecdysone accumulates in the greatest amount in leaves, where its content reaches 2.5%, while the roots and inflorescences contain 0.34 and respectively 1.7%.

A strategy has been developed for screening flora for the presence of ecdysteroids based on preliminary chromatographic analysis of seed extracts, which allows us to draw a conclusion about the ability of plants to synthesize them and to conduct a large-scale study of the distribution of ecdysteroids in the world flora. More than 450 species of the order *Caryophyllaceae* were analyzed, including 266 species of the families *Caryophyllaceae* belonging to 28 genera (Zibareva 2003; Dinan 1998).

The reliability of the results on the identification of ecdysteroid-containing species was confirmed by radioimmunoassay (RIA), biotesting on cell cultures of Drosophila melanogaster and high-performance liquid chromatography. Thus, 95 new ecdysteroid-containing species of this family were found. The species of *Caryophyllaceae*, including species of the genus *Lychnis* L. and 75 species of *Silene* L were found. For the first time, the genera *Melandrium* Roehl., *Petrocoptis* A. Braun were proposed as sources of ecdysteroids. The chemical composition of ecdysteroids of 14 species of the families of *Caryophyllaceae* was determined: 12 species of the genus *Silene* and 2 species of *Lychnis* L.

The main patterns of the distribution of ecdysteroids in the genus *Silene* were revealed and chemotaxonomic studies of sections of this genus were carried out (Vasilchenko 1936).

It has been established that various *Silene* species differ both in qualitative composition and in quantitative content of ecdysteroids (Yusupova et al. 2020b; Ramazonov et al. 2020).

4.2 Phytoecdysteroids of the Plant *Ajuga Turkestanica* (Rgl.) Brig

Phytoecdysteroids isolated from *Ajuga turkestanica* (Fig. 4.1) are shown in Table 4.1. It should be noted that the content of turkesterone in the aerial parts significantly exceeds its content in the roots (Saatov et al. 1990).

Below we are listing the isolated compounds and some details related to them.

22-Acetate-cyasterone (35), $C_{31}H_{46}O_9$, mp 212–215 °C (acetone-hexane), $[\alpha]D_{20}$ + 96.2 (methanol).

The IR spectrum of this substance (35) contains absorption maxima corresponding to hydroxyl absorption (3440 cm^{-1}), 1745 (δ-lactone), α, β-unsaturated keto grouping (1655 cm^{-1}) and bands at 1755 and 1255 cm are observed −1, indicating the presence of an ester group. This is also evidenced by the presence of a three-proton singlet at 1.94 ppm in the ^{1}H NMR spectrum of the ecdysteroid, 22-O-acetate-cyasterone, at 1.94 ppm. In the mass spectrum of ecdysteroid 35, there are ion peaks with m/z 363, 345, and 327, characteristic of cyasterone (Hikino et al. 1968; Takemoto et al. 1967).

When comparing the ^{1}H NMR spectra of cyasterone and 22-O-acetate-cyasterone (35), a significant difference is revealed only in the position of the resonance lines of the proton at C-22, which shift to a weak field and appear at 5.32 ppm The values of chemical proton shifts in the ^{1}H NMR spectrum of ecdysterone were in good agreement with those for 22-O-acetate-cyasterone. The data presented allowed us to identify ecdysteroid 35 with 22-O-acetate-cyasterone (Usmanov et al. 1978).

Ajugalactone (36), $C_{29}H_{40}O_8$, mp 225–236 °C (acetone-hexane), (chloroform-ethanol) (Saatov et al. 1977). For ajugalactone ecdysteroid, absorption typical for ecdysteroids was observed in the IR spectrum at 3350 cm^{-1} (OH), 1713 cm^{-1} (C = O at C-I2), 1693 cm^{-1} (δ-lactone) and 1675 cm^{-1} (cyclohexelone). Peaks of ions with m/z 516 (M+), 498, 480, 377, 360, 359, 341, 316, 315, 314, 183, 139, and 111, the mass numbers of which were convened with fragments, were present in the mass spectrum of ecdysteroid 36 ajugalactone (Koreeda et al. 1970a). Based on the above physicochemical constants and spectral data, ecdysteroid 36 is ajugalactone (Koreeda et al. 1970).

Fig. 4.1 *Jivuchka turkestanica—Ajuga turkestanica* (Rgl.) Brig

Table 4.1 The content of ecdysteroids in the aerial organs and roots of the plant *Ajuga turkestanica*

Name	Yield, %	
	Aerial part	Roots
22-Acetalcyasterone	0,12	–
Ajugalactone	0,001	–
AjugasteroneB	–	0,003
α-Ecdysone	0,001	–
Cyasterone	0,030	–
Hydroxyecdysone	0,20	0,03
Turkesterone	0,17	0,043

Ajugasterone B (37). $C_{29}H_{46}O_7$, m.p. 238–240 °C (Usmanov et al. 1977). The infrared spectrum of ajugasterone B (37) has maxima corresponding to hydroxyl absorption (3420 cm^{-1}) and α, β-unsaturated keto grouping (1660 cm^{-1}). Its mass spectrum contains ion peaks with m/z 506 (M+), 488, 427, 426, 363, 362, 345, 327, 301, 300, 143, 125, 107, 97 and 79. Physicochemical constants and spectral data of ecdysteroid 37 indicate the identity of the substance with ajugasterone B.

Cyasterone (38). $C_{29}H_{44}O_8$, mp 158–160 °C (methanol), [α]D_{20} + 60.0 (pyridine). The infrared spectrum of ecdysteroid 38 showed an absorption characteristic of ecdysteroids at 3450 (OH), 1752 (lac-lactone), and 1660 cm^{-1} (7-en-6-keto grouping). In the mass spectrum of ecdysteroid, there are ion peaks with m/z 363, 345, and 327, characteristic of ecdysteroids with an ecdysterone nucleus, including cyasterone (Takemoto et al. 1967). The chemical shifts of protons in the ^1H NMR spectrum of the ecdysteroid are in good agreement with those for cyasterone (Yusupova et al. 2019). Acetylation of cyasterone with acetic anhydride in pyridine resulted in the production of cyasterone triacetate.

Turkesterone (32), $C_{27}H_{44}O_8$, amorph. [α] D_{20} + 52.0 (methanol) (Saatov et al. 1990).

In the IR spectrum, an ecdysteroid showed an absorption characteristic of ecdysteroids at 3450 (OH) and 1660 cm^{-1} (7-en-6-keto grouping). Peaks of ions with m/z 379 ($C_{21}H_{31}O_6$), 361 ($C_{21}H_{24}O_5$), 343, and 325 inherent in ecdysteroids containing four hydroxyl groups in the steroid nucleus were noted in the mass spectrum of ecdysteroid 32. The decomposition of the side chain of this compound proceeds with the formation of intense ion peaks with m/z 99 and 81, typical for fragmentation of the ecdysterone side chain. Turkesterone (32) forms diacetonide, in the mass spectrum of which there is a peak of a molecular ion with m/z 576, as well as peaks of ions with m/z 201, 143, 125, 102, 99, and 81, characteristic of the decomposition of the side chain of ecdysterone diacetonide.

Based on the data, the authors concluded that turkestone contains seven hydroxyl groups—four in the steroid nucleus and three in the side chain (Usmanov et al. 1975).

The chemical shifts of protons in the ^1H NMR and ^{13}C NMR spectra of the ecdysteroid coincided with the corresponding values for Turkesterone (Usmanov et al. 1975).

From this clear-headed—*Lamiaceae*, a small-branched shrub with a woody caudex of 25–28 cm in height. Ecdysteroids—ecdysterone—0.05%, turkesterone from roots—0.052% were isolated from the aerial part. They have anabolic, tonic, and stimulating effects (Mamarasulov et al. 2020).

Ajuga turkestanica (Rgl.) is characterized by a narrow range. It grows in areas of variegated rocks on the southwestern slopes of the Gissar Range. The raw materials stocks and the role of *Ajuga turkestanica* in the addition of vegetation cover in Kashkadarya and Surkhandarya regions of Uzbekistan were studied (Mamarasulov et al. 2020).

4.3 Phytoecdysteroids of the Plant *Rhaponticum Carthamoides* (Wild) Iljin

In total, 17 species belong to the genus *Leuzea*, 12 grow in the CIS, and 9 species of this plant grow in Central Asia (Fig. 4.2), (Borisov 1963).

Rhaponticum carthamoides (wild) Iljin—Sophloroid Leuzea (family *Asteraceae*)—a perennial herb. Underground organs with a specific smell are represented by a horizontal, woody, branched, dark brown rhizome with numerous thin, hard, short roots 17–20 cm long. The total length of the branches of an individual's rhizome is often about 1 m, reaching 2 m or more. About 5–50 shoots develop on the rhizome. At the top of generative shoots, single inflorescences of a basket, 4–8 cm in diameter, are formed. The basket contains five-membered, tubular purple-pink flowers in an amount of 320–430 pieces. Fruits of the achene, ellipsoidal, grayish-brown, tetrahedral or almost flat, ribbed, 6–8 mm long, 3–4 mm wide, in the amount of 220–360 (Sridhar et al. 2016).

Sterols, inulin polysaccharide, triterpene glycosides, antraglycosides, flavonoids, tannins, and ecdysteroids were found in the plant roots (Sridhar et al. 2016; Baltaev 1979).

Fig. 4.2 Leuzea—*Rhaponticum carthamoides* (Willd.)

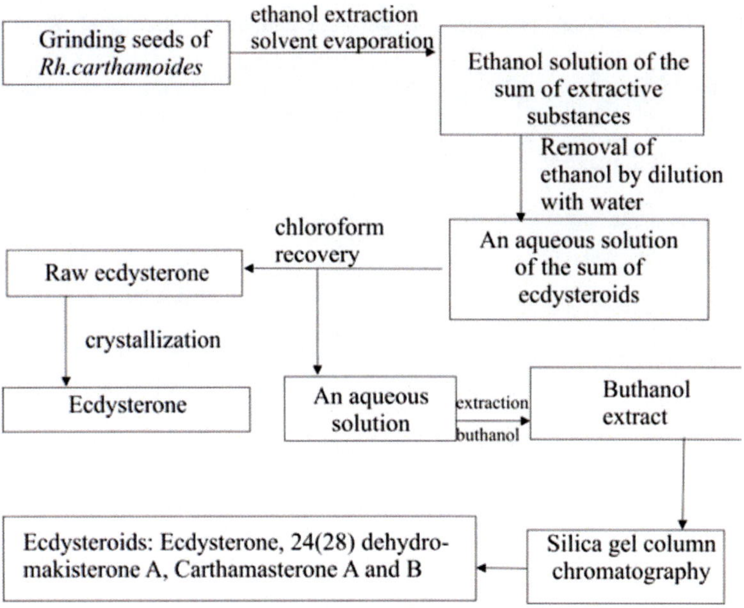

Fig. 4.3 Scheme for the isolation of ecdysteroids from the seeds of *Rhaponticum carthamoides*

Chromatographic analysis of the roots and seeds of the Salfloroid leuzea showed that they differ in the content of ecdysteroids in both qualitative and quantitative terms.

A study was made of the seeds of the Leuzea sofloroid plant. Ecdysteroids were isolated according to Fig. 4.3.

The crushed seeds of the plant were extracted with ethanol. The ethanol extract was concentrated, diluted with water. The aqueous solution was extracted first with chloroform. When standing for 24 h, ecdysterone crystallized out of the solution. Then phytoecdysteroids from the aqueous residue were exhaustively extracted with butanol. Butanol was distilled off, the residue was chromatographed on a silica gel column and individual ecdysteroids were isolated (Fig. 4.3). Physicochemical constants of isolated ecdysteroids are given in Table 4.2.

4.3.1 *The Structure of Carthamasterone A*

Carthamasterone A (39) (Fig. 4.4) has the composition $C_{28}H_{44}O_{10}$, so pl. 170–172 °C (methanol–ethyl acetate) (Ramazanov et al. 1997). In the UV spectrum, there is maximum absorption at 245 nm (lgε 4.01).

In the IR spectrum, there is a wide absorption band at 3405 cm^{-1} (OH group), as well as at 1674 cm^{-1}, corresponding to the absorption frequency of the keto

Table 4.2 Physicochemical properties of phytoecdysteroids isolated from *Rhaponticum carthamoides*

Name of the compounds	Gross formula	Temperature melting, °C	$[\alpha]_D$°, (CH_3OH)	Content, %
Carthamasterone A	$C_{28}H_{44}O_{10}$	170–172		0,0007
Carthamasterone B	$C_{28}H_{44}O_8$	179–182		0,0008
24(28)-Dehydromakisterone A	$C_{28}H_{44}O_7$	243–245	62.0 ± 2	0,0004
22-O-Benzoat-polypodine B	$C_{34}H_{48}O_9$	196–198	60.0 ± 2	0.003
Ecdysterone	$C_{27}H_{44}O_7$	240–242	61.2 ± 2	1,5
α-Ecdysone	$C_{27}H_{44}O_6$	236–238	63.5 ± 2	0.01

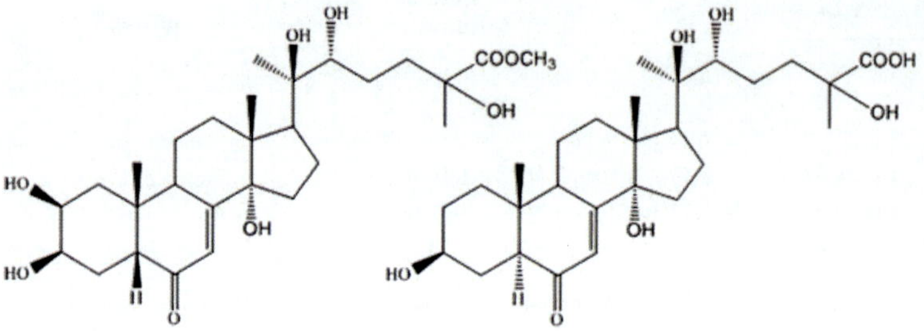

Carthamasterone A (39) Carthamasterone B (40)

Fig. 4.4 The structure of carthamasterone A and B

group conjugated to the double bond. In addition, at 1710 cm^{-1}, an inflection band is observed corresponding to the carbonyl group of the ester function. This is also evidenced by the presence of a three-proton singlet of the carbomethoxyl group in the ^{1}H NMR spectrum of carthamosterone A (39) at 3.88 ppm (Table 4.3).

In the mass spectrum of carthamasterone A (39), the peak of the molecular ion is absent. At the same time, there are ion peaks with m/z 379, 361, 343, 325, 317, and 316, indicating the presence of four hydroxy groups in the steroid nucleus (Scheme 4.3.2) (Galbraith and Horn 1966; Rashkes and Abubakirov 1980).

The ^{1}H NMR spectrum of carthamasterone A (39) reveals singlet signals of protons of four methyl groups at 1.15, 1.20, 1.49, and 1.59 ppm, as well as resonant lines of single protons at 3.58 (H-9), 3.88 (H-22), 4.0 -4.3 (H-2.3), and 6.26 ppm (H-7). The assignment of these signals are given in Table 4.3.

A comparison of the characteristics of these signals with those of ecdysteroids known in the literature revealed the closeness of the chemical shifts of most protons of carthamasterone A (39) and polypodine B (41), with the exception of CH_3-27.

The data presented above, in total, allow us to note that carthamasterone A (39) is a natural derivative of polypodine B containing a carbomethoxyl moiety in the side chain.

As can be seen from Table 4.3, when switching from polypodine B (41) to carthamasterone A(39), the values of the chemical shifts of H-22 and CH_3-18 do

Table 4.3 The values of chemical shifts of protons in the ^{1}H NMR spectra of polypodine B (41), ecdysterone (2), makisterones A (39), B (40), and 22-O-benzoate polypodine B (42) (δ, ppm, C_5D_5N, O-TMS)*

Proton position										
H-2	H-3	H-7	H-9	H-22	CH_3-18	CH_3-19	CH_3-21	CH_3-26	CH_3-27	$COOCH_3$
Polypodine B										
4.0	4.3	6.23	3.59	3.85	1.20	1.15	1.58	1.37	1.37	–
Ecdysterone										
4.0	4.3	6.26	3.58	3.88	1.21	1.06	1.58	1.36	1.36	–
Carthamasterone A										
4.0	4.3	6.26	3.58	3.88	1.20	1.15	1.59	–	1.49	3.88
Carthamasterone B										
4.0	4.3	6.24	3.58	3.88	1.20	1.07	1.59	–	1.49	3.88
22-O-Benzoate-polypodine B										(Ar–H)
4.15	4.15	6.17	3.50	5.70, d J = 7.5 Hz	1.07	1.00	1.65	1.17		7.35(3H) 8.24(2H), d J = 7.5 Hz

*Signals of protons of methyl groups are singlet in nature; the H-7 proton appears as a broadened singlet, the remaining signals as broadened multiplets

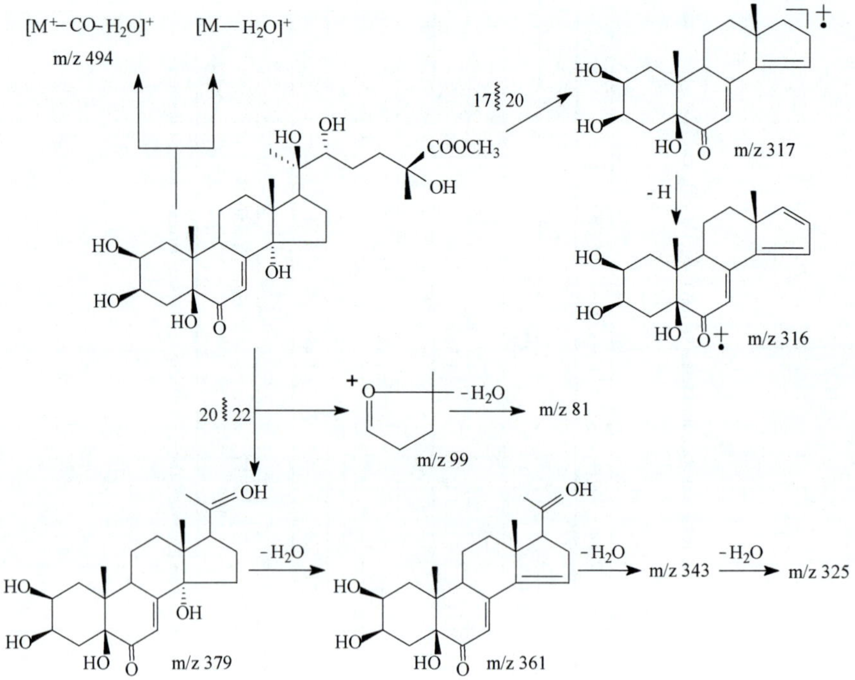

Fig. 4.5 The main ways of mass spectrometric decay of carthamasterone A (39)

not change, and besides, the signals of only four methyl groups are present in the ^{1}H NMR spectrum of carthamasterone A(39). Moreover, one of the singlet signals of the methyl groups experiences a paramagnetic shift of 0.12 ppm. A similar change is found in the literature for ecdysteronic acid (Chou and Lu 1980). These data unambiguously indicate the formation of a carbomethoxyl group in the molecule of carthamasterone A (68) due to one of the methyl groups at C-25 (Fig. 4.5).

Thus, carthamasterone A(39) is a natural derivative of polypodine B (41), which contains a carbomethoxyl group at C-25.

4.3.2 The Structure of Carthamasterone B

Carthamasterone B (40) (Fig. 4.2), $C_{28}H_{44}O_9$, m.p. 179–182 °C (from methanol-ethyl acetate) (Ramazonov et al. 1997). The UV spectrum of the described compound is characterized by a maximum at 245 nm (lgε 4.00), characteristic of α, β-unsaturated ketones.

In the IR spectrum of carthamasterone B, in addition to absorption bands due to hydroxyl groups (3400 cm^{-1}) and a keto group conjugated to a double bond

(1670 cm^{-1}), there is a band at 1710 cm^{-1} corresponding to the absorption of the carbonyl group of the ester residue.

In the mass spectrum of carthamasterone B, the peak of the molecular ion is absent and the following main peaks of fragment ions with m/z are found: 460, 363, 345, 327, 301, and 300. The first peak of the fragment is formed due to the removal of two water molecules from the molecular ion carbon, and the rest are characteristic of the decay of the steroid part of the ecdysterone molecule (Fig. 4.6) (Rashkes and Abubakirov 1980).

In the resonance region of methyl groups, the ^1H NMR spectrum of carthamasterone B, in contrast to ecdysterone, contains singlet signals of protons of only four methyl groups at 1.07 (CH_3-19), 1.20 (CH_3-18), 1.49 (CH_3-27), and 1.59 (CH_3- 21) ppm. At 3.88 ppm, the methyl singlet of the carbomethoxy group is located. The values of chemical shifts of the remaining single protons are identical to those in the spectrum of ecdysterone (Table 4.3).

It is noteworthy that upon transition from the ^1H NMR spectrum of ecdysterone to the NMR spectrum of cardosterone B, a paramagnetic shift of CH_3-27 by 0.13 ppm takes place, as it is observed when comparing the spectral characteristics of ecdysterone and ecdysteronic acid (Baltaev 2000).

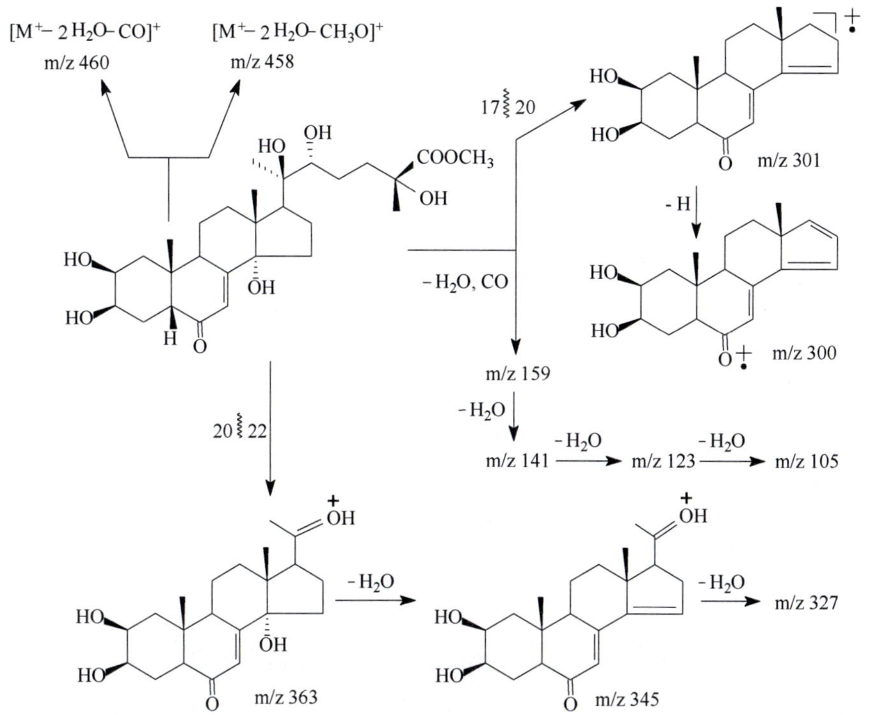

Fig. 4.6 The main ways of mass spectrometric decay of carthamasterone B (40)

Fig. 4.7 24 (28)-dehydromakisterone A (20)

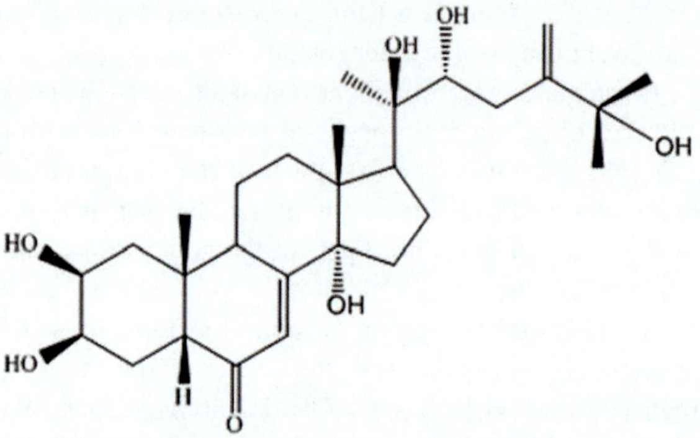

24 (28)-dehydromakisterone A (20)

All the above data made it possible to unequivocally state that the basis of the carthamasterone B molecule (40) is ecdysterone with only one difference in the side-chain structure—a carbomethoxyl group is formed due to one of the methyl groups at C-25 in the case of carthamasterone B. Thus, carthamasterone B is a naturally occurring 25-carbomethoxy-ecdysterone.

4.3.3 The Structure of 24 (28)-Dehydromakisterone A

24(28)-dehydromakisterone A (20), (Fig. 4.7) $C_{28}H_{44}O_7$, after crystallization from a methanol-acetone mixture, 243–245 °C, $[\alpha]_D + 62.0°$ (methanol) (Ramazonov et al. 1997). In the IR spectrum, there was an absorption typical for most ecdysteroids at 3400–3500 (OH), 1664 cm^{-1} (conjugated keto group).

In the mass spectrum of ecdysteroid 24 (28) -dehydromakisterone A (20), the peaks of ions with m/z 363, 345, 327, 301, and 300 indicated the presence of three hydroxyl groups in the tetracyclic core.

Fragmentation of the side chain of 24(28) -dehydromakisterone A (20), as well as the decay of the corresponding portion of the molecule of 24(28) -dehydromakisterone A (20) (Baltaev et al. 1978), is characterized by high-intensity peaks with m/z 113, 111, 97, 95, 93, 83, and 71.

These data, as well as a direct comparison on TLC, led to the conclusion that ecdysteroid is 24(28) -dehydromakisterone A (Baltaev et al. 1978).

4.3.4 The Structure of 22-O-Benzoate-Polypodine B

22-O-Benzoate-polypodine B (42) (Fig. 4.8), $C_{34}H_{48}O_9$.

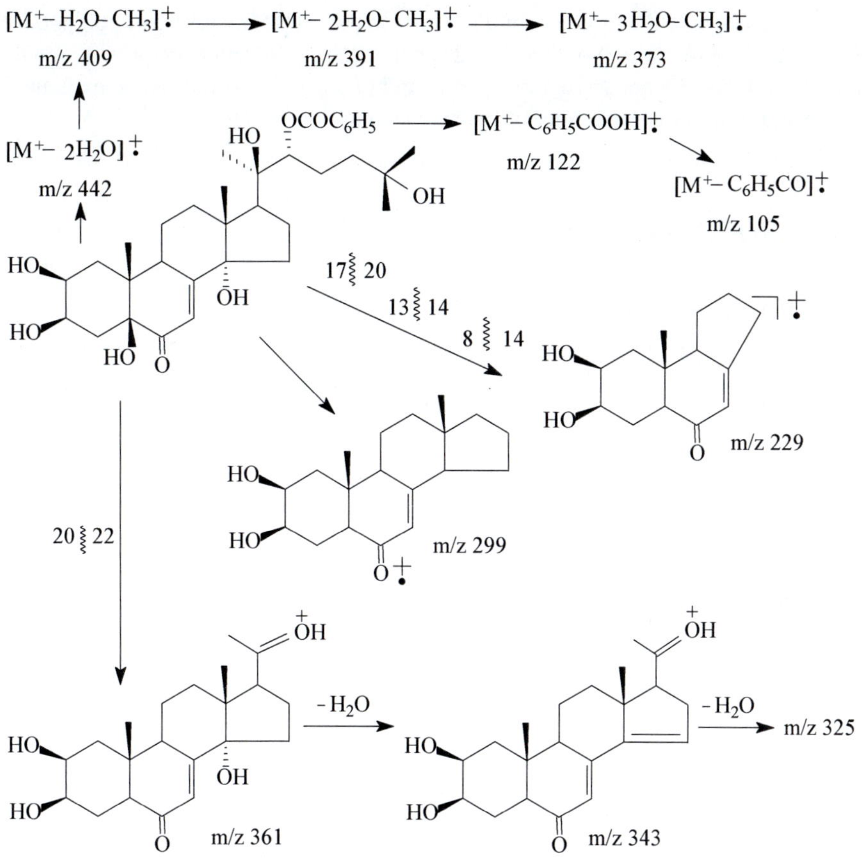

Fig. 4.8 The main ways of mass spectrometric decay of 22-O-benzoate-polypodine B (42)

The IR spectrum of 22-O-benzoate-polypodine B contains absorption bands due to hydroxyl groups (3432 cm^{-1}) and α, β-unsaturated keto grouping (1654 cm^{-1}). In addition, there are absorption bands at 1710 and 1281 cm^{-1} (ester), 1610, 1587, 714 cm^{-1} (aromatic ring), suggesting that there is an ester group in 22-O-benzoate-polypodine B (39) aromatic in nature.

The signals of five aromatic protons at 7.35 (3H) and 8.24 (2H) ppm also testify to this. In the ^1H NMR spectrum, as well as intense peaks of ions with m/z 122 ($C_7H_6O_2$), 105 (C_7H_5O), and 77 (C_6H_5) inherent to the decomposition fragments of benzoic acid.

The peak of the ion with m/z 379 (gap C-20 20 C-22) and its derivatives with m/z 361, 343, and 325, observed in the mass spectrum of 22-O-benzoate-polypodine B (42), suggest that benzoic acid is not bound to the steroid core, but attached to the side chain (Sadykov 2005), (Fig. 4.8).

A comparison of the proton chemical shifts in the ^1H NMR spectra of polypodine B and 22-O-benzoate polypodine B (42) shows that in the latter the proton signals at C-22 and 21-CH_3 are shifted to a weak field (Table 4.3), which can be explained by the addition of the remainder of benzoic acid to the oxy group at C-22.

4.3.5 *α-Ecdysone Identification*

α-ecdysone (1), $C_{27}H_{44}O_6$, with m.p. 236–238 °C.

In the IR spectrum of α-ecdysone, absorption bands are observed at 3440–3460 cm^{-1} (OH), 1657 cm^{-1} (keto group conjugated with a double bond), characteristic of ecdysteroids. In the mass spectrum of α-ecdysone, peaks of ions with m/z 446 (M^+ -H2O), 428, 413, 410, 395, 348, 330, 300, 99, and 81, which coincide with fragments of the α-ecdysone, are observed (1).

A comparison of the constants and chemical proton shifts in the ^1H NMR and ^{13}C NMR spectra coincided with the corresponding indices for α-ecdysone given in the literature (Saatov et al. 1986a).

The Leuzea—*Rhaponticum carthamoides* is a perennial herbaceous plant from the family *Compositae*, 50–180 cm high, with a ligneous, thickened, branching rhizome and numerous thin, elastic, and elastic roots. Rhizomes and roots from brown to dark brown in color, with a specific smell. The stem is finely furrowed, simple, cobweb-like, especially under inflorescences. Leaves are bare or irregularly cobweb-shaped, deeply pinnately dissected with a larger final lobe; the lower leaves are petiolate plates with a length of 12–40 cm and a width of 5–20 cm; the upper leaves are sessile, the most—the upper whole, coarse. Inflorescence is a single basket at the top of the stem, 3–6 cm in diameter. Violet-violet flowers. Seeds (achenes) are brownish, longitudinally ribbed, slightly widened upward 5–8 mm and 3–4 mm across, with a crest of cirrus hairs, fused at the base into a continuous ring. All 1000 seeds 11–19 g.

Endemic plant is found only in the Sayan Mountains and Altai, Kuznetsk Alatau (Russia). It lives mainly at an altitude of 1700–2000 m above sea level. It rarely penetrates alpine meadows. In places, especially near the upper border of the forest and among the cedar woodlands, it forms continuous thickets. For the culture of Leuzea, the most favorable regions of the middle zone of the European part of the former Union, as well as some regions of Siberia. In places of natural growth of leuzea, it blooms in July-early August, in culture (Moscow region) in June.

4.4 Phytoecdysteroids of the Plant *Rhaponticum Integrifolium* C. Winkl

Rh. integrifolium C. Winkl (family *Asteraceae*) (Fig. 4.9) is found in vivo in Central Asia, in particular in the Surkhandarya region, on the Ferghana Range, and on the

Fig. 4.9 *Rhaponticum integrifolium* C. Winkl. (family *Asteraceae*)

Pamir-Alai. This is a perennial herbaceous plant 40–150 cm high. Achenes are ovate, bare, light, truncated at the apex with a crown-like margin. The weight of 1000 pieces of seeds is 19.5 g.

TLC analysis using vanillin sulfuric acid and column chromatography revealed a number of major ecdysteroids with different structures. Plant extracts of various organs showed that they differ in the content of ecdysteroids. The most promising in quantitative and qualitative terms were seeds containing phytoecdysteroids. Phytoecdysteroids were studied—α-ecdysone, 24 (28) -dehydromakisterone A (20), ecdysterone (2), viticosterone E (43) (Table 4.4). α-Ecdysone (1) was first discovered in this type of plant (Baltaev et al. 1978; Sadikov 2005).

Viticosterone E (43), $C_{29}H_{46}O_8$, m.p. 195–196 °C, $[\alpha]_D + 59.2 \pm 2$ °C (c 0.50; methanol) (Ramazanov et al. 2007).

The mass spectrum of viticosterone E (34) contains ion peaks with m/z 462 $(M\text{-}60)^+$, 444 $(M\text{-}60\text{–}18)^+$, 426, 411, 408, 393, 363, 345, 327, 309, 301, 300, 99, and 81, which are also characteristic for the fragmentation of ecdysterone. This indicates that the basis of viticosterone E and ecdysterone is the same hydrocarbon skeleton and functional groups.

Table 4.4 Physicochemical constants and data on the quantitative content of ecdysteroids isolated from the plant *Rhaponticum integrifolium* are given

Compounds	Gross formula	Temperature melting, °C	$[\alpha]_D$, (CH_3OH)	Content, %
α-Ecdysone	$C_{27}H_{44}O_6$	235–237	63.5 ± 2^0	0,02
24(28)-Dehydromakisterone A	$C_{28}H_{44}O_7$	244–246	54.4 ± 2^0	0,0031
Ecdysterone	$C_{27}H_{44}O_7$	239–241	62.3 ± 2^0	1,43
Viticosterone E	$C_{29}H_{46}O_8$	195–197	59.7 ± 2^0	0,0036

In the infrared spectrum of viticosterone E (43), in addition to absorption bands characteristic for this class of compounds at 3430–3450 cm^{-1} (OH) and 1670 cm^{-1} (7-en-6-keto grouping), maxima were observed at 1275 cm^{-1} corresponding to the ester group. Accordingly, in the ^1H NMR spectrum, viticosterone E (43) at 1.92 ppm the signal of the methyl group of the acetyl residue appears.

When comparing the ^1H NMR spectra of viticosterone E (43) and ecdysterone (2), the shift of the resonance signals of C-26 and C-27 methyl groups in the former to a weaker field indicates the presence of the acetyl group at C-25.

This assumption is confirmed by the practical coincidence of the mass and ^1H NMR spectra for viticosterone E (43), as well as by their direct comparison on thin-layer chromatography (Ramazonov et al. 2007).

Rh. Integrifolium (Fig. 4.9) natural conditions, Integrifolium is found on gravelly slopes in the middle zone of the mountains of Central Asia, in particular in the Surkhandarya region, on the Ferghana Range and on the Pamir-Alai. This is a perennial herbaceous plant 40–150 cm high. The leaves are leathery, whole-edge, oblong elliptical, stem, ovate, drawn to both ends. The basket is spherical, 4.5–6 cm. Achene is ovate, glabrous, bright, truncated at the apex, with a crown-shaped margin.

Botanists at ICPS (Academy of Sciences, RUz) made the plant propagated by sowing seeds in October–November. In the first year of life, it formed a rosette of basal leaves. Flowering and fruiting occurred in the second year of life. Perennial specimens began to grow in the third decade of March, budding was observed in late April-early May, flowering in mid-May, fruit ripening in June. Intensive plant growth occurs in April–May, its height at this time reaches 100–120 cm. The weight of 1000 pieces of seeds is 19.8 g. From 1, it will be collected 41.7 kg of seeds.

The results can be used to optimize the growing conditions of a plant in culture and the recommended raw materials can be successfully cultivated and used to obtain an ecologically safe tonic preparation "ecdysten".

4.5 Phytoecdysteroids *Silene Brahuica* Boiss

Chromatographic analysis of the above-ground organs and the roots of the Brahuy'smelter showed (Saatov 1983). That in the content of ecdysteroids they differ both in qualitative and quantitative terms.

The underground organs of the plant were studied and phytoecdysteroids—ecdysterone (2), 25-O-acetate-integristerone A (44), sileneoside C (26), sileneoside H (33), and sileneoside G (34) were isolated. Ecdysteroids were isolated according to (Fig. 4.10). Physicochemical constants of selected ecdysteroids are given in Table 4.5.

Compounds ecdysterone and sileneoside C (26) (Fig. 4.11), according to physicochemical constants, spectral data (IR, ^1H and ^{13}C NMR) and according to the results of chemical transformations are identified, respectively, with known ecdysteroids.

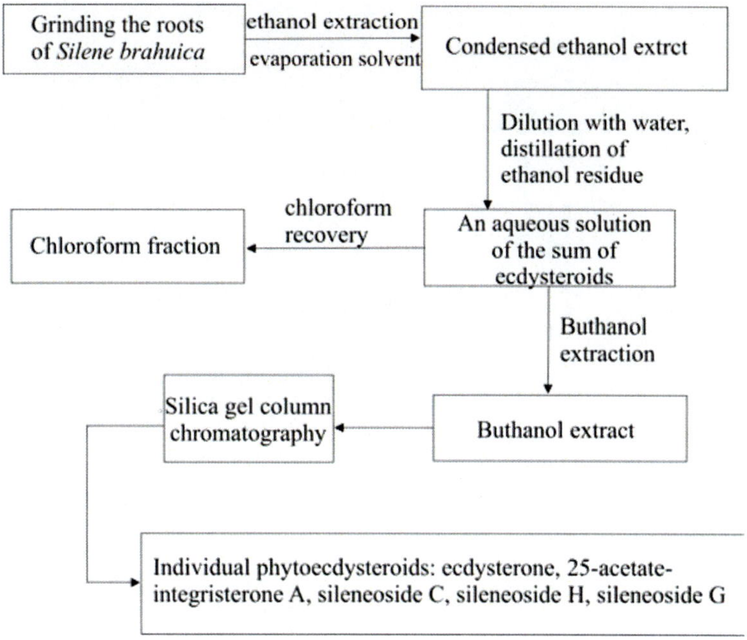

Fig. 4.10 Scheme for isolating phytoecdysteroids from the roots of *Silene brahuica*

Table 4.5 Physicochemical properties of phytoecdysteroids isolated from *Silene brahuica*

Name of the compounds	Formula	M.P., °C	$[\alpha]_D^{20}$ CH_3OH	Yield, in %
Ecdysterone	$C_{27}H_{44}O_7$	240–242	60.7 ± 2	0.0940
IntegristeroneA-25-acetate	$C_{29}H_{46}O_9$	198–200	–	0.001
Sileneoside C	$C_{33}H_{54}O_{13}$	232–234	90.0 ± 2	0.0032
Sileneoside H	$C_{35}H_{56}O_{14}$	210–212	–	0.0005
Sileneoside G	$C_{39}H_{64}O_{17}$	225–227	121 ± 2	0.0016

4.5.1 Identification of Ecdysterone and Sileneoside C

Ecdysterone (115) (Fig. 4.11), $C_{27}H_{44}O_7$, m.p. 241–242 °C, $[\alpha]_D$ 22 + 63.2° (in methanol).

In the IR spectrum, absorption bands characteristic of ecdysteroid ecdysterone are observed at 3470 (OH) and 1665 cm^{-1} (7-en-6-keto grouping).

The peaks of ions with m/z 462 (M + H_2O)+, 444, 426, 408, 363, 345, 327, 309, 300, 99, 81, and 69 characteristic of ecdysteron are present in the mass spectrum (Zatsny et al. 1975).

The above data, as well as a direct comparison of the ^1H NMR spectra (Table 4.6) and R_f on TLC, showed that the compound is ecdysterone (Saatov 1983).

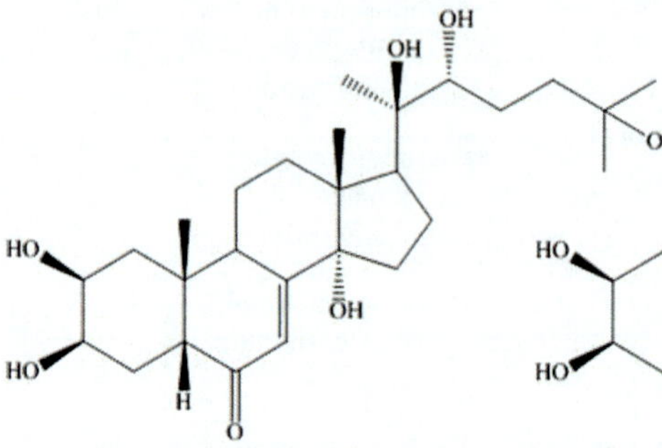

Ecdysterone (2)

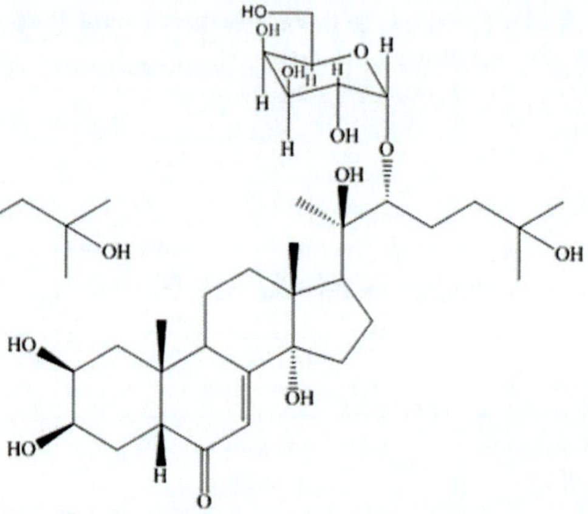

Sileneoside (26)

Fig. 4.11 Ecdysteroids isolated from *Silene brahuica*

Table 4.6 Chemical shifts of protons of ecdysterone (2) and sileneoside C (26) (C_5D_5N, δ. Ppm, TMS)

Position of proton									
H-2	H-3	H-7	H-9	H-22	CH_3-18	CH_3-19	CH_3-21	CH_3-26	CH_3-27
Ecdysterone									
4.01	4.07	6.12	3.42	3.70	1.06	0.94	1.44	1.22	1.22
Sileneoside C									
4.13	4.13	6.06	3.41	3.58	1.11	1.26	1.48	1.24	1.29

Sileneoside C (26), $C_{33}H_{54}O_{13}$, m.p. 232-234^{0}C, $[\alpha]_D + 90.0 \pm 2^0$, (with 1.02; methanol), in the IR spectrum, had a wide band at 3370–3440 cm^{-1}, corresponding to stretching vibrations of hydroxyl groups, and also a maximum at 1658 cm^{-1}, which corresponds to the absorption frequency of the keto group conjugated to a double bond.

In the low and medium mass spectra of sileneoside C, peaks of ions with m/z 163, 145, corresponding to the sugar residue and fragments with m/z 143, 99, 81, and 69, corresponding to the side chain of ecdysteroids were observed. Also, peaks of ions with m/z 361, 343, 325, and 316, characteristic of ecdysteroids with four hydroxyl groups in the steroid nucleus, were also present in the mass spectrum of sileneoside C.

Physicochemical constants and spectral data testified to the identity of the phytoecdysteroid with sileneoside C (Saatov 1983), (Table 4.6).

4.5.2 The Structure of 25-O-Acetate-Integristerone A

25-O-Acetate-Integristerone A (44), $C_{29}H_{46}O_9$, m.p. 198–200 °C (methanol). In the IR spectrum of 25-O-acetate-integristerone A (44), in addition to absorption due to hydroxyl groups (3200 cm^{-1}) and α, β-unsaturated keto grouping (1655 cm^{-1}), there are absorption bands at 1730 and 1244 cm^{-1} indicating the presence of an ester residue. The presence of a three-proton singlet at 1.96 ppm in the ^1H NMR spectrum of 25-O-acetate-integristerone A (44) indicates the presence of one acetyl group in this compound.

In the mass spectrum of 25-O-acetate-integristerone A (44), the peak of the ion with m/z 379 (bond cleavage C-20, C-22) and its derivatives with m/z 361, 343, and 325 suggest that acetyl the group is located in the side chain, and not in the steroid core (Sadykov and Saatov 1999), (Fig. 4.12).

When comparing the ^1H NMR spectra of 25-O-acetate—integristerone A (44) and integristerone A (45), a significant difference is revealed only in the position of the signals of methyl groups at C-26 and C-27, which are shifted to a weak field (Table 4.7). A similar shift to a weak field is also observed in the case of viticosterone E (43) (Sadykov and Saatov 1999; Zatsny et al. 1975) also having a 25-acetyl group.

This fact suggests that in 25-O-acetate—integristerone A (44), the acetyl group is attached to the oxy group at C-25.

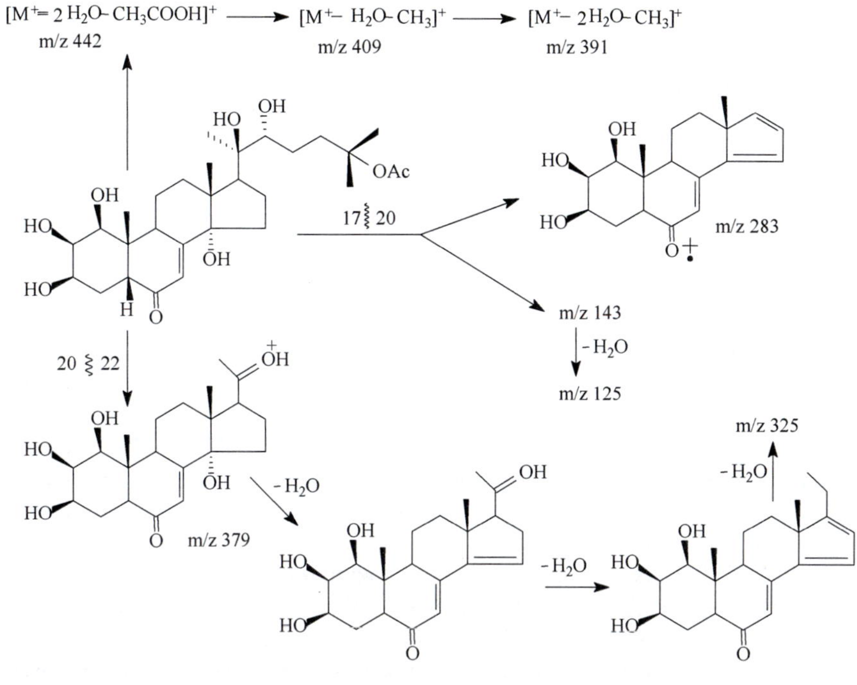

Fig. 4.12 The main ways of mass spectrometric decay of 25-O-acetate-integristerone A

Table 4.7 Chemical shifts of protons of 25-O-acetate-integristerone A (44), integristerone A (45) and 2,3,20,22,25-O-pentaacetate-integristerone A (δ,, ppm, C_5D_5N, TMS)

Position of proton											
H-1	H-2	H-3	H-7	H-9	H-22	CH_3-18	CH_3-19	CH_3-21	CH_3-26	CH_3-27	OAc
25-O-acetate-integristerone A											
4.30	–	4.38	6.30	3.60	3.87 d 3 J = 9.4 Hz	1.26	1.43	1.63	1.44	1.50	1.96
Integristerone A											
4.30	4.30	4.30	6.17	3.57	3.77	1.19	1.40	1.58	1.39	1.39	
2,3,20,22,25-O-penta acetate-integristerone A											
5.85	5.60	5.60	6.12	3.62	5.03	1.46	1.46	1.62	1.41	1.46	1.98 (6H) 2.08 (9H)

Signals of protons of methyl group H-7 appear as a singlet; the rest of the signals (with the exception of H-22) are extended multiplets

Acetylation of 25-O-acetate-integristerone A (44) gave pentaacetate. In the ^1H NMR spectrum of pentaacetate, the CH_3-26 and CH_3-27 signals remained virtually unchanged.

Based on the foregoing, it can be concluded that ecdysteroid is 25-O-acetate—integristerone A (Sadykov and Saatov 1999).

4.5.3 *The Structure of Sileneoside H*

Sileneoside H (33), $C_{35}H_{56}O_{14}$, m.p. 210–212 °C, has in the UV spectrum a maximum absorption at 242 nm, characteristic of the 7-en-6-on part of ecdysteroids.

In the IR spectrum of sileneoside H (33), absorption bands are observed in the region of 3398 cm^{-1} (OH groups), 1655 (cyclohexanone), and 1707 and 1282 (ester group).

The use of ammonia in the mass spectrum as a reagent gas led to the formation of an ion with m/z 700, which corresponds to the molecular formula $C_{35}H_{56}O_{14}$.

The presence of 25-O-acetate-integristerone A (44) (Sadykov and Saatov 1999), as genin, was confirmed by its identification in the products of enzymatic cleavage of sileneoside N (104), carried out by the sum of enzymes isolated from sweet almond, which contains β- galactosidase (Sadykov et al. 2000).

Acid hydrolysis gave sugar identified by paper chromatography as D-galactose.

The location of the sugar residue in the molecule of sileneoside H (33) was established on the basis of ^1H and ^{13}C NMR spectra.

The proton and carbon signals of the steroid skeleton of sileneoside H (33) were identical to those of integristeron A (45), except for signals of the side chain. The location of the acetyl group at C-25 was confirmed by a shift in the signals of the methyl groups at C-26 and C-27 (by +0.2 ppm for protons and +1.8–2.6 ppm for carbons) compared to integristerone A (Table 4.7).

The presence of a hexose residue was proven by ^{1} H NMR peaks in the region of δ 3.3–5.1 ppm and ^{13}C NMR spectrum, where the signals of six additional oxygenated carbons are observed in the region of δ 61.8–102.9 ppm.

The C-22 signal (δ 90.6) was shifted to a weak field (by 12.3 ppm compared with Integristerone A) and this suggested the location of the hexose residue at C-22, which was finally confirmed by ^{1}H-, ^{13}C-correlations H-1' → C-22 and H-22 → C-1 'observed in the PFG-HMBC spectrum of sileneoside H.

The identity of the sugar residue with α-D-galactopyranoside was established by equipping the characteristic signal of the anomeric proton at δ 5.10 (d, J = 3.3 Hz) (Jones et al. 1993), and chemical shifts C-2 ', C -3 ', C-4' and C-5 'in the ^{13}C NMR spectrum (Schneider et al. 2003).

4.6 Phytoecdysteroids *Silene Tomentella* Schischk

Silene tomentella Schischk plant is shown in Fig. 4.13, the stems are numerous, woody at the very base, straight or ascending at the base, simple or with few branches, very dense—finely fluffy at the bottom, bare at the top, and sometimes sticky. Blossoms and bears fruit in May–June. It grows on rocky and gravelly slopes in the outlier slides. Height 20–30 cm. To extract the sum of ecdysteroids used above-ground plant organs.

The aerial part of the *S. tomentella* plant was extracted with ethanol (Fig. 4.14). The extract was concentrated, diluted with water. Hydrophobic impurities precipitated were removed. Ethanol was evaporated *in vacuo*, the aqueous residue was treated with hexane. From the purified aqueous fraction, ecdysteroids were recovered first with ethyl acetate and then with butanol. The dry residue obtained after distillation of ethyl acetate was chromatographed on an alumina column and on SiO_2. At the same time, individual substances were isolated—tomentesterone A (46) and B (47), viticosterone E (43), 2-deoxy-α-ecdysone (22), polypodine B (41), ecdysterone (2) (Yusupova et al. 2021) (Table 4.8).

The amount of substances obtained after distillation of butanol was chromatographed on a silica gel column. An additional amount of ecdysterone (2), integristerone A (45), and sileneoside D (27) were isolated. Table 4.8 shows the physicochemical constants and the yield of ecdysteroids from plant materials.

Fig. 4.13 *Silene tomentella* Schischk

4.6.1 The Structure of Tomentesterone A

Tomentesterone A (46), $C_{36}H_{50}O_7$, m.p. 139–140 °C (methanol–water) is the least polar ecdysteroid from all isolated compounds from the aerial part of *Silene tomentella* (Ramazanov et al. 1995).

The IR spectrum has absorption bands at 3420 cm^{-1} (OH) and 1660 cm^{-1}, corresponding to the absorption frequency of the keto group conjugated to the double bond. In addition, the IR spectrum of the studied tomentesterone A (46) contains absorption bands at 1720, 1730 cm^{-1} (ester group), 1610, 1590 cm^{-1} (benzene ring). The signals of five aromatic protons at 7.48 (3H) and 8.18 (2H) ppm also testify to this. in the ^1H NMR spectrum (Table 4.6) and intense peaks of ions in the mass spectrum with m/z 122 ($C_7H_6O_2$) and 105 (C_7H_5O), characteristic of benzoic acid (Fig. 4.15).

In addition, in the ^1H NMR spectrum, there is a signal of a three-proton singlet at 2.14 ppm, indicating the presence of one acetyl group in tomentesterone A (46) (Table 4.9).

A comparative analysis of the characteristics of the NMR spectra of the ecdysteroids of tomentesterone A (108) and 5β-2-deoxy-ecdysteroid indicates their proximity, with the exception of chemical shifts of the resonance lines of CH_3-19, 21, 26/27, and H-22. From the values of the chemical shifts of protons and groups (Table

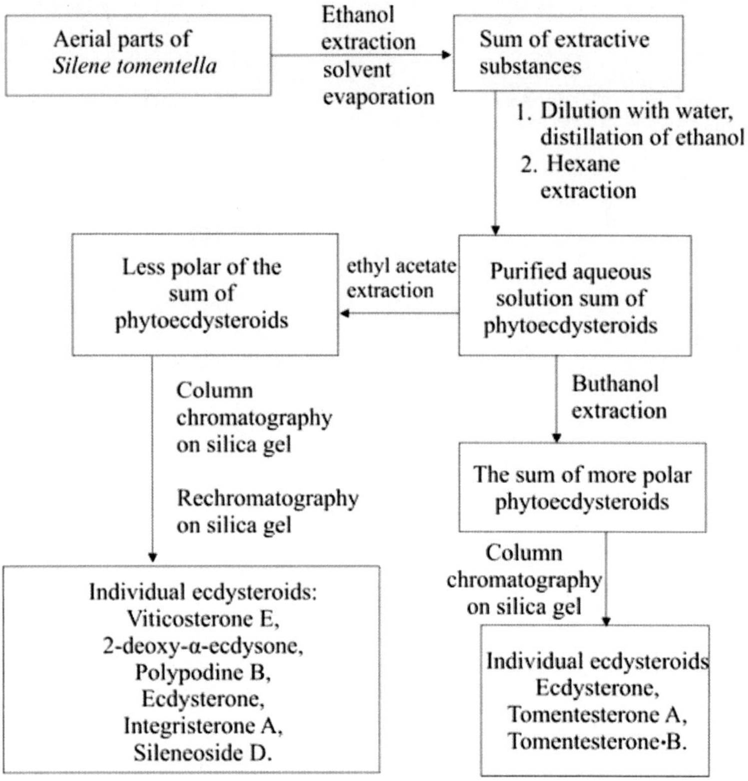

Fig. 4.14 Isolation of phytoecdysteroids from the aerial parts of *Silene tomentella*

Table 4.8 Physicochemical properties of phytoecdysteroids isolated from *Silene tomentella*

Name of the compounds	Formula	M.P. °C	[α]D CH_3OH	Yield, (%)
Tomentesterone A	$C_{36}H_{50}O_7$	139–140	–	0,0007
Tomentesterone B	$C_{34}H_{48}O_6$	145–147	–	0,0004
Viticosterone E	$C_{29}H_{46}O_8$	195–196	+59.2 ± 2	0,0014
2-Deoxy-α-ecdysone	$C_{27}H_{44}O_5$	234–235	+93.2 ± 2	0,116
Polypodine B	$C_{27}H_{44}O_8$	252–254	+94.2 ± 2	0,0022
Ecdysterone	$C_{27}H_{44}O_7$	241–242	+63.2 ± 2	0,0340
Integristerone A	$C_{27}H_{44}O_8$	246–248	+36.1 ± 2	0,0200
Sileneoside D	$C_{39}H_{54}O_{12}$	240–242	+91.4 ± 2	0,018

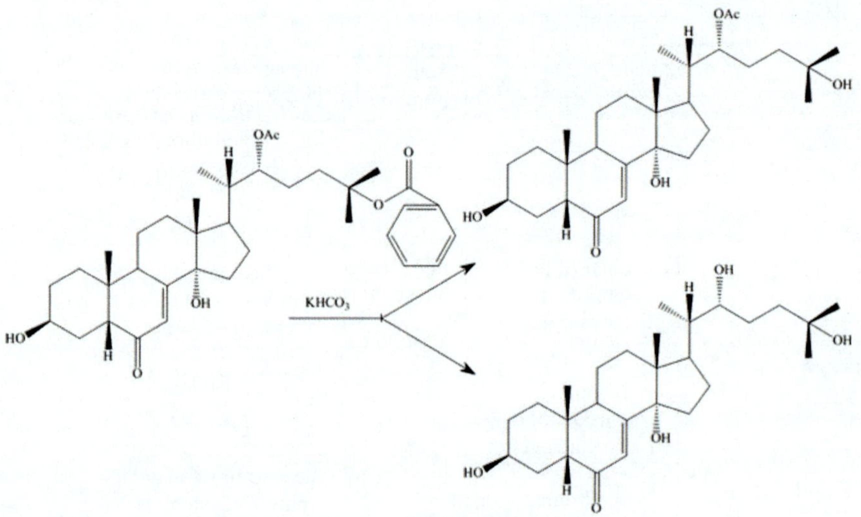

Fig. 4.15 Chemical transformations of tomentesterone A

4.9), it is seen that in the case of tomentesterone A (46), in comparison with 5β-2-deoxy-ecdysteroid, the CH_3-19 signal undergoes a diamagnetic shift of 0.18 ppm. A similar phenomenon is observed in the transition from 5β-2-deoxy-ecdysteroid to its 5α-isomer of 2-deoxy-ecdysteroid (Dzhukharova et al. 1995). Hence, it was concluded that the rings A and B in the steroid part of the ecdysteroid of tomentesterone A are mutually trans-linked, with H-5 having an α orientation. Indeed, 22-O-acetate-2-deoxy-α-ecdysone and 5α-2-deoxy-α-ecdysone (22) were isolated from the products of the alkaline hydrolysis of tomentesterone A (46). In the IR spectrum of another hydrolysis product, 22-O-acetate-2-deoxy-α-ecdysone, there are characteristic absorption bands at 1735 and 1260 cm^{-1} and in the PMR spectrum there are resonance lines of one acetyl group, H-22, CH_3-21, and CH_3-26/27 occur at 2.08, 5.33, 1.13, and 1.35 ppm, respectively. If we compare these characteristics with those shown in Table 4.9 for tomentesterone A (46), we find the absence of aromatic proton signals in the spectrum of 5α-2-deoxy-α-ecdysone, the conservation of the chemical shift of H-22 and CH_3-21, and the diamagnetic displacement singlet from CH_3-26/27 at 0.25 ppm These data are taking into account the values of chemical shifts of protons of the corresponding groups of 5α-2-deoxy-α-ecdysone (22) and 22-O-acetate-2-deoxy-α-ecdysone (Dzhukharova et al. 1995) and clearly indicate that the compound is 5α-2-deoxy-α-ecdysone-22-acetate.

A comparative analysis of all spectral characteristics given in Table 4.6.2 allowed us to conclude that tomentesterone A (46) is 22-O-acetate-25-O-benzoate-5α-2-deoxy-α-ecdysone (Ramazonov et al. 1995).

Mass spectrum: m/z (%): 576 (M-H_2O; 0.8), 566 (0.5), 474 (46), 566 (0.5), 472 (10), 452 (46), 444 (46), 439 (9), 412 (46), 397 (46), 379 (11), 332 (30), 331 (35), 284 (100), 234 (13), 231 (12), 122 (50), 105 (45), 99 (46), 81 (47) (Fig. 4.16).

Table 4.9 Chemical shifts of the protons of tomentesterone A, 5α-2-deoxy-α-ecdysone-22-acetate, 5α-2-deoxy-α-ecdysone, tomentesterone B, 3,22-diacetate of tomentesterone B and 2-deoxy-α-ecdysone (δ, ppm, C_5D_5N, 0-TMC)

Position of proton									
H-3	H-22	H-7	CH_3-18	CH_3-19	CH_3-21	CH_3-26	CH_3-27	H_{Ar}	OAc
Tomentesterone A									
4.10	5.34	6.19 J = 3 Hz	0.74	0.89	1.13 J = 6.5 Hz	1.60	1.60	7.48 (3H) 8.18 (2H)	2.14 (3H)
5α-2-Deoxy-α-ecdysone-22-acetate									
4.12	5.33	6.14 J = 3 Hz	0.72	0.90	1.13 J = 6 Hz	1.35	1.35	–	2.08 (3H)
5α-2- Deoxy-α-ecdysone									
4.05	4.05	6.18 J = 3 Hz	0.73	0.93	1.27 J = 6 Hz	1.41	1.41	–	–
Tomentesterone B									
4.09	4.09	6.16 J = 3 Hz	0.71	0.89	1.30 J = 6 Hz	1.65	1.65	7.49 (3H) 8.22 (2H)	–
3,22-Diacetate tomentesterone B									
4.71	4.91	5.91 J = 3 Hz	0.67	0.87	0.84 J = 6 Hz	1.59	1.61	7.47 (3H) 8.01 (2H)	2.04 (3H) 2.06 (3H)
2-Deoxy-α-ecdysone									
4.13	4.13	6.16	0.75	1.07	1.31 J = 6.5 Hz	1.43	1.43	–	–

The signal of CH_3-21 group is a doublet, of other methyl groups—singlets. The proton H-7 in compound 4 appears as a broadened singlet; in other cases, in the form of a doublet, the remaining signals are broadened multiplets

4.6.2 The Structure of Tomentesterone B

Tomentesterone B (47), $C_{34}H_{48}O_6$, m.p. 145–147 °C (methanol–water), isolated from the aerial organs of *Silene tomentella* with a very small yield (Ramazanov et al. 1996).

In the IR spectrum, a wide absorption band is observed at 3429 cm^{-1} (OH), as well as the absorption frequency at 1662 cm^{-1}, which corresponds to the vibration of the carbonyl of the keto group conjugated to the double bond. The absorption bands at 1712, 1292 and 1643, 1584, 713 cm^{-1} correspond, respectively, to the ester group and the benzene ring. These data, as well as intense ion peaks in the mass spectrum with m/z 122 [$C_7H_6O_2$ (100%)], 105 (C_7H_5O), and 77 (C_6H_5) indicate that tomentesterone B contains the remainder of benzoic acid (Fig. 4.17) Signals of

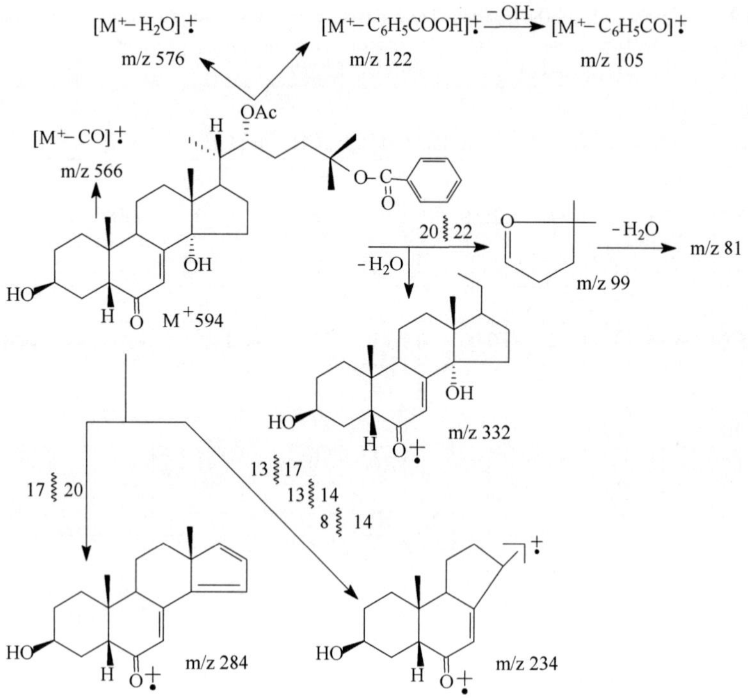

Fig. 4.16 The main ways of the mass spectrometric decay of tomentesterone A

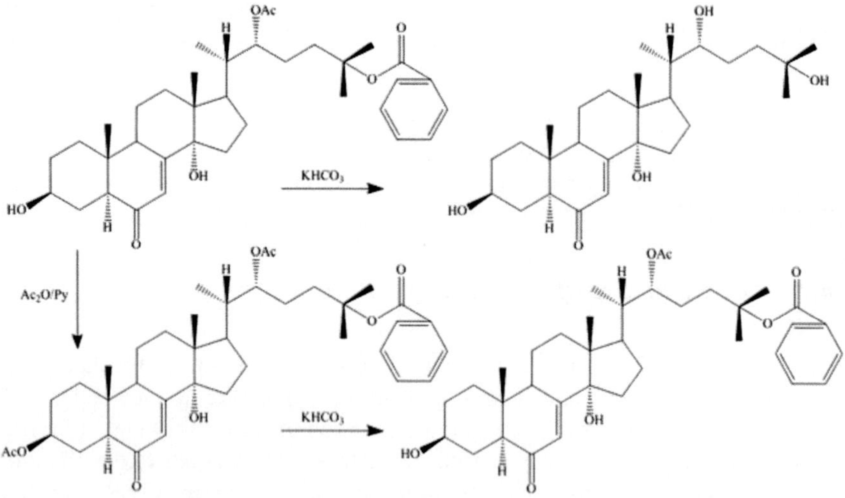

Fig. 4.17 Chemical transformations of tomentesterone B

five aromatic protons at 8.22 (2H) and 7.49 ppm (3H) in the NMR spectrum indicate the presence of one benzoyl group (Table 4.9).

5α-2-deoxy-α-ecdysone was identified in the neutral fraction of the products of alkaline saponification of tomentesterone B (22) (Dzhukharova et al. 1995). and in the acidic part of the hydrolyzate, the presence of benzoic acid was detected.

Mass spectrum: m/z (%): 430 (M + 122; 24), 415 (24), 412 (44), 397 (24), 379 (24), 332 (44), 284 (84), 234 (82), 122 (100), 105 (66), 99 (32), 77 (42) (Fig. 4.18).

Acetylation of tomentesterone B (47) yielded 3,22-diacetate of tomentesterone B. In its mass spectrum, intense peaks of ions with m/z 374 ($C_{23}H_{34}O_4$) and 276 ($C_{17}H_{24}O_3$) are observed (Rashkes and Abubakirov 1980), indicating that the remainder of benzoic acid is in the side chain of the ecdysteroid. Indeed, when comparing the characteristics of the ^{1}H-NMR spectra of tomentesterone B (47) and 5α-2-deoxy-α-ecdysone (22), a significant weak-field shift of the resonance lines of 26/27 methyl groups in the spectrum of the former is observed (Table 4.9).

A similar shift was observed earlier in the spectrum of tomentesterone A (Ramazanov et al. 1995). This fact suggests that the oxy group is esterified at tomentesterone B in benzoic acid.

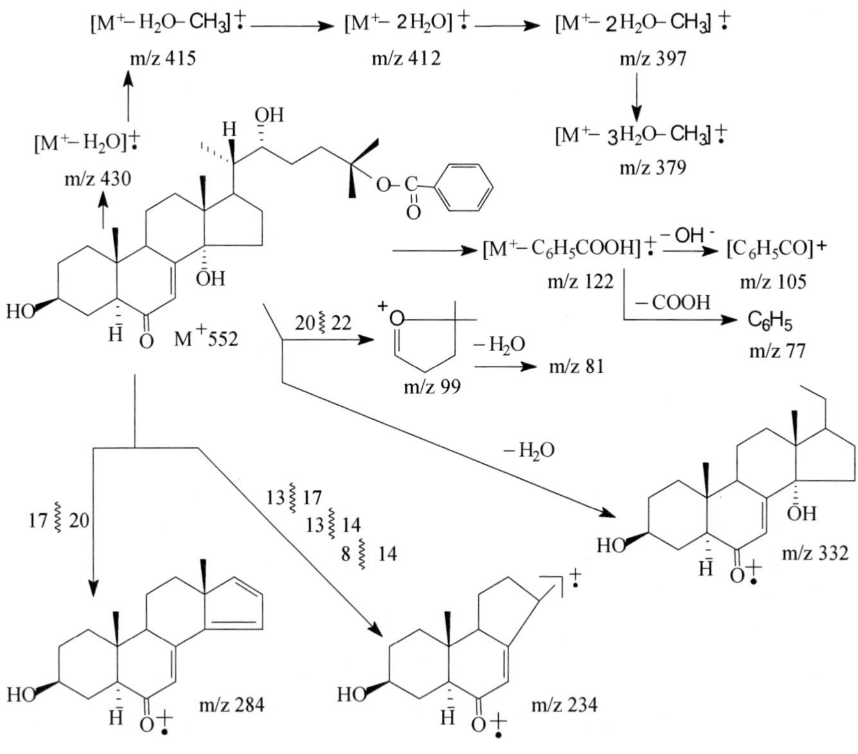

Fig. 4.18 The main ways of mass spectrometric decay of tomentesterone B

To confirm the structure of timentesterone B, the diacetate of tomentesterone B (109) was saponified with a 0.5% aqueous methanol solution of potassium bicarbonate. As one would expect, under these conditions, the acetyl group at C-22 remained unaffected and the isolated reaction product in its physicochemical constants (R_f, IR, and mass spectra) coincided with tomentesterone A (46).

These data indicate that the remainder of benzoic acid is indeed attached to the oxy group at C-25.

Thus, the ecdysteroid of tomentesterone B (47) is 25-O-benzoate-5α-2-deoxyecdysone (Ramazanov et al. 1996).

4.6.3 *Identification of Viticosterone E, 2-Deoxy-A-Ecdysone, Polypodine B, Ecdysterone, Integristerone A, Sileneoside D*

Isolation of known ecdysteroids from the aerial part of *S. tomentella* and their identification with previously described ones.

Viticosterone E (43), $C_{29}H_{46}O_8$, after crystallization from acetone, had m.p. 195–196 °C, $[\alpha]_D + 59.2 \pm 2$ °C (c 0.50; methanol) (Ramazanov et al. 1995). In the infrared spectrum of viticosterone E (43), in addition to absorption bands characteristic for this class of compounds, at 3430–3450 cm^{-1} (OH) and 1670 cm^{-1} (7-en-6-keto grouping), a maximum was observed at 1275 cm^{-1}. corresponding to the ester group. In the ^1H NMR spectrum of viticosterone E (43) at 1.92 ppm, the signal of the methyl group of the acetyl residue is observed.

2-Deoxy-α-ecdysone (22) (Fig. 4.19), $C_{27}H_{44}O_5$, after recrystallization from aqueous ethanol, had m.p. 234–235 °C, $[\alpha]_D + 93.2 + 2°$ (s 0.50; methanol) (Yusupova et al. 2020a).

In the IR spectrum of 2-deoxy-α-ecdysone (22), the absorption band at 3440 (OH) and 1645 cm^{-1} (α, β-unsaturated keto grouping), as well as the maximum in the UV spectrum at 245 nm (lgε 4.10) gave the ability to attribute this substance to the class of ecdysteroids.

Analysis of the mass spectrum of 2-deoxy-α-ecdysone (22), the key fragments of which had m/z 448 (M+), 430, 412, 332, 314, 284, 99, and 81, confirmed this assumption and at the same time indicated that the compound in question contains one hydroxy group in the tetracyclic core less than α-ecdysone (1) (Karlson et al. 1965).

Based on the above physicochemical constants, spectral data (Table 4.10), as well as direct comparison on TLC shows a very good identity with an authentic sample (Saatov et al. 1979) substance 52 which was identified as 2-deoxy-α-ecdysone. Polypodine B (85), $C_{27}H_{44}O_8$, m.p. 250–251 °C, $[\alpha]_D + 94.2 \pm 2°$ (c 0.50; methanol) (Yusupova et al. 2019).

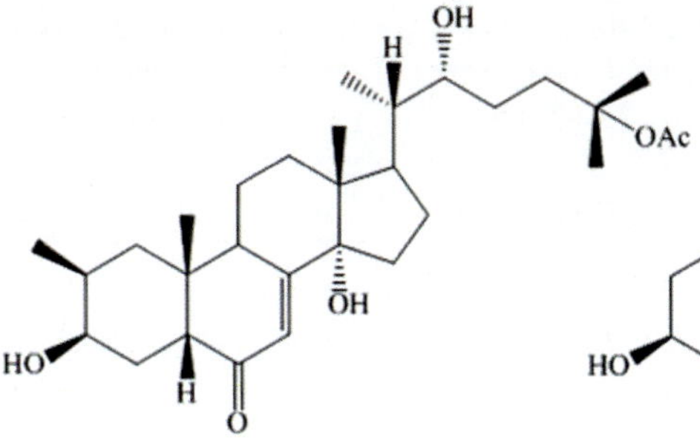

Viticosterone E (43)

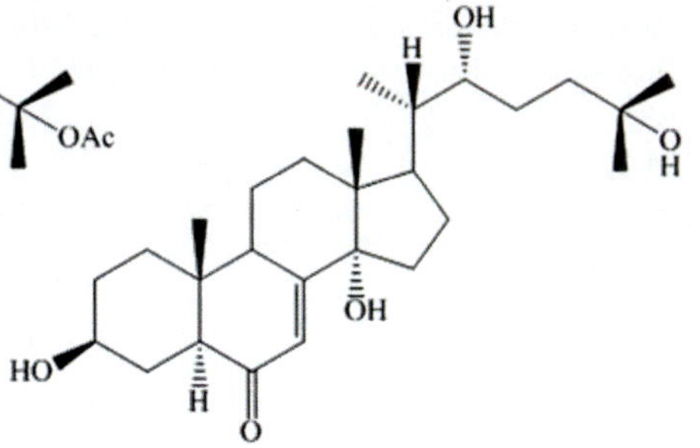

2-Deaxy-α-acdysone (22)

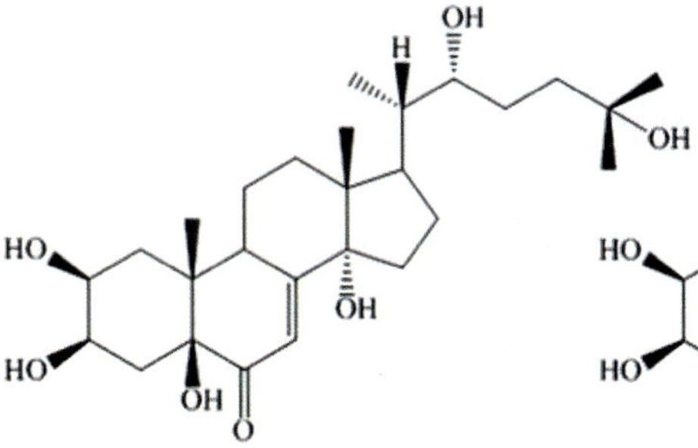

Polypodine B (41)

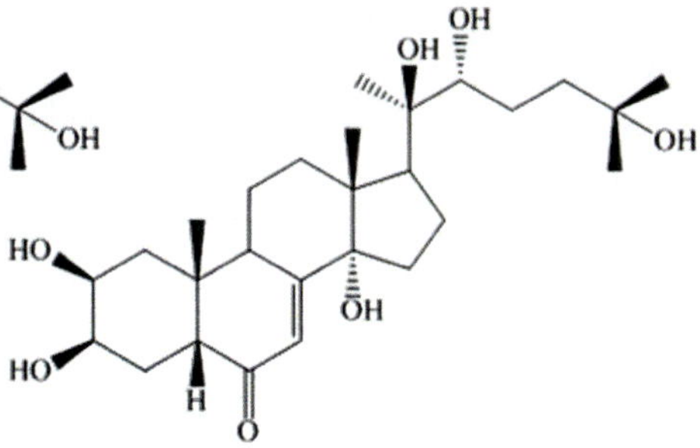

Ecdysterone (2)

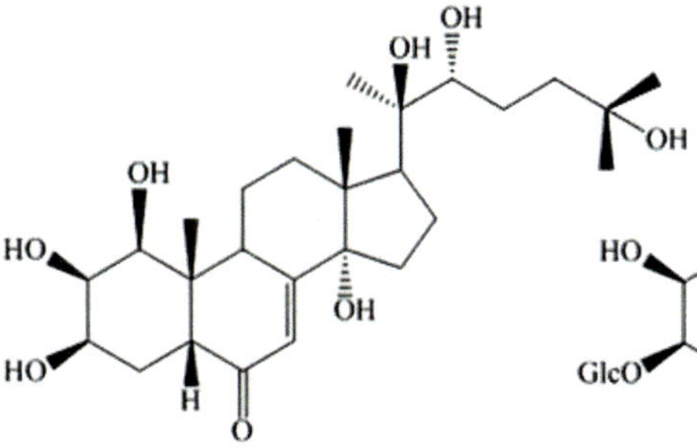

Integristerone A (45)

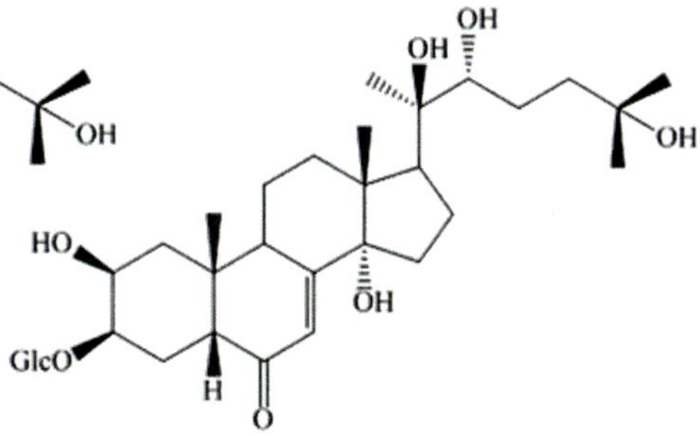

Ecdysterone (2)

Fig. 4.19 Ecdysteroids isolated from *S. tomentella*

The IR spectrum of this ecdysteroid contained maxima corresponding to hydroxyl absorption (3400 cm^{-1}) and α, β-unsaturated keto grouping (1673 cm^{-1}). In the UV spectrum of polypodine B, a maximum was observed at 244 nm (log $\varepsilon = 3.9$).

In the mass spectrum of polypodine B (41), peaks of ions with m/z 478 (M + H_2O), 460, 442, 424, 379, 361, 343, 325 were observed, indicating the presence of four hydroxy groups in the steroid nucleus (Usmanov et al. 1975).

The values of chemical proton shifts in the ^1H NMR spectrum of polypodine B (41) coincided with similar values for polypodine B (41) (Yusupova et al. 2019), (Table 4.10).

Table 4.10 Chemical shifts of protons of viticosterone E (43), 2-deoxy-α-ecdysone (22), polypodine B (41), ecdysterone (2), integristerone A (45), sileneoside D (27) (δ, ppm, C_5D_5N, 0-TMS)[a]

Position proton											
H-1	H-2	H-3	H-7	H-9	H-22	CH_3-18	CH_3-19	CH_3-21	CH_3-26/27	H'-	OAc
Viticosterone E											
	–	4.18	6.15	3.65	3.81	1.21	1.07	1.58	1.42;1.49	–	1.92
2-Deoxy-α-ecdysone											
	–	4.13	6.16	–	4.13	0.75	1.07	1.13;d J = 6.5 Hz	1.43	–	–
Polypodine B											
	–	4.23	6.20	3.61	3.83	1.19	1.16	1.56	1.37	–	–
Ecdysterone											
–	–	4.23	6.20	–	–	1.20	1.07	1.57	1.38	–	–
Integristerone A											
4.30	4.30	4.30	6.17	3.57	3.77	1.19	1.40	1.58	1.39	–	–
Sileneoside D											
		4.09	6.23	3.50	3.88	1.22	0.99	1.61	1.38	5.61 J = 3.9 Hz	–

[a]Signals of protons of methyl groups are singlet in nature. In all cases, the H-7 proton appears as a broadened singlet; the rest of the signals (with the exception of H-11) are extended multiplets

The data presented, as well as a direct comparison with a reliable sample (TLC), allowed us to identify it as polypodine B.

Ecdysterone (2), $C_{27}H_{44}O_7$ obtained after crystallization from acetone with a melting point of 241–243 °C, $[\alpha]_D + 63.2 + 2°$ (c 0.30; methanol) (Ramazonov et al. 1995).

The infrared spectrum of ecdysterone showed a characteristic absorption at 3400 (OH) and 1665 cm^{-1} (7-en-6-keto grouping). In the UV spectrum of ecdysterone, a maximum was observed at 245 nm (log $\varepsilon = 4.01$). The main pathways for the mass spectrometric decay of ecdysterone were shown in Fig. 4.20.

Integristerone A (45), $C_{27}H_{44}O_8$ with m.p. 246–248° C, $[\alpha]_D + 36.1 \pm 2°$ (from a mixture of ethyl acetate–methanol), (c 0.43; methanol) (Ramazonov et al. 1995). UV spectrum—245 nm (lgε 4.00). In the IR spectrum, there is an absorption typical for most ecdysteroids at 3400 and 1660 cm^{-1}.

In the mass spectrum of integristerone A (45), ions with m/z 478 $(M\text{-}H_2O)^+$, 460, 442, 424, 379, 361, 343, and 325 indicated the presence of four hydroxyl groups in the tetracyclic core (Zibareva 2000).

The broadened singlet at 4.28 ppm in the ^1H NMR spectrum is due to three protons geminally located to the secondary OH groups. The coincidence of the value of the indicated chemical shift of the compound with those for integristone A indicates that

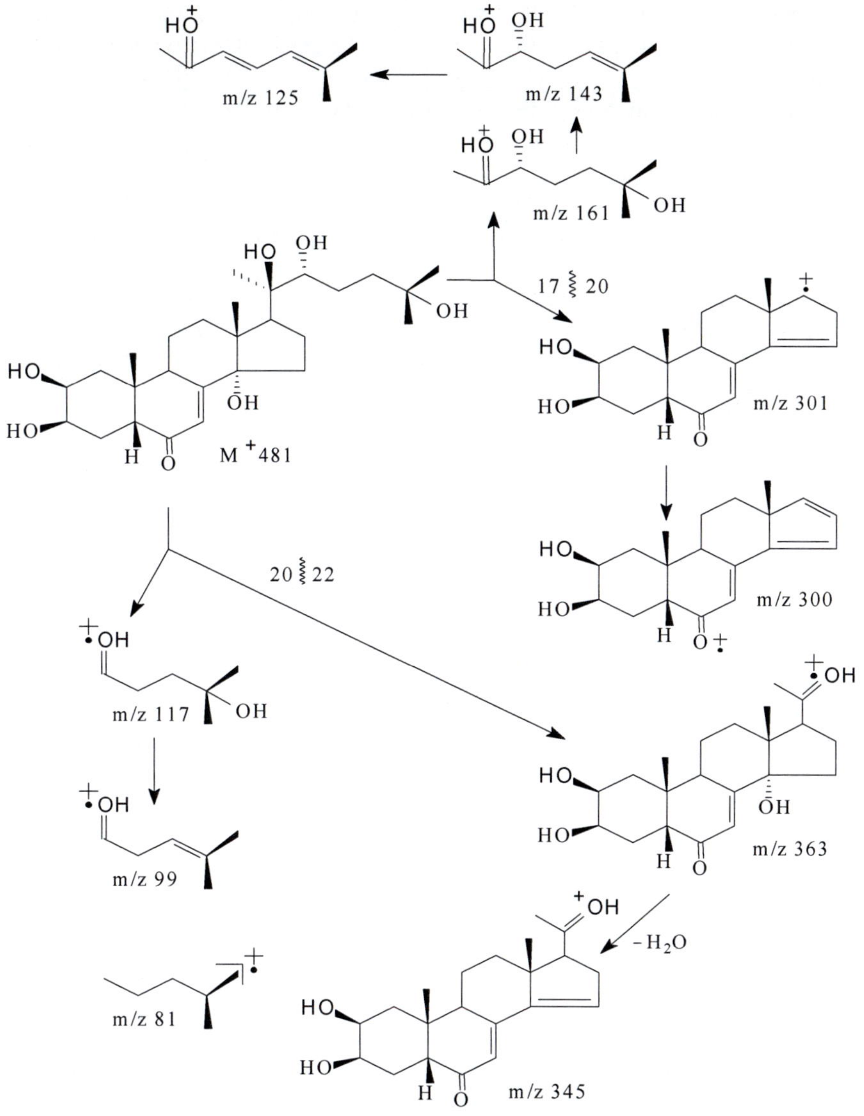

Fig. 4.20 The main pathways for the mass spectrometric decay of ecdysterone

the additional hydroxyl group in comparison with ecdysterone is located at C-1. The values of the signals of the protons of the methyl groups coincide with the values of the corresponding signals of the integristerone A (45) (Table 4.10). The identification of integristerone A (45) was carried out on the basis of physicochemical constants, spectral data and was identified by direct comparison with the reference sample (Saatov et al. 1981).

Sileneoside D (27), $C_{39}H_{54}O_{12}$, after recrystallization from a methanol-acetone mixture, had m.p. 240–242° C, $[\alpha]_D + 91.4 \pm 2°$ (methanol) (Ramazonov et al. 1995).

In the UV spectrum, a maximum characteristic of ecdysteroids was observed at 247 nm (log ε 4.00).

In the IR spectrum of silenoside D (27), there is a wide absorption band at 3380–3430 cm^{-1}, corresponding to stretching vibrations of hydroxyl groups. The band at 1650 cm^{-1} corresponds to the absorption of a keto group conjugated to a double bond.

The peak of the molecular ion in the mass spectrum of sileneoside D (27) is absent. The presence in the mass spectrum of sileneoside D (27) of products of sequential dehydration of a molecular ion with m/z 624, 606, 588, and 570 in combination with fragments with m/z 363, 345, 327, 99, 81, and 69 inherent in ecdysteroids (Usmanov et al. 1973), as well as fragments with m/z 163 and 145 corresponding to fragmented hexose, suggested that the compound 27 is an ecdysterone glycoside.

The chemical shifts of protons in the ^1H NMR spectrum of sileneoside D (27) coincided with similar parameters (Table 4.10) (Saatov et al. 1984).

The data presented, as well as a direct comparison with a reliable sample (TLC), allowed us to identify sileneoside D.

Silene tomentella is a perennial herbaceous plant from the clove family, 20–30 cm high. The stems are numerous, woody at the base, straight or ascending at the base, simple or with few branches, very densely small-leaved underneath, bare above, and sometimes sticky. Flowers are few in racemose inflorescences, on relatively short, densely pedicel stalks. Blossoms and bears fruit in May–June. It grows on rocky and gravelly slopes in the outlier slides. Endemic plant, found in the Bukhara region and in Karakalpakistan.

4.7 Phytoecdysteroids of Plants of the Genus *Silene Wallichiana* Klotzsch

S. Wallichiana is a perennial herb from the clove family. The stems are several, straight, or ascending below, branched, bare above, 40–100 cm in height. Leaves are oblong-lanceolate, oblong or reverse-oblong, at the apex of the b.m. drawn, very sharp, bare, bluish-green above. The box is almost spherical, 7–9 mm long. Seeds are brown, sharp-tuberous, about 1 mm wide. Blossoms and bears fruit in June-July. It is found in the shade of rocks and trees in the middle zone of mountains. It is widespread in Tashkent, Ferghana, Samarkand, Kashkadarya, and Surkhandarya regions of Uzbekistan (Fig. 4.21).

A study of the ecdysteroids of the plant *Silene wallichiana* Klotzch (family *Caryopyllaceae*) allowed the authors (Saatov et al. 1988a) to isolate the known

Fig. 4.21 Wallich Smile—*Silene Wallichiana* Klotzsch

phytoecdysteroids, 2-deoxy-α-ecdysone, from the aerial part and roots (22), 2-deoxyecdysterone (23), 22-O-benzoate-ecdysterone (48), ecdysterone (2), integristerone A (45), and one new phytoecdysteroid 49—3-O-benzoate-2-deoxy-ecdysterone.

The IR spectrum data (1733, 1290, 1610, 720 cm^{-1}), as well as the peaks of ions with m/z 122 ($C_7H_6O_2$), 105 (C_7H_5O), and 77 (C_6H_5) in the mass spectrum, indicate that 3-O-benzoate-2-deoxyecdysterone (49) contains the remainder of benzoic acid. In its ^{1}H NMR spectrum, the signals of 2 aromatic protons at 8.12 (3H) and 7.37 ppm (3H) 2 indicate the presence of one benzoyl group.

In the mass spectrum of 3-O-benzoate-2-deoxyecdysterone (49), in addition to the above, peaks of ions with m/z 347, 329, 303, 234, 233 are observed, which makes it possible to attribute it to 2-deoxyecdysteroid (23) (Rashkes and Abubakirov 1980).

When comparing the NMR spectra of 2-deoxyecdysterone (23) and 3-O-benzoate-2-deoxyecdysterone (49), a significant difference is revealed only in the position of the resonance lines of the proton at C-3. The spectrum of 3-O-benzoate-2-deoxyecdysterone (49) has a peak with m/z 338, which is formed due to the splitting of C/D rings (Koreeda et al. 1969). The presence of this fragment indicates the connection of the remainder of benzoic acid with the steroid part of the molecule.

Among the products of alkaline hydrolysis of 3-O-benzoate-2-deoxyecdysterone (49), in addition to benzoic acid, 2-deoxyedysterone (23) and 5α-2-deoxyecdysterone were isolated.

Table 4.11 Chemical shifts of protons of 2-deoxyecdysterone (23), 3-O-benzoate-2-deoxyecdysterone (49) and 5α-2-deoxyecdysterone (δ, ppm O-HMDS)[a]

Position proton								
H-3	H-7	H-9	H-22	CH_3-18	CH_3-19	CH_3-21	CH_3-26/27	Aromatic proton
2-Deoxyecdysterone								
3.99	6.10	3.40	3.73	1.10	0.93	1.46	1.26	–
2-Deoxyecdysterone-3-O-benzoate								
5.22	6.12	3.40	3.80	1.12	0.92	1.50	1.25	7.37(3H), 8.12(2H)
5α-2-Deoxyecdysterone								
3.708	5.98	–	3.70	1.07	0.79	1.45	1.25	–

[a]The signals of the protons of methyl groups are singlet-like, in all cases the H-7 proton appears in the form of a broadened singlet, other signals are broadened multiplets

In the ^{1}H-NMR spectrum of 5α-2-deoxyecdysterone (Table 4.11), a noticeable shift of the signal of the 19-methyl group to a strong field is observed compared with that of ecdysteroid 12 ($\Delta\delta = 0.14$ ppm). The magnitude of the chemical shifts of the remaining signals is virtually unchanged. The indicated diamagnetic shift suggests (Saatov et al. 1987b) that the compound obtained by alkaline isomerization of ecdysteroid 49 is 3-O-benzoate-2-deoxyecdysterone. Thus, the new ecdysteroid 49 is 3-O-benzoate-2-deoxyecdysterone.

References

Baltaev UA (1979) Phytoecdysteroids of plants of the genus *Rhaponticum*: Dis.. cand. Chem Sci 106

Baltaev UA (2000) Phytoecdysteroids: structure, sources and pathways of biosynthesis in plants. Russ J Bio Chem 26:892–925. https://doi.org/10.1023/a:1026662505403

Baltaev UA, Gorovits MB, Abdullaev ND et al (1978) Phytoecdysones of *Rhaponticum* integrifolium III. Integristerone B. Chem Nat Compd 14:388–393. https://doi.org/10.1007/BF00565240

Borisov (1963) Flora of the USSR: Ed. V.L. Komarova. Leningrad 28:308

Dinan L (1998) A strategy towards the elucidation of the contribution made by phytoecdysteroids to the deterrence of invertebrate predators on plants. Rus J Plant Physiol 45(3):347–359

Dinan L, Harmatha J, Volodin V, Lafont R (2009) Phytoecdysteroids: diversity, biosynthesis and distribution. In: Smagghe G (eds) Ecdysone: structures and functions. Springer, Dordrecht. https://doi.org/10.1007/978-1-4020-9112-4_1

Dragana ZJ, Sava MV, Milan SS, Ljiljana RČ, Marina DT (2015) Secondary metabolite content and in vitro biological effects of Ajuga chamaepitys (L.) Schreb. SuBSp. Chamaepitys. Arch. Biol. Sci., Belgrade 67(4):1195–1202. https://doi.org/10.2298/ABS150225095J

Dzhukharova MK, Saatov Z, Abdullaev ND (1995) Phytoecdysteroids of plants of the genus *Silene* XVI. 5α-sileneoside E from *Silene brahuica*. Chem Nat Compd 31:207–210. https://doi.org/10.1007/BF01170207

Galbraith MN, Horn DHS (1966) An insect-molding hormone from a plant. J Chem Soc Chem Commun 24:905–906. https://doi.org/10.1039/C19660000905

Hikino H, Hikino Y, Nomoto K, Takemoto T (1968) Cyasterone, an insect metamorphosing substance from *Cyathulacapitata*: structure. Tetrahedron 24:4895–4906. https://doi.org/10.1016/S0040-4020(01)98686-7

Imai S, Fujioka S, Murata E, Otsuka K, Nakanishi K (1969) Structure of the phytoecdysone, ajugasterone B. Chem Commun 82. https://doi.org/10.1039/C29690000082

Jakovljević DZ, Vasić SM, Stanković MS, Čomić LR, Topuzović MD (2015) Secondary metabolite content and *in vitro* biological effects of *Ajuga chamaepitys* (L.) Schreb. SuBSp. *Chamaepitys*. Arch Biol Sci, Belgrade 67(4):1195–1202. https://doi.org/10.2298/ABS150225095J

Jones C, Mulloy B, Thomas AH (1993) Methods in molecular biology. Spectroscopic methods and analyses. Humana Press, Totowa 17(7):366

Karlson P, Hoffmeister H, Hoummel H et al (1965) Reaktionen des Ecdysonmolekus. Chem Ber 98(7):2394–2402. https://doi.org/10.1002/cber.19650980743

Koreeda M, Narada N, Nakanishi K (1969) Interaction between non-adjacent Benzoate Groups: an Extension of the Dibenzoate cherally Rule. J Chem Soc Chem Commun 548–549. https://doi.org/10.1039/C29690000548

Koreeda M, Nakanishi K, Goto M (1970) Ajugalactone, an insect molding inhibitor as tested by the chile dipping method. J Am Chem Soc 92(25):7512–7513. https://doi.org/10.1021/ja00728a068

Mamarasulov B, Davranov K, Jabborova D (2020) Phytochemical, pharmacological and biological properties of *Ajuga turkestanica* (Rgl.) Brig (Lamiaceae). Ann Phytomedicine 9(1):44–57. https://doi.org/10.21276/ap.2020.9.1.6

Politova NK, Punegov VV, Volodin VV et al (1997) Synthesis of 20-hydroxyecdysone 25-acetate. Chem Nat Compd 33:57–60. https://doi.org/10.1007/BF02273925

Pongrácz Z, Báthori M, Simon A, Tóth G, Mák M (2003) Máthé I. 9,20-Dihydroxyecdysone, a new natural ecdysteroid with an unique structure from *Silene italica* ssp. *nemoralis*. J Nat Prod 66(3):450–1. https://doi.org/10.1021/np0205194

Ramazanov NS (2005) Phytoecdysteroids and other biologically active compounds from plants of the genus *Ajuga*. Chem Nat Compd 41:361–369. https://doi.org/10.1007/s10600-005-0153-4

Ramazanov NS, Maksimov ES, Saatov Z et al (1995) Phytoecdysteroids of plants of the genus *Silene*. XVII. Tomentesterone from *Silene tomentella*. Chem Nat Compd 31:600–603. https://doi.org/10.1007/BF01164888

Ramazanov NS, Maksimov ES, Saatov Z et al (1996) Phytoecdysteroids of plants of the *Silene* genus. XVIII. Tomentesterone from *Silene tomentella*. Chem Nat Compd 32:47–49. https://doi.org/10.1007/BF01373789

Ramazanov NS, Maksimov ES, Saatov Z et al (1997) Phytoecdysteroids of plants of the genus *Rhaponticum* I. Carthamosterone A from *Rh. carthamoides*. Chem Nat Compd 33:301–302. https://doi.org/10.1007/BF02234880

Ramazonov NS, Mamadalieva NZ, Bobaev ID (2007) Phytoecdysteroids from five species of the genus *Silene*. Chem Nat Compd 43:117–118. https://doi.org/10.1007/s10600-007-0049-6

Ramazonov N, Yusupova U, Egamova F, Syrov V (2020) Anti stress activity of phytoecdysteroids isolated from aerial part of *Silene claviformis*. Chem Sci Inter J 29(7):9–17. https://doi.org/10.9734/csji/2020/v29i730190

Rashkes YV, Abubakirov NK (1980) Characteristics of the mass spectra of ecdysteroids with different numbers of –OR groups. Chem Nat Compd 16:372–381. https://doi.org/10.1007/BF00568371

Saatov Z, Usmanov BZ, Abubakirov NK (1979) Phytoecdysones of *Silene praemixta* I. Silenosterone. Chem Nat Compd 15:700–703. https://doi.org/10.1007/BF00565565

Saatov Z, Gorovits MB, Abdullaev ND et al (1981) Phytoecdysteroids of plants of the genus Silene. III. Sileneoside A—A new glycosidic ecdysteroid of Silene brachuica. Chem Nat Compd 17:534–539. https://doi.org/10.1007/BF00574372

Saatov Z, Gorovits MB, Abdullaev ND et al (1982) Phytoecdysteroids of plants of the genus *Silene*. IV. Sileneoside C—A galactoside of intergristerone A from *Silene brahuica*. Chem Nat Compd 18:193–196. https://doi.org/10.1007/BF00577192

Saatov Z, Abdullaev ND, Gorovits MB et al (1984) Phytoecdysteroids of plants of the genus *Silene*. VII. Sileneoside D—ecdysterone 3-O-α-D-galactopyranoside from *Silene brahuica*. Chem Nat Compd 20:700–703. https://doi.org/10.1007/BF00580027

Saatov Z, Gorovits MB, Melibaev S et al (1986a) Phytoecdysteroids of plants of the genus *Silene*. IX. Ecdysterone 22-O-benzoate from *Silene scabrifolia*. Chem Nat Compd 22:71–73. https://doi.org/10.1007/BF00574584

Saatov Z, Gorovits MB, Abubakirov NK (1987b) Phytoecdysteroids of plants of the genus *Silene*. XV. 2-Deoxy-α-ecdysone 22-O-benzoate from *Silene wallichiana*. Chem Nat Compd 23:708–711. https://doi.org/10.1007/BF00596648

Saatov Z, Mamathanov AU, Syrov VN, Gorovits MB, Yakubova MR, Khushbaktova ZA, Shakirov TT, Abubakirov NK (1990) A method of obtaining a substance with anabolic activity "Turkesterone". Academy Science. 1571822. Bulletin 22

Saatov Z (1983) Phytoecdysteroids of plants of the genus *Silene*: Dis. Cand. Chem. Sciences. Tashkent 139

Sadikov ZT, Saatov Z (1999) Phytoecdysteroids of plants of the genus *Silene* XX. Integristerone A 25-acetate from *Silene brahuica*. Chem Nat Compd 35:440–441. https://doi.org/10.1007/BF02282513

Sadykov Z, Saatov Z, Girault JP, Lafont R (2000) Silenoside H, a New phytoecdysteroid from *Silene brachuica*. J Nat Prod 63:987–988. https://doi.org/10.1021/np990609h

Sadykov ZT (2005) Phytoecdysteroids of plants *Silene brahuica, S. claviformis* and *Rhaponticum carthamoides*. Dis. cand. Chem. Sciences. Tashkent 105

Schneider B, Gershenzon J, Graser G et al (2003) One-dimensional ^{13}C NMR and HPLC-^{1}H NMR techniques for observing carbon-13 and deuterium labelling in biosynthetic studies. Phytochem Rev 2:31–43. https://doi.org/10.1023/B:PHYT.0000004196.73829.4e

Sridhar V, Vinesh LS, Mani M (2016) Mealybugs and their management in agricultural and horticultural crops. In: Mani M, Shivaraju C (eds) Medicinal Plants. Springer, New Delhi. https://doi.org/10.1007/978-81-322-2677-2_57

Takemoto T, Ogawa S, Nishimoto N (1967) Isolation of the moulting hormones of insects from Achyranthes radix. Yakugaku Zasshi 87(12):1469–1473. https://doi.org/10.1248/yakushi1947.87.12_1469

Usmanov VZ, Gorovits MB, Abubakirov NK (1971) Phytoecdysones of *Ajuga turkestanica*. Chem Nat Compd 7:520. https://doi.org/10.1007/BF00564772

Usmanov BZ, Gorovits MB, Abubakirov NK (1973) Phytoecdysones of *Ajuga turkestanicaII*. Chem Nat Compd 9:125–126. https://doi.org/10.1007/BF00580923

Usmanov BZ, Gorovits MB, Abubakirov NK (1975) Phytoecdysones of *Ajuga turkestanica*. III. The structure of turkesterone. Chem Nat Compd 11:484–487. https://doi.org/10.1007/BF00566790

Usmanov BZ, Saatov Z, Abubakirov N (1977) Phytoecdysones of *Ajuga turkestanica*. V. Chem Nat Compd 13:595. https://doi.org/10.1007/BF00569612

Usmanov BZ, Rashkes YV, Abubakirov NK (1978) Phytoecdysones of *Ajuga turkestanica* VI. 22-Acetylcyasterone. Chem Nat Compd 14:175–178. https://doi.org/10.1007/BF01134622

Vasilchenko (1936) Flora of the USSR: Ed. V.L. Komarova. Leningrad 6:577

Yakubova MR, Genkina GL, Shakirov TT et al (1978) Chromato-spectrophotometric method of determining ecdysterone in plant raw material. Chem Nat Compd 14:627–629. https://doi.org/10.1007/BF00937614

Yusupova UY, Usmanov DA, Ramazonov NS (2019) Phytoecdysteroids from the Plant *Dianthus helenae*. Chem Nat Compd 55(2):393–394. https://doi.org/10.1007/s10600-019-02701-y

Yusupova U, Usmanov D, Azamatov A, Ramazonov N, Rejepov J (2020a) Phytochemical constituents and biological activities of *Dianthus helenae* Vved., growing in Uzbekistan. Nat Prod Res. https://doi.org/10.1080/14786419.2020.1862834

Yusupova UY, Usmanov DA, Ramazonov NS (2020b) Phytoecdysteroids from the aerial part of *Silene popovii*. Chem Nat Compd 56(3):562–563. https://doi.org/10.1007/s10600-020-03092-1

Yusupova UY, Ramazonov NS, Bobakulov KM, Egamova FR, Syrov VN, Usmanov DA (2021) Biological active components of *Silene tomentella* and their pharmacological properties. Chem Plant Source 1:197–202. https://doi.org/10.14258/jcprm.2021018323

Zatsny IL, Gorovits MB, Rashkes YV et al (1975) Phytoecdysones of *Serratula* III. Mass spectrometric study of the acetates and acetonides of ecdysterone and viticosterone E. Chem Nat Compd 11:170–172. https://doi.org/10.1007/BF00570661

Zibareva L (2000) Distrbution and levels of phytoecdysteroids in plants of the genus silene during development. Arch Insect Biochem Phisiol 43(1):1–8. https://doi.org/10.1002/(SICI)1520-6327(200001)43:1

Zibareva LN (2003) Phytoecdysteroids of plants of the family *Caryophyllaceae*. Abstract. dis .doctor Chem. sciences. Novosibirsk 31

Chapter 5
Reactions of Phytoecdisteroids and Their Chemical Modification

Abstract In this chapter, it is shown the main experimental methods for studying phytoecdysteroids, their isolation, and studies of the reactivity of various functional groups of phytoecdysteroids, as well as obtaining modified derivatives. New derivatives of 3-O-p-bromobenzoyltiocarbamate-2-deoxyecdysterone and 3,22-O-di(p-bromobenzoylthiocarbamate) -2-deoxy-ecdysterone were obtained. Based on the interaction of cyasterone (obtained from Ajuga turkestanica) with paraiodobenzoylisothiocyanate, new derivatives of cyasteronoyl-thiocarbamates, 2,3,22-O-triacetate-cyasterone were obtained. Furthermore, synthesis of 2,3,20,22-O-diacetonide-20-hydroxyecdysone, 2,3-O-monoacetonide-20-hydroxyecdysone, and 20,22-O-monoacetonide-20- hydroxyecdysone, 2,3-O-monoacetonide-22-O-benzoate-20-hydroxyecdysone, synthesis of 22-O-benzoate-20-hydroxyecdysone. In this regard, the synthesis of monoesters and 3-keto-derivative ecdysteroids with the aim of creating a new type of biologically active compounds based on them is promising. Carrying out chemical transformation using available phytoecdysteroids as the starting compounds is the simplest and cheapest way to obtain rare natural phytoecdysteroids. The chemical transformation of 20-hydroxyecdysone has been developed with the aim of further synthesis of the conjugate of 22-O-benzoate-20-hydroxyecdysone. One of the promising directions for the transformation of ecdysteroids is the use of the acylation reaction. By the interaction of 20-hydroxyecdysone with ethyl chloroformate, 2,3–0-di- (ethyl carbonate) -20,22–0-carbonate-20-hydroxyecdysone were obtained. The purpose of such transformations is to enhance the biological effect of natural compounds and the search for new types of activity.

Keywords Chemical modifications · Derivatives of phytoecdysteroids · Synthesis · Reaction of phytoecdysteroids

U. Y. Yusupova et al., *Phytoecdysteroids*,
https://doi.org/10.1007/978-981-16-6711-4_5

5.1 Synthesis of Thiocarbamate Derivatives of 2-Deoxyecdysterone

The aerial part of *Silene wallichiana* was extracted with 86% aqueous alcohol. The extraction was repeated five more times, steering for 8 h. The extract was concentrated, diluted with water, the precipitate was removed. Ethanol was evaporated *in vacuo*. The aqueous residue was treated with butanol. The solvent was distilled off *in vacuo*. The residue was crystallized from an alcohol-acetone mixture. Individual ecdysteroids were isolated-2-deoxy-α-ecdysone (22), 2-deoxyecdysterone (23), and ecdysterone (2).

Based on the interaction of 2-deoxyecdysterone (23) with para-bromobenzoyl-isotiocyanate, new derivatives of 3-O-p-bromobenzoyltiocarbamate-2-deoxyecdysterone and 3,22-O-di(p-bromobenzoylthiocarbamate) -2-deoxy-ecdysterone were obtained (Fig. 5.1).

The reaction was carried out in pyridine medium at a temperature of 100–110 °C for 4–5 h. It should be noted that new derivatives of thiocarbamates were obtained with rather high yields (71.2%). The signals of benzene protons appearing in the spectrum to the right of the emission signal are

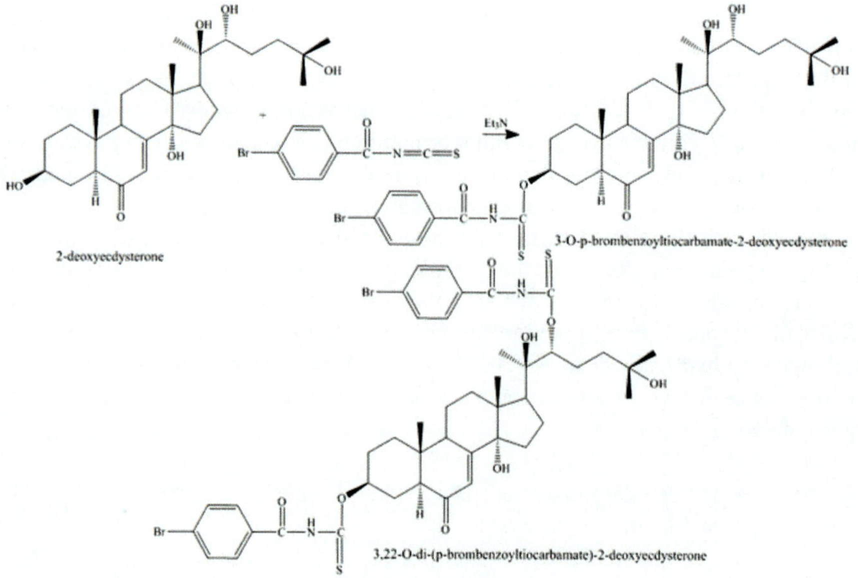

Fig. 5.1 Obtaining thiocarbamate derivatives of 2-deoxyecdysterone

Table 5.1 Chemical proton shifts 3-O-p-bromobenzoylthiocarbamate - 2-deoxyecdysterone, 3.22-O-di (p-bromobenzoyl-thiocarbamate)-2-deoxyecdysterone (δ, ppm O-HMDS)

Position proton								
H-3	H-7	H-9	H-22	CH_3-18	CH_3-19	CH_3-21	CH_3-26/27	Aromatic proton
3-O-p-bromobenzoylthiocarbamate- 2-deoxyecdysterone								
3,81	6,14	3,38	3,67	1,06	0,77	1,43	1,22	7.89 (H); 7.37 (H): 7.11 (2H)
3.22-O-di (p-bromobenzoyl-thiocarbamate)-2-deoxyecdysterone								
3,86	6,09	3,18	3,44	1,02	0,69	1,39	1,26	8.0(2H); 7.5(2H): 7.23 (4H)

Synthesis of 3-O-p-bromobenzoylthiocarbamate-2-deoxyecdysterone, 3,22-O-di (p-bromobenzoyl-thiocarbamate) -2-deoxyecdysterone

possibly related to the reaction positively polarized into the product. The synthesized substances are water-insoluble brown crystals. The structure of derivatives -3-O-p-bromobenzoylthiocarbamate–2-deoxyecdysterone and 3.22-O-di(p-bromobenzoylthiocarbamate)–2-deoxyecdysterone was confirmed by ^{1}H NMR spectroscopy (see Table 5.1).

In a three-necked flask equipped with a reflux condenser with a calcium chloride tube, a stirrer, a thermometer, 0.8 g is placed. (0.0017 mol) of 2-deoxyecdysterone (23), to which 2 ml of freshly distilled triethylamine are added. Under heating (80–90 °C.) and stirring, 4.0 g (0.163 mol) of p-bromobenzoyl-isothiocyanate dissolved in 10 ml of pyridine are added dropwise. The reaction mixture is stirred for 4 h at 100–110 °C. After the reaction, the mass is transferred into a 0.5-L glass, to which 100 ml of water is added. The yellow precipitate formed was dissolved in ethyl acetate and chromatographed on an Al_2O_3 column. A stain was obtained on TLC with $R_f = 0.83$, $R_f = 0.71$ (system—chloroform: ethanol = 5: 0.5).

The yield of 3-O-p-bromobenzoylthiocarbamate-2-deoxyecdysterone 71.2%, m.p. 229–230 °C. 3.22–O–di (p-bromobenzoylthio-carbamate)-2-deoxyecdysterone 15.1%, m.p. 244–246 °C. 3-O-p-bromo-benzoylthiocarbamate-2-deoxyecdysterone and 3,22-O-di(p-bromo-benzoylthiocarbamate) -2-deoxyecdysterone are dissolved in DMF, trifluoroacetic acid, pyridine. IR spectrum (3-O-p-bromobenzoylthiocarbamate-2-deoxyecdysterone) (KBr, ν, cm^{-1}): 3362 (N–H), 1673 (C = O), 1603 (aromatic ring), 1268 (C–O–C), 1160 (C = S).

IR spectrum of 3.22-O-di (p-bromobenzoylthiocarbamate) -2-deoxy-ecdysterone (KBr, ν, cm^{-1}); 1237 (C–O–C), 1681 (C = O), 1175 (C = C); 1597 (aroma. Rings), 3345 (N–H) (Ramazanov and Bobaev 2006).

5.2 Synthesis of Derivatives of Cyasterone Thiocarbamates

Based on the interaction of cyasterone (38) (obtained from *Ajuga turkestanica*) with para-iodobenzoylisothiocyanate, new derivatives of cyasteronoyl-thiocarbamates were obtained (Fig. 5.2).

The reaction was carried out in dry pyridine at a temperature of 100–110 °C for 6 h. It should be noted that new derivatives of cyasteronyl thiocarbamates were obtained with high yields.

All synthesized substances are yellow and white crystals, insoluble in water.

The structure of derivatives of 2,3,22-tri-(O-p-iodobenzoylthiocarbamate)-cyasterone was confirmed by PMR spectroscopy (Table 5.2).

Synthesis of 2,3,22-three O- and 2-O-para-iodobenzoylthiocarbamate-cyasterone

0.52 g are placed in a three-necked flask equipped with a reflux condenser with a calcium chloride tube, a stirrer, and a thermometer. (0.001 mol) of cyasterone (38) to which 2 ml of freshly distilled dry triethylamine is added. While heating (70–80 °C.) and stirring, 1.18 g (0.0041 mol) of p-iodobenzoylisothiocyanate dissolved in 10 ml of pyridine are added dropwise. The reaction mixture is stirred for 6 h at 100–110 °C. After the reaction, the mass is transferred into a 0.5 L beaker and 80 ml of water is added. The yellow crystals precipitated were dissolved in butanol and chromatographed on an Al_2O_3 column in an ethanol: chloroform (50: 2) system. After distillation of the solvent and drying, a yellow and white powder is obtained. Yield 2,3,22 - O-tri (para-iodobenzoylthiocarbamate)—cyasterone,

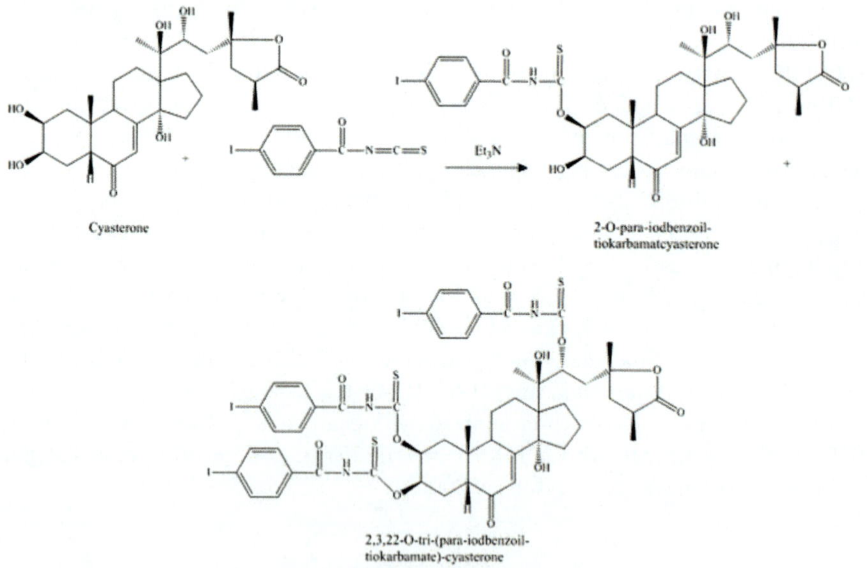

Fig. 5.2 Obtaining derivatives of cyasteronyl thiocarbamates

Table 5.2 Chemical shifts of protons of cyasterone, 2,3,22-triacetate–cyasterone, 2,3,22—O-tri (p-iodobenzoylthiocarbamate) -cyasterone, 2—O-p-iodobenzoyl thiocarbamate cyasterone (δ, ppm O-HMDS)[a]

Position proton									
H-2	H-3	H-7	H-9	H-22	CH_3-18	CH_3-19	CH_3-21	CH_3-27/29	R-radical proton
Cyasterone									
3,93	4,10	6,13	3,42	3,80	1,08	0,94	1,46	1,25	-
2,3,22-Triacetatecyasterone									
5,10	5,21	5,94	3,06	4,94	0,80	1,0	1,44	1,21–1,37	CH_3-COO 1.8;2.0;2.10
2,3,22 – O-Tri(para-iodbenzoilthiocarbamate)-cyasterone									
5,15	5,27	6,13	3,33	4,83	0,83	0,98	1,42	1,18	8.9(2H);7.7(3H):8.27 (7H)
2 –O-para- iodbenzoilthiocarbamate)-cyasterone									
5,07	4,31	6,09	3,48	3,87	0,90	0,93	1,45	1,22	8.6 (H); 7.81 (H):8.43 (2H)

[a]Signals of protons of methyl groups are of a singlet character, in all cases, the H-7 proton appears as a broadened singlet, other signals are broadened multiplets

69.4% (0.96 g), m.p. 267–269 °C. 2-O-para-iodobenzoyl-thiocarbamatcyasterone, 12.4% (0.17 g), mp 284–287 °C. A stain appears on TLC with R_f = 0.78 2,3.22-O-tri (para-iodobenzoylthiocarbamate) -cyasterone, R_f = 0.69 2-O-para-iodobenzoylthiocarbamate-cyasterone, (ethanol system: chloroform = 5:0.5).

The products 2,3,22–O-tri(para-iodobenzoylthiocarbamate)-cyasterone and 2-O-para-iodobenzoyl-thiocarbamate-cyasterone are dissolved in DMF, trifluoroacetic acid, pyridine, DMSO (Ramazonov and Syrov 2006).

IR spectrum of 2,3,22–O-tri (para-iodobenzoyl-thiocarbamate)-cyasterone (KBr, ν, cm -1); 3363 (N–H), 1685 (C = O), 1600 (aromatic ring), 1219 (C–O–C), 1177 (C = S).

IR spectrum of 2–O-para-iodobenzoylthiocarbamate-cyasterone (KBr, ν, cm-1); 3339 (N–H), 1667 (C = O), 1596 (aromat. rings), 1243 (C–O–C), 1182 (C = S).

Synthesis of 2,3,22-O-triacetate-cyasterone

2,3,22-O-triacetate-cyasterone. A solution of 100 mg of cyasterone (38) in 1 ml of pyridine was acetylated with 1 ml of acetic anhydride at room temperature for 24 h. Excess reagents were removed in vacuo. The residue was chromatographed on silica gel. Elution of the column with the system (chloroform–methanol (50: 1, 25: 1, 10: 1)) gave 85 mg of 2,3,22-O-triacetate-cyasterone.

Recrystallized from methanol 2,3,22-O-triacetate-cyasterone, $C_{35}H_{50}O_{11}$ m.p. 252–254 °C (methanol), $[\alpha]_D^{20} + 68.2 \pm 2°$ (c 0.37; methanol), IR spectrum of 2,3,22-O-triacetate-cyasterone—KBr, ν, cm^{-1}: 3460–3480 (OH), 1778 (γ-lactone), 1742–1243 (ester group), 1673 (C = O), 1657 (7-en-6-keto group), 1268 (C–O–C) (Baltaev 2000; Usmanov et al. 1977).

Literature data for triacetate-cyasterone: m.p. 251–252 °C, $[\alpha]_D^{20}$ + 69.0 °C (pyridine) (Hikino et al. 1968).

Mass spectrum: m/z (%): 646 (M+; 0.3), 628 (0.5), 618 (0.8), 610 (2), 568 (5), 550 (55), 539 (25), 455 (5), 385 (10), 334 (100), 232 (56), 201 (5), 183 (7), 113 (25), 69 (97).

5.3 Synthesis of Silenosterone

Ecdysteroids are found in plant objects both in free form and in the form of conjugates: esters of acetic, benzoic, sulfuric acids, and isopropylidene derivatives. Silenosterone (21), an ecdysteroid with a ketone group at C-3, was first found in the aerial organs of *Silene praemixta* (Saatov et al. 1979).

3-Dehydroecdysteroids are known as metabolites of animal origin (Karlson et al. 1972; Karlson and Koolman 1973). K-3 ecdysteroids were found only in Calliphora vicina.

In this regard, the synthesis of monoesters and 3-keto-derivative ecdysteroids with the aim of creating a new type of biologically active compounds based on them is promising.

Carrying out chemical transformation using available phytoecdysteroids as the starting compounds is the simplest and cheapest way to obtain rare natural phytoecdysteroids. One of the practically available natural ecdysteroids is 2–deoxy-α-ecdysone (22). Based on the structural features of its molecule, the acetylation of 2-deoxy-α-ecdysone (22) was carried out with acetic anhydride in pyridine, which made it possible to obtain 3.22-O-diacetate-2-deoxy-α-ecdysone (Fig. 5.3) (Karlson et al. 1965).

Saponification of 3.22-O-diacetate-2-deoxy-α-ecdysone gave 22-monoacetate-2-deoxy-α-ecdysone. The IR spectrum of 22-O-monoacetate-2-deoxy-α-ecdysone contains absorption bands characteristic of the ester group. The NMR spectrum of 22-O-monoacetate-2-deoxy-α-ecdysone contains a three-proton singlet at 1.95 ppm, indicating that there is one in the 3.22-O-diacetate-2-deoxy-α-ecdysone molecule acetyl group. Indeed, when compared the NMR spectra of 2-deoxy-α-ecdysone and 22-O-monoacetate-2-deoxy-α-ecdysone, a significant difference is revealed only in the position of the resonance lines of the proton at C-22.

In the spectrum of the compound 22-O-monoacetate-2-deoxy-α-ecdysone, the proton signal at C-22 is shifted to a weak field by 1.14 ppm and resonates at 5.14 ppm in the form of a multiplet. The data presented indicate the esterification of the hydroxy group at C-22 with the acetyl group. Analysis of the mass spectrum of 22-O-monoacetate-2-deoxy-α-ecdysone, the key fragments of which had m/ z 331, 314, 303, 285, 284, confirmed this assumption and at the same time indicated that the 22-O- monoacetate-2-deoxy-α-ecdysone contains two oxy groups in the steroid core and is similar to 2-deoxy-α-ecdysone (22) (Chong et al. 1970). Oxidation of 22-O-monoacetate-2-deoxy-α-ecdysone with chromic anhydride at room temperature resulted in 3-keto-22-acetyl-2-deoxy-α-ecdysone. In the IR spectrum of the

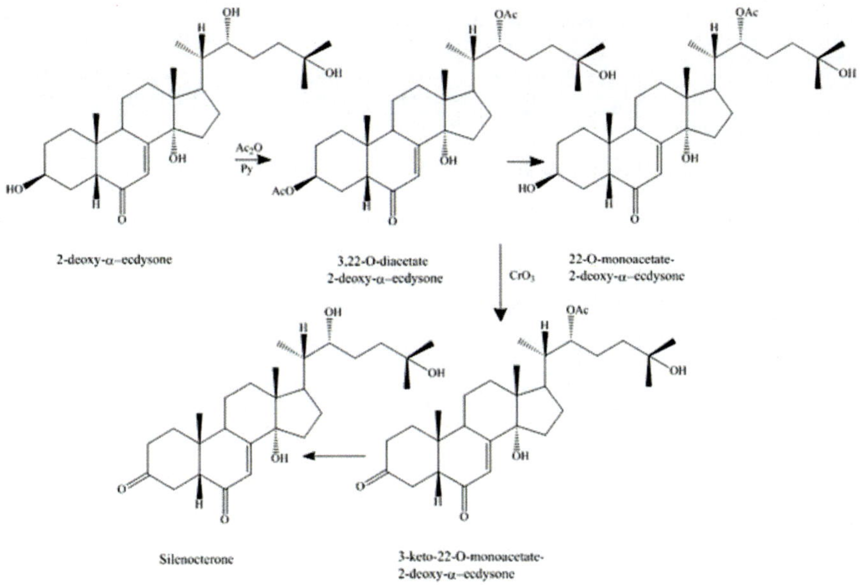

Fig. 5.3 For the preparation of 22-monoacetate-2-deoxy-α-ecdysone

obtained 3-keto-22-acetyl-2-deoxy-α-ecdysone, an absorption band at 1710 cm^{-1} is observed, which is characteristic of an additional non-conjugated carbonyl function in the six-membered ring.

When comparing the molecular ions of 22-monoacetate-2-deoxy-α-ecdysone (M + 490) and 3-keto-22-monoacetate-2-deoxy-α-ecdysone (M + 488), it turned out that the molecular weight of the latter is two units less than 22-monoacetate-2-deoxy-α-ecdysone. The fragmentation of the side chain of both compounds is the same, which is subject to the coincidence in the magnetic resonance spectra of these substances of the chemical shifts of the protons of methyl groups at C-21, C-26, and C-27 (Table 5.3).

Comparison of the mass spectra numbers of the steroid core's main fragments of 22-O-monoacetate-2-deoxy-α-ecdysone (m/z 332, 314, 284, 234) with the ions of the part of the 3-keto-22-monoacetate-2-deoxy-α -ecdysone (m/z 330, 312, 282, 232) shows that the elements causing the difference in the structure of 22-O-monoacetate-2-deoxy-α-ecdysone and 3-keto-22-monoacetate-2-deoxy-α -ecdysone are in the tetracyclic skeleton.

In the mass spectra, the decay ions of 3-keto-22-monoacetate-2-deoxy-α-ecdysone have two a.m.u. smaller than the corresponding fragments of 22-O-monoacetate-2-deoxy-α-ecdysone. This difference indicates that the hydroxyl group at C-3 of 22-O-monoacetate-2-deoxy-α-ecdysone was oxidized to a ketone. These data are also exposed when comparing the ^{1}H NMR spectra of 22-O-monoacetate-2-deoxy-α-ecdysone and 3-keto-22-monoacetate-2-deoxy-α-ecdysone: in the spectrum

Table 5.3 Chemical shifts of protons (δ-C_5D_5N, O-HMDS)[a]

Position of proton								
H-3	H-22	H-7	H-9	CH_3-18	CH_3-19	CH_3-21	CH_3-26/27	-OAc
2–Deoxy-α-ecdysone								
4.00	4.00	6.13	3.42	0.61	0.95	1.17 3 J = 6 Hz	1.26	–
3,22-O-diacetate-2-deoxy-α-ecdysone								
4.82	5.19	6.06	3.20	0.60	0.85	1.00 3 J = 6 Hz	1.20	1.85; 1.95
22-O-Monoacetate 2-deoxy-α-ecdysone								
3.93	5.14	5.59	3.30	0.59	0.76	1.01 3 J = 6 Hz	1.22	1.95
3-keto-22-monoacetate-2-deoxy-α-ecdysone								
–	5.13	5.94	–	0.60	0.89	0.99 3 J = 6 Hz	1.21	1.92
Silenosterone								
–	3.90	5.96	3.45	0.62	0.94	1.163 J = 6 Hz	1.27	–

[a]Signal of CH_3-21 groups—doublet, of other groups—singlets. Proton H-7 is manifested in the form of a broadened singlet, astral—like broadened multiplets

of 3-keto-22-O-monoacetate- 2-deoxy-α-ecdysone there is no proton signal at C-3 (Table 5.3).

3-Keto-22-O-monoacetate was subjected to alkaline hydrolysis. A product was isolated from the reaction products, which, according to spectral data and physicochemical constants, was identified with silenosterone (Saatov et al. 1979).

Synthesis of 3,22-diacetate-2-deoxy-α-ecdysone

7 g of 2-deoxy-α-ecdysone (22) (obtained from *Silene praemixta* (Nakanishi et al. 1968). 61 ml of acetic anhydride were acetylated in 142 ml of pyridine at room temperature for 24 h. Water was poured into the reaction mixture and left overnight in a refrigerator. Precipitated crystals were filtered off and chromatographed on a silica gel column. Elution of the column with chloroform gave 6.14 g of 2,3-O-diacetate-2-deoxy-α-ecdysone, $C_{31}H_{48}O_7$, m.p. 245–246 °C (from methanol–water mixtures), $[\alpha]_D^{24} + 39.0 + 3°$ (c 0.25; methanol). IR spectrum (KBr, ν, cm^{-1}): 3460 (OH), 1740, 1250 (ester group), 1660 (7-en-6-keto grouping) (Sing and Thakur 1982).

Synthesis of 22-O-acetate-2-deoxy-α-ecdysone

1.71 g of 3,22-O-diacetate-2-deoxy-α-ecdysone was dissolved in 48 ml of a 0.5% methanol solution of KOH. The reaction was allowed to stand for 6 h, then neutralized with acetic acid and diluted with water. The precipitate was filtered off, washed with water, and chromatographed on a silica gel column. Elution with a chloroform–methanol system (100:1) gave 1.25 g of 22-O-acetate-2-deoxy-α-ecdysone (50), $C_{29}H_{46}O_6$, which after recrystallization from aqueous methanol had a melting point of 158–160 °C $[\alpha]_D^{22} + 51.3 + 2°$ (c 0.20; methanol). IR spectrum (KBr, ν, cm^{-1}): 3445 (OH), 1730, 1255 (ester group), 1665 (7-en-6-keto group).

Mass spectrum: m/z (%): 490 (M + ; 6), 472 (12), 462 (12), 444 (27), 412 (22), 397 (41), 394 (26), 384 (47), 379 (16), 332 (12), 331 (14), 303 (4), 302 (14), 285 (46), 284 (100), 235 (94), 234 (85), 99 (46), 81 (50).

Synthesis of 5α-2-deoxyecdysterone from 2-deoxyecdysterone

To 150 mg of 2-deoxyecdysterone (23) in 15 ml of methanol was added 120 mg of potassium bicarbonate in 3 ml of water. The reaction mixture was left for three days at room temperature. Then it was neutralized with acetic acid, methanol was evaporated to half the volume, diluted with water, and extracted with ethyl acetate. The ethyl acetate extract was evaporated to dryness. The residue was chromatographed on a silica gel column. Elution with a chloroform–methanol 15:1 system gave 53 mg of 5α-2-deoxyecdysterone, $C_{27}H_{44}O_6$ mp. 235–237 °C (from aqueous methanol).

5.4 Synthesis of 22-O-Benzoate-20-Hydroxyecdysone

A scheme for the chemical transformation of 20-hydroxyecdysone has been developed with the aim of further synthesis of the conjugate of 22-O-benzoate-20-hydroxyecdysone (48).

By acetalization of 20-hydroxyecdysone with acetone in the presence of phosphoromolybdic acid, 2,3,20,22-O-diacetonide-20-hydroxyecdysone is formed. In the mass spectrum of 2,3,20,22-O-diacetonide-20-hydroxyecdysone, there are ion peaks characteristic of 2,3,20,22-O-diacetonide-20-hydroxyecdysone (Saatov et al. 1987a), which is confirmed by the practical coincidence of the mass spectra and the ^{1}H NMR spectra (Table 5.4), as well as their direct comparison on a thin-layer chromatogram (TLC).

Treatment of 2,3,20,22-O-diacetonide-20-hydroxyecdysone with diluted acetic acid (Ves'kina and Odinokov 2012) led to 2,3-O-monoacetonid-20-hydroxyecdysone (Dzhukharova et al. 1991). identified according to the mass and NMR spectra (Table 5.4). In the mass spectrum of the ecdysteroid 2,3,20,22-O-diacetonide-20-hydroxyecdysone, there is an ion peak with m/z 404, characteristic of monoacetonide 20-hydroxyecdysone (Galbraith et al. 1973; Zatsny et al. 1973, 1975). This indicates that the isopropylidene group is associated with the steroid part of the molecule and replaces the diol group at C-2 and C-3.

A compound identified with 20,22-O-monoacetonide-20-hydroxyecdysone was isolated from the reaction mixture along with 2,3-O-monoacetonide-20-hydroxyecdysone (Table 5.4) (Dzhukharova et al. 1991). Further processing of 2,3-O-monoacetonide-20-hydroxyecdysone with benzoyl chloride in a dry pyridine solution led to the production of 2,3-O-monoacetonide-20-hydroxyecdysone-22-benzoate (Fig. 5.4) [248].

2,3-O-Monoacetonide-22-O-benzoate-20-hydroxyecdysone had a maximum absorption in the UV spectrum at 235 nm (logε = 4.02). In the IR spectrum of 2,3-O-monoacetonidide-22-O-benzoate-20-hydroxyecdysone, absorption bands at 3460 cm^{-1} (OH groups) and at 1650 cm^{-1}, corresponding to the absorption of the

Table 5.4 Chemical shifts of protons of ecdysteroid derivatives (CD_3 OD, O-TSP-d4 δ, m.g., J/Hz)[a]

Proton	20-hydroxyecdysone	2,3,20,22-*O*- diacetonide 20-hydroxyecdysone	2,3-*O*- monoacetonide -20- hydroxyecdysone	2,3-*O*- monoacetonide -20- hydroxyecdysone	2,3-O-monoacetonide -22-O-benzoate-20-hydroxyecdysone	22-O-benzoate 20-hydroxyecdysone
1	**2**	**3**	**4**	**5**	**6**	**7**
1 a	1.43					
1 e	1.78					
1 a	3.83	4.27 m,$W_{1/2} = 23$	4.27 m,$W_{1/2} = 23$	3.84 m,$W_{1/2} = 23$	4.27 m,$W_{1/2} = 23$	3.83 m,$W_{1/2} = 23$
J-12,3,3						
3 e	3.94	4.27 m,$W_{1/2} = 9$	4.30 m,$W_{1/2} = 9$	3.95 m,$W_{1/2} = 9$	4.31 m,$W_{1/2} = 9$	3.95 m,$W_{1/2} = 8$
J-3,3,3						
4 a	1.65					
4 e	1.75					
5	2.38 d,J = 12.5	2.24 t, J = 8.6	2.24 t, J = 8.6	2.39 d, J-12.5	2.24 t, J = 8.6	2.39 d, J-12.7, 4.4
7	5.80 d, J = 2.5	5. 80 d, J = 2.4	5.78 d, J = 2.3	5.81 d, J = 2.5	5.81 d, J = 2.2	5.82 d, J = 2.5
9 e	3.14 m,$W_{1/2} = 22$	2.93 m,$W_{1/2} = 23$	2.94 m,$W_{1/2} = 23$	3.15 m,$W_{1/2} = 23$	2.95 m,$W_{1/2} = 23$	3.16 m,$W_{1/2} = 23$
11 a	1.65					
11 e	1.78					
12 a	2.13	2.10	2.12	2.12	2.16	2.18

(continued)

Table 5.4 (continued)

Proton	20-hydroxyecdysone	2,3,20,22-*O*- diacetonide 20-hydroxyecdysone	2,3-*O*- monoacetonide -20- hydroxyecdysone	2,3-*O*- monoacetonide -20- hydroxyecdysone	2,3-O-monoacetonide -22-O-benzoate-20-hydroxyecdysone	22-O-benzoate 20-hydroxyecdysone
	J = 13,13.5	J = 13,13.5	J = 13,13.5	J = 13,13.5	J = 13,13.5	J = 13,13.5
12 e	1.85					
15 a	2.00					
15 b	1.55					
16 a	1.95					
16 b	1.75					
17	2.39 m	2.39 t, J = 8.7	2.39 m, $W_{1/2}$ = 20	2.32 t, J = 8.7	2.48 t, J = 9.2	2.48 t, J = 9.0
22	33.33 d, J = 11,2	3.68 m,$W_{1/2}$ = 15	3.30 d, J = 11.2	3.69 m,$W_{1/2}$ = 15	5.15 d, J = 11.2)	5.15 d, J = 11.2
23 a	1.30					
23 b	1.65					

[a]s-singlet, d-doublet, t-triplet, m-multiplet

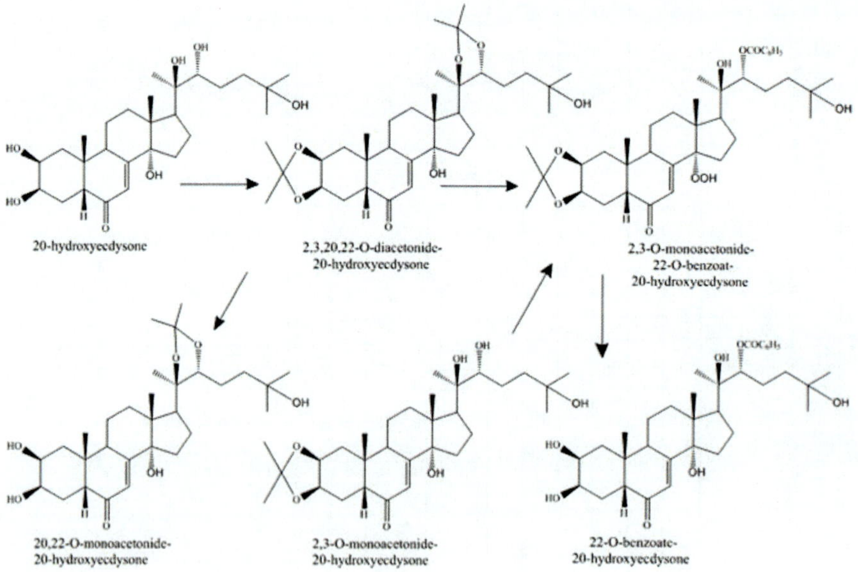

Fig. 5.4 Obtaining 2,3-O-monoacetonide-20-hydroxyecdysone-22-benzoate

keto group conjugated with a double bond, are noted. In addition, in the IR spectrum, absorption bands are observed at 1725, 1710, 1255 (ester groups) and 1610, 1558, 720 cm^{-1} (benzene ring).

In the region of high mass spectra numbers in the mass spectrum of 2,3-O-monoacetonide-22-O-benzoate-20-hydroxyecdysone, a peak of the m/z 624 molecular ion and a number of peaks formed due to the cleavage of several water and benzoic acid molecules appear: m/z 642 ($M + NH_4$), 625 (NH), 607, 589, 571, 520, 503, 502, 485, 467.

^{1}H NMR, IR, and mass spectra were identified with 2,3-O-monoacetonide-22-O-benzoate-20-hydroxyecdysone previously synthesized from 22-O-benzoate-20-hydroxyecdysone (48) (Saatov et al. 1988a). Hydrolysis of 2,3-O-monoacetonide-22-O-benzoate-20-hydroxy-ecdysone with diluted acetic acid gave 22-O-benzoate-20-hydroxyecdysone. The IR spectrum of this compound has absorption bands at 1710 and 1285 cm^{-1} (ester group), 1610, 1587, 720 cm^{-1} (benzene ring). This is also evidenced by the signals of five aromatic protons in the ^{1}H NMR spectrum (Table 5.4).

The data presented together indicate the addition of the remainder of benzoic acid to the hydroxyl group at C-22. Hydrolysis of 2,3-O-monoacetonide-22-O-benzoate-20-hydroxyecdysone with dilute acetic acid gave 2,3-O-monoacetonid-22-O-benzoate-20-hydroxy-ecdysone. The IR spectrum of this compound contains absorption bands at 1710–1285 cm^{-1} (ester), 1610, 1587, 720 cm^{-1} (benzene ring).

This is also evidenced by the signals of five aromatic protons in the ^{1}H NMR spectrum (Table 5.4). The data presented collectively indicate the addition of the remainder of benzoic acid to the hydroxyl group at C-22. Thus, ecdysteroid is 22-O-benzoate-20-hydroxyecdysone (48).

Synthesis of 2,3,20,22-O-diacetonide-20-hydroxyecdysone

To a suspension of 1.2 g of 20-hydroxyecdysone (2) in 170 ml of anhydrous acetone was added 24 mg of phosphoromolybdic acid, the reaction mixture was shaken at room temperature until 20-hydroxyecdysone was completely dissolved (2). The solution was concentrated *in vacuo* to a small volume, diluted with water, neutralized with sodium bicarbonate solution. The neutral solution was extracted with ethyl acetate, the extract after evaporation was chromatographed on a silica gel column. Elution with the system (chloroform–methanol 12:1) gave 0.95 g (70% yield) 2,3,20,22-O-diacetonide-20-hydroxyecdysone, $C_{33}H_{52}O_7$, m.p. 230–232 °C (ether-acetone), $[\alpha]_D{}^{20}$ + 36.8 + 2 °C (c 0.30; methanol). IR spectrum (KBr, ν, cm^{-1}): 3460 (OH), 1670 (7-en-6-keto grouping). Mass spectrum m/z: 578 (M + NH_4) 561 (M + H), 543 (M + H-H_2O), 538 (M + NH_4-C_2H_6O), 521 (M + H-C_2H_6O), 503 (M + H-C_2H_6O-H_2O), 480, 447, 445, 380, 364, 304, 279, 234, 193, 160, 132, 78.

Synthesis of 2,3-O-monoacetonide-20-hydroxyecdysone and 20,22-O-monoacetonide-20-hydroxyecdysone

2,3,20,22-O-diacetonide-20-hydroxyecdysone 0.9 g was hydrolyzed with acetic acid (70 ml) in aqueous methanol (12:19) at room temperature for 4 h. The reaction mixture was diluted with water and neutralized with a bicarbonate solution of sodium. The neutral solution was extracted with ethyl acetate, the extract was chromatographed on a column using a system (chloroform–methanol 15: 1), and 135 mg (yield 15%) of 20-hydroxyecdysone 2,3-O-monoacetonide, $C_{30}H_{48}O_7$ m.p. 242–244 °C (ethyl acetate), $[\alpha]_D{}^{20}$ + 59.2 + 2 °C (c 0.32; methanol). UV spectrum (C_2H_5OH λmax, nm): 246 (lgε 4.00).

IR spectrum (KBr, ν, cm^{-1}): 3400–3450 (OH), 1650 (7-en-6-keto grouping). Mass spectrum m/z: 538 (M + NH4) 521 (NH)+, 503 (M + H-H_2O)+, 485 (M + H-2H_2O)+, 467 (M + H-3H_2O)+, 404, 387, 330, 318, 279, 232, 186, 156, 130, 101, 99, 84.

Upon further washing of the column with the system (chloroform–methanol 15: 1), 155 mg (yield 17%) of 20.22-O-monoacetonide-20-hydroxyecdysone were isolated (Saatov et al. 1987c; Dzhukharova et al. 1991). Mass spectrum. m/z: 505, 502, 487, 469, 462, 445, 429, 427, 411, 409, 363, 345, 329, 327, 301, 300, 201, 161, 143, 125, 102, 99, 81.

Synthesis of 2,3-O-monoacetonide-22-O-benzoate-20-hydroxyecdysone

2,3-O-monoacetonide-20-hydroxyecdysone (125 mg), dissolved in 20 ml of dry pyridine, was benzoylated with benzoyl chloride (3 ml) at room temperature for 24 h. After removal of the solvents, the residue was chromatographed on a silica gel column. Eluting with a system (chloroform–methanol 15:1), 55 mg (yield 37%) of 2,3-O-monoacetonide-22-O-benzoate-20-hydroxyecdysone, $C_{37}H_{52}O_8$ m.p. 242–244 °C (chloroform) were obtained. Mass spectrum, m/z: 642 (M^+ NH_4), 625 (NH)+, 607, 589, 571, 520, 503, 502, 485, 467, 420, 409, 362, 282, 247, 203, 163, 99, 78.

Synthesis of 22-O-benzoate-20-hydroxyecdysone

2,3-O-Monoacetonide-22-O-benzoate-20-hydroxyecdysone 45 mg was hydrolyzed with 6 ml of acetic acid in aqueous methanol (2.6 ml) at room temperature for 3 h. The reaction mixture was treated as described above and after chromatography on column system (chloroform–methanol, 12:1) received 24 mg (yield 56%) of 22-O-benzoate-20-hydroxyecdysone (48), $C_{34}H_{48}O_8$, m.p. 202–205 °C. Mass spectrum m/z: 602 (M $^+$NH4), 585 $(NH)^+$, 567 $(M\,^+H\text{-}H_2O)^+$, 549 $(M\,^+H\text{-}2H_2O)^+$, 531 $(M^+H\,\text{-}3H_2O)^+$, 520 $(M\,^+NH_4\text{-}C_7H_6O_2)^+$, 502, 486, 467, 445, 444, 427, 409, 353, 346, 282, 247, 240, 166, 140, 105, 99, 78 (Saatov et al. 1986).

5.4.1 *New Acetonide Derivatives of 20-Hydroxyecdysone*

Currently, a promising direction in the use of natural compounds is the transformation of the structure of secondary plant metabolites, leading to previously unknown derivatives. The purpose of such transformations is to enhance the biological effect of natural compounds and the search for new types of activity. One of the promising directions for the transformation of ecdysteroids is the use of the acylation reaction. As derivatives of 20-hydroxyecdysone (2)-protected by 2,3-hydroxyl groups, 2,3-O-monoacetonide is convenient for the following selective transformations of one of the most representative phytoecdysteroids. We have studied the transformations of 2,3-O-monoacetonide-20-hydroxyecdysone (121), 2,3,20,22-di-O-acetonide-20-hydroxyecdysone (123) of one of the most common phytoecdysteroids—20-hydroxyecdysone (2) under the action of phosphoric-molybdenum acid (PMA). Obtained by a known method (Afonkina et al. 2006), 2,3-O-monoacetonide-20-hydroxyecdysone (121) with a yield of 32%, m.p. 136–137 °C, 20.22-O-monoacetonide-20-hydroxyecdysone (122) with a yield of 27%, m.p. 223–224 °C and 2,3,20,22-di-O-acetonide-20-hydroxyecdysone (123) with a yield of 19°/o, m.p. 134–135° C (Fig. 5.5).

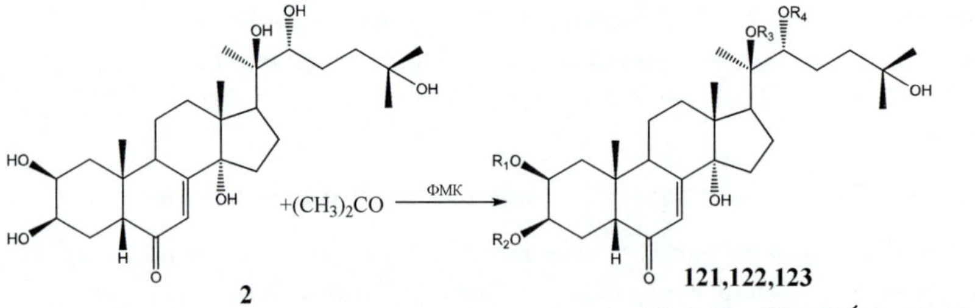

2

121,122,123

121. $R_1+R_2=(CH_3)_2C<$; $R_3+R_4=H$

122. $R_1+R_2=H$; $R_3+R_4=(CH_3)_2C<$

123. $R_1+R_2=(CH_3)C<$; $R_3+R_4=(CH_3)_2C<$

Fig. 5.5 Obtaining acetonide derivatives of 20-hydroxyecdysone

A mixture of compounds (121, 122, 123) was separated by column chromatography. Their structure was confirmed by the ^{13}C-, ^{1}H NMR spectra (Table 5.5), as well as by their direct comparison with the sample materials in a thin-layer chromatogram.

A method for the chemical transformation of 2,3-O-monoacetonide-20-hydroxyecdysone (121) and 2,3,20,22-di-O-acetonide-20-hydroxyecdysone (123) was developed with the aim of further synthesis of conjugates of compounds 124, 125, 126, 127, 128, 129, 130, 131. Further processing of 2,3-O-monoacetonide-20-hydroxyecdysone (70) with acetic anhydride in a solution of dry pyridine leads to 2,3-O-monoacetonide-O-22,25-O-diacetate-20-hydroxyecdysone (124) (Fig. 5.6).

As a result of acylation with acetic anhydride 121, derivatives of 2,3-O-monoacetonide-22.25-di-O-acetyl-20-hydroxyecdysone were obtained (124). Using the acid hydrolysis of ecdysteroid 124, the acetonide protective group of the compound was removed under the action of 10% HC1 and acetate fragments were obtained at the C-22 and C-25 positions of the 20-hydroxyecdysone (125).

The chemical structure of compound 125 was determined according to ^{1}H and ^{13}C NMR spectra. In the ^{1}H NMR spectrum, a weak-field shift of the H-22 signal (δ 5.32) by 1.56 ppm is observed and compared to that in molecule 2. In the strong-field part of the spectrum at 1.83 and 1.95 ppm, two additional singlet signals of methyl groups from acetate fragments resonate clearly. In the ^{13}C NMR spectrum, the signals of the C-22 and C-25 carbon atoms are shifted to a weaker field (80.56 and 82.35 ppm, respectively). In addition, in the aliphatic part of the spectrum, signals of an additional two carbon atoms are observed at 26.52 and 26.66 ppm, related to metal–carbon atoms. In a weaker field in the region at 170.74 and 171.76 ppm, signals of two carbon atoms of the carbonyl group of the acetate group are manifested. The above results indicate that acetate fragments in compound 125 are located at the C-22 and C-25 positions of the 20-hydroxyecdysone molecule (Table 5.6).

Further, by the interaction of 2,3-(3-monoacetonide-20-hydroxyecdysone (70) with iropionic anhydride, 2,3-O-monoacetonide-22-O-propionyl-20-hydroxyecdysone (159) was obtained with a yield of 57%. in 126, in the presence of 10% HC1, 22-O-propionyl-20-hydroxyecdysone (127) was synthesized with a yield of 69% (Fig. 5.7).

In the ^{1}H NMR spectrum of compound 127, the H-22 proton signal appears as a doublet ($J = 10.8$) at 5.39 ppm, which is shifted by 1.63 ppm into a weaker field. Characteristic signals of the ethyl group are observed in the spectrum at 0.96 (t, $J = 7.5$ Hz) and 2.26 (q, $J = 7.5$ Hz) ppm Analysis of ^{13}C NMR and DEPT spectra of compound 127 showed the presence of signals of 30 carbon atoms. In the ^{13}C NMR spectrum of compound 127, a weak field shift of the C-22 signal (5 80.74 ppm) by 2.7 ppm is observed in comparison with that in compound 2. In addition, in the spectrum at 10.13, 28.63, and 175.21 ppm signals of an additional three carbon atoms—propionic acid residues, are observed. These data indicate that the ester moiety in compound 127 is at position C-22.

The acylation of 2,3-O-monoacetonide-20-hydroxyecdysone (121) with acetic anhydride proceeds predominantly at the hydroxyl group at the secondary 22nd carbon atom and the tertiary 25th carbon atom to form 22.25-di-O-acetyl-2,3-O-monoacetonide-20-hydroxyecdysone (124). With propionic anhydride, the reaction

Table 5.5 Chemical shifts of ^{1}H and ^{13}C NMR compounds 121, 123, (C_5D_5N, δ, ppm, J/Hz) and DEPT experiment data

Atom C	DEPT	121		123	
		δC	δH (J/Hz)	δC	δH (J/Hz)
1	CH_2	38.9		37.5	
2	C	73.6	4.19,m	72.0	4.23 (m)
3	C	73.3	4.08	71.5	4.26
4	CH_2	27.8		31.3	
5	CH	52.6	2.55, m	50.7	2.36 (dd, 12.5, 4.7) 2.36 (dd, 12.5, 4.7)
6	C	205.7		203.0	
7	CH	121.9	6.20 (d, 2)	121.1	5.83 (d. 2.2)
8	C	167.3		163.8	
9	CH	35.9	3.15 (m)	34.4	2.80 (m)
10	C	39.8		37.7	
11	CH_2	21,7		20.4	
12	CH_2	32.6		30.8	
13	C	49.0		47.4	
14	C	85.3		84.6	
15	CH_2	31.8		26.3	
16	CH_2	21.6		21.1	
17	CH	50.7	3.00 (t, 8.9)	48.9	2.22 (t, 8)
18	CH_3	18.1	1.19(s)	16.9	0.78 (s)
19	CH_3	24.1	0.99 (s)	23.3	0.98 (s)
20	C	78.0		84.3	
21	CH_3	21.2	1.60 (s)	21.5	1.16(s)
22	CH	78.5	3.87 (d, 8.1)	81.9	3.65 (dd. 9.5, 2.5)3.65 (dd. 9.5,2.5)
23	CH_2	27.5		23.5	
24	CH_2	42.5		41.3	
25	C	71.4		70.3	
26	CH_3	29.1	1.35 (s)	28.9	1.24(s)
27	CH_3	29.8	1.35 (s)	28.4	1.25 (s)
2,3-$(CH_3)_2$-	CH_3	109.6;	1.3, 1.55	108.2	1.33, 1.41(s)
C =		28.9: 26.7	(s)	26.4: 29.2	
20,22-	CH_3			106.9;	1.33, 1.49 (s)
$(CH_3)_2$-C =				26.8: 29.5	
20,22-CH_3	CH_3				

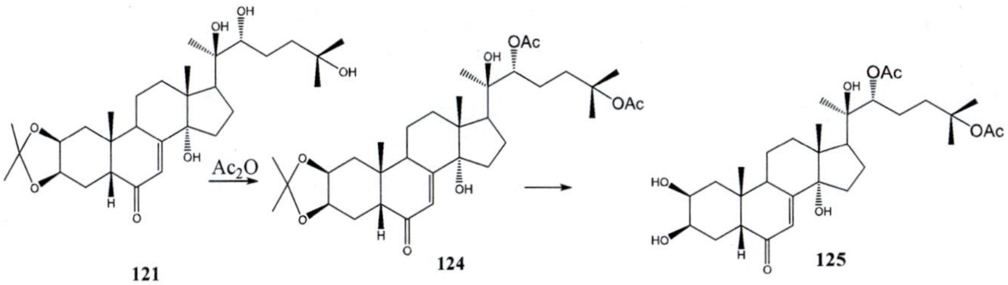

121 124 125

Fig. 5.6 Preparation of 22.25-O-diacetate-20-hydroxyecdysone (125)

proceeds only at the secondary 22nd carbon atom and 2,3-O-monoacetonid-22-O-propionyl-20-hydroxyecdysone is formed (126).

In the esterification of 2,3-(9-monoacetonide-20-hydroxyecdysone (121), the reaction ability depends on the radical of the carbon atom of the carboxylic acid anhydride, if the CH_3-C(O) group joins easily, a di-ether is formed. If the CH_3-CH_2-C(O) group is attached only to the hydroxyl group, an ether is formed at the secondary 22nd carbon atom.

The interaction of 2,3,20,22-diacetononide-20-hydroxyecdysone (123) with butyric anhydride gave the 25-O-butanoyl-20-hydroxyecdysone (129) (Fig. 5.8).

The structure of molecule 129 was established by analyzing the spectra of ^{1}H, ^{13}C NMR, and DEPT experiment (Table 5.6). In the ^{1}H NMR spectrum in a stronger field at 0.68 ppm, a signal appears in the form of a triplet (J = 7.4 Hz) of the methyl group of the butyric acid residue. In the ^{13}C NMR spectrum of compound 129, the signal of the C-25 carbon atom is shifted to a weaker field (80.76 ppm) compared with that in molecule 2. Moreover, the chemical shifts of the remaining hydroxyl-containing carbon atoms do not change significantly. In the aliphatic part of the spectrum at 12.08, 27.55, and 40.48 ppm, additional carbon atoms of the butyric acid residue appear. In a weaker field, along with the signals of the steroid molecule at 173.36 ppm, a signal of the carbon atom of the carbonyl group appears. All of the above results indicate that the butyric acid residue is in the C-25 position of the 20-hydroxyedisone molecule. Acylates of 2,2-dimethyl succinic acid are very promising anti-HIV agents, which stimulated the development of methods for the preparation of individual compounds (Afonkina et al. 2006; Shafikov et al. 2007; Levina et al. 2008).

Based on the interaction of 2,3-monoacetonide-20-hydroxyecdysone (121) with succinic anhydride, 2,3-O-monoacetonide-22-O-succinyl-20-hydroxyecdysone (130) were obtained and by hydrolysis (10% HCl) of monoacetonide 130 was obtained ecdysteroid 22-O-succinyl-20-hydroxyecdysone (131) (Fig. 5.9).

In this case, a mixture of monoacetonide 70 and 2,3-monoacetonide-22-O-succinyl-20-hydroxyecdysone (130) was obtained, which was isolated by column chromatography. The reaction of 2,3-O-monoacetonide-20-hydroxyecdysone (121) with succinic anhydride proceeds only at the secondary 22nd carbon atom and 2,3-O-monoacetonid-22-O-succinyl-20-hydroxyecdysone is formed (130). Hydrolysis of

Table 5.6 Chemical shifts of ^{1}H and ^{13}C NMR compounds 125, 127, 129, and 131 (C_5D_5N, δ, ppm, J/Hz)

Atom C	125	127	129	131	125	127	129	131
1	38.46	38.47	38.47	38.47				
2	68.65	68.68	68.68	68.67	4.05, d, (11.0)	4.08, d(yu.9)	4.07, dd, (11.4; 3.4)	4.05, dd, (12.0; 3.6)
3	68.56	68.57	68.57	68.56	4.12, s	4.12, s	4.11, s	4.11, s
4	32.98	32.97	32.98	32.95				
5	51.89	51.91	51.91	51.90	2.89, dd,	2.89, m,		2.89, dd,
					(13.1; 3.6)			(13.2; 3.6)
6	203.97	204.03	204.05	204.02				
7	122.29	122.25	122.24	122.17	6.10, d, (2.3)	6.12,d, (2.4)	6.11,d, (2.3)	6.13, d, (2.0)
8	166.36	166.45	166.47	166.62				
9	34.87	34.89	34.89	34.92	3.46, t (8.5)	3.46, t (8.9)	3.45, t (8.6)	3.46, t (8.6)
10	39.18	39.18	39.18	39.17				
11	21.57	21.58	21.57	21.62				
12	32.52	32.50	32.51	32.52				
13	48.57	48.64	48.65	48.60				
14	84.60	84.64	84.64	84.66				
15	32.18	32.21	32.20	32.28				
16	22.06	22.10	22.05	21.99				
17	51.01	50.96	50.89	50.60	2.82, t, (9.0)	2.87, t, (9.1)		2.84, t, (8.6)
18	18.39	18.40	18.42	18.44	1.04, s	1.04, s	1.03, s	1.09, s
19	24.93	24.95	24.96	24.97	0.93, s	0.93, s	0.92, s	0.94, s
20	76.62	76.84	76.88	77.34				
21	21.77	22.88	22.96	22.19	1.52, s	1.52, s	1.52, s	1.45, s
22	80.56	80.74	81.21	77.77	5.32, d, (9.9)	5.39, d, (10.8)	5.45, dd, (10.8; 1.9)	3.73, dd, (9.9; 4.0)
23	25.93	26.62	26.58	24.23				
24	38.54	42.24	42.07	43.22				
25	82.35	69.77	69.82	80.76				
26	22.74	30.61	30.67	27.28	1.28, s	1.22, s	1.23, s	1.17, s
27	22.74	30.18	30.13	27.14	1.30, s	1.23, s	1.21,s	1.19, s
1'	170.74 171.76	175.21	173.82	173.36				

(continued)

Table 5.6 (continued)

Atom C	125	127	129	131	125	127	129	131
2’	26.52 26.66	28.63	31.02	40.48	1.83,s 1.95,s	2.26, k, (7.5)		
3’		10.13	30.85	27.55		0.96, t, (7.5)		
4’			176.13	12.08				0.68, t,
								(7.4)
OH					5.42, s	5.25, s		5.95, s
OH					5.94, s	5.59, s		6.25, s
OH					6.05, s	5.97, s		
OH					6.33, s	6.04, s		
OH						6.33, s		

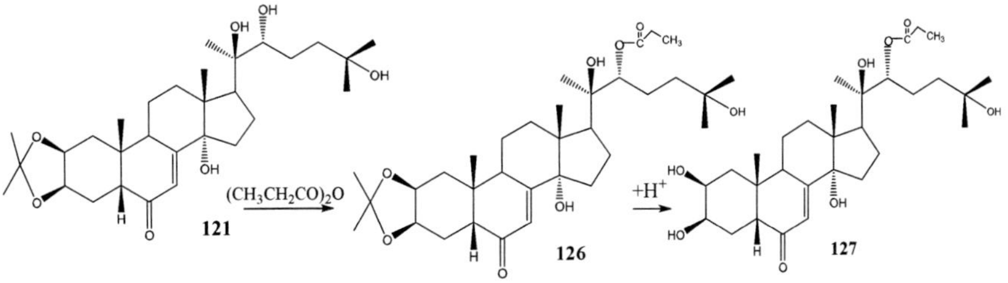

Fig. 5.7 Preparation of 22-O-propionyl-20-hydroxyecdysone (127)

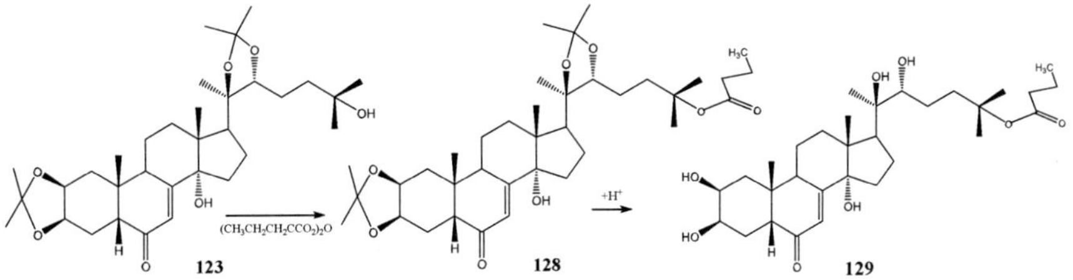

Fig. 5.8 Preparation of 25-O-butanoyl-20-hydroxyecdysone (129)

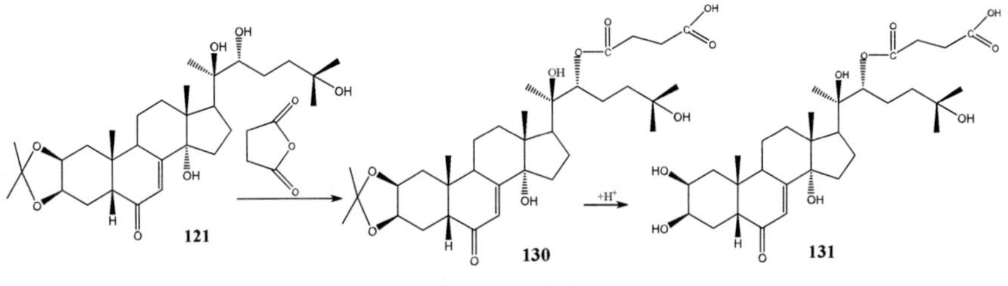

Fig. 5.9 Obtaining 22-O-succinyl-20-hydroxyecdysone

2,3-O-monoacetonide-22-O-succinyl-20-hydroxyecdysone (130) in 10% HC1 (acidification to pH 4) led to compound 131. In the ^{1}H NMR spectrum of molecule 131, the signals H-2, H-3, and H-22 have chemical shifts that are approximately the same as in compound 127 (4.07 ppm, 4.11 ppm, and 5.45 ppm, respectively). Analysis of the ^{13}C NMR spectra of compound 131 showed a paramagnetic shift in the weak field of the signal of the C-22 carbon atom (81.21 ppm). In addition, in the spectrum along with the signals of the steroid molecule in the aliphatic part of the spectrum at 31.02 and 30.85 ppm the signals of two carbon atoms resonate—the residue of succinic acid. In a weaker field at 173.82 and 176.13 ppm, signals of carbon atoms of two carbonyl groups appear. These data allow us to establish the attachment of the ester residue at C-22.

Thus, new ecdysteroid derivatives were obtained—22.25-O-diacetyl-20-hydroxyecdysone (125), 22-O-propionyl-20-hydroxyecdysone (127), 25-O-butyryl-20-hydroxyecdysone (129), 22-O-succinyl-20-hydroxyecdysone (131).

5.4.2 *Synthesis of 2,3-O-Di-(Ethyl Carbonate) -20,22-O-Carbonate-20-Hydroxyecdysone*

By the interaction of compound 2 with ethyl chloroformate, 2,3–0-di- (ethyl carbonate)-20,22–0-carbonate-20-hydroxyecdysone were obtained (132) (Fig. 5.10). An individual compound 132 was isolated from a mixture of compounds 2 and 132 using column chromatography (Hardman 1987). Purified 2,3–0-di- (ethyl carbonate)-20,22–0-carbonate-20-hydroxy-ecdysone (132) was analyzed using ^{1}H, ^{13}C, NMR spectra and experiments by Dept, HetCor.

The structure of 2,3-O-di-(ethyl carbonate)-20,22-O-carbonate-20-hydroxyecdysone (132) follows from the spectral data: the structure of O-tetra-substituted 20-hydroxyecdysone (132) as O-ethyl carbonate (132) confirmed by ^{1}H NMR. The proton signal H-2 manifests itself in the form of a doublet—triplet (J = 12.6; 3.9) at 4.84 ppm, which is shifted by 0.79 ppm weak field, proton H-3 signal observed at

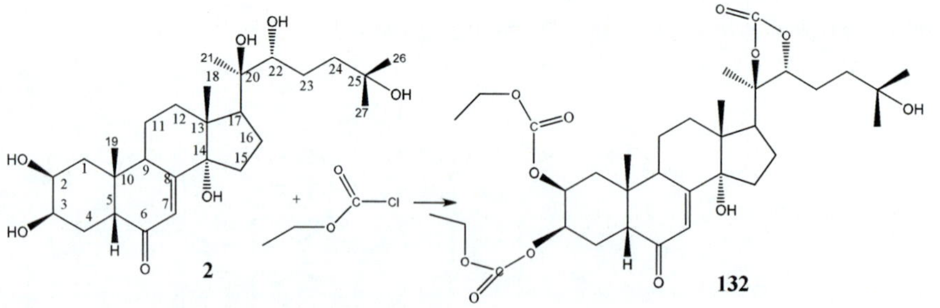

Fig. 5.10 Reaction of 2,3-O-di- (ethyl carbonate)-20,22-O-carbonate-20-hydroxyecdysone

5.09 ppm (br.s., J = 2.4), the chemical shift has an H-22 signal at 4.33 ppm (dd, J = 8.9; 4.0).

An analysis of the ^{13}C NMR spectra of compound 132 showed a strong shift in the weak field of the signal of the carbon atom C-25 (68.35 ppm) as well as along with the carbon atoms C-2 (71.87 ppm), C-3 (70.83 m. d.) and S-22 (84.22 ppm).

Thus, compound 132 has the structure of 2,3-O-di-(ethyl carbonate)-20,22-O-carbonate-20-hydroxyecdysone (Hardman 1987; Karlson 1994).

5.4.3 *Synthesis of 20-Hydroxyecdysone Hexanesulfonic Acid Derivatives*

In order to obtain biologically active compounds, we carried out the reaction of 20-hydroxyecdysone with a hexanesulfonic acid chloride (Fig. 5.11). The reaction was carried out according to the method in dry pyridine at 35 °C for 0.5 h according to the following scheme:

The compounds obtained were: 17.46% 2-O-(hexanesulfo) -20-hydroxyecdysone (133), 23.07%) 2.22-O-di-(hexanesulfo) -20-hydroxyecdysone (134). On TLC, the individual compounds showed two spots with $R_f = 0.29$ (133), $R_f = 0.48$ (134), respectively (system chloroform: methanol (6: 1)). The structure was established as a result of the analysis of ^{1}H, ^{13}C, NMR spectra.

Studies have shown that free 20-hydroxyecdysone is inactive with respect to most test cultures of microorganisms. The introduction of the hexanesulfo group at the C-2, C-22 position in the 20-hydroxyecdysone molecule leads to a sharp increase in antimicrobial activity.

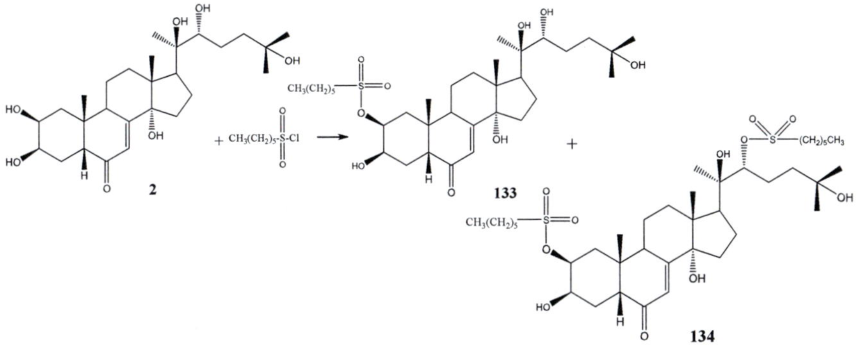

Fig. 5.11 Preparation of 2-O, 2,22-O-di- (hexanesulfo) -20-hydroxyecdysone

5.4.4 *Synthesis of O-Chlorobenzoyl Derivatives of 20-Hydroxyecdysone*

A study of the chemical modification of 20-hydroxyecdysone (2) made it possible to find out the reactivity of the hydroxy groups for conjugation and rank them according to the decrease in reactivity in the following order: C2 > C22 > C3 > C25 > C20 > C14 (Akhrem et al. 1973).

The acylation reaction of 20-hydroxyecdysone was carried out by the interaction of o-chlorobenzoic acid chloride in pyridine (Silva et al. 2005; Odinokov et al. 2005). Therefore, with direct esterification with a large excess of o-chlorobenzoic acid chloride, a mixture of 20- hydroxyecdysone mono-, di-, tri-, and tetra-ethers is rather easily obtained (Fig. 5.12).

With the subsequent separation of the reaction products by column chromatography (silica gel), new derivatives were obtained- 2-O-(O-chlorobenzoyl)-20-hydroxyecdysone (135), 2,3-di-O-(o-chlorobenzoyl)-20-hydroxyecdysone (136), 2,3,22-tri-O-(o-chlorobenzoyl)-20-hydroxyecdysone (137), 2,3,22,25-tetra-O-(O-chlorobenzoyl)-20-hydroxyecdysone (138). Their structure was established as a result of analysis of H, ^{13}C NMR spectra, and 2D experiments (Table 5.7).

The structure of the mono-aroylation product of 20-hydroxyecdysone (2) as compound 135 was proved by the data of ^{1}H and ^{13}C NMR spectra. In the spectrum, the signals of methane protons H-3α and H-22 are present at 4.45 and 3.92 ppm accordingly, the proton signal H-2α appears in the form of a doublet of triplets and at 5.54 ppm. In spectrum 2, the H-2a signal resonates at 4.07 ppm, i.e., at 1.47 ppm in a stronger field. In the spectrum of monoester 135, the signal of the equatorial proton H-3α is shifted to a weak field by only 0.34 ppm. In the aromatic part of the spectrum, the signals of the four protons of the chlorobenzoyl residue resonate clearly. The ^{13}C NMR spectrum of compound 135 shows a weak-field shift of the C-2 signal (8 74.64 ppm) by 6.0 ppm compared with that in molecule 2. Moreover, the chemical shifts of the remaining hydroxyl-containing carbon atoms have not changed significantly. Therefore, compound 135 is 2-O-(o-chlorobenzoyl)-20-hydroxyecdysone.

In the ^{1}H NMR spectrum of di-O-ether 136, weak-field shifts of the signals of the H-2 and H-3 protons are observed (5, 5.73, 5.99, respectively) compared with those of compound 2. The chemical shift of H-22 slightly changed. In the ^{13}C NMR

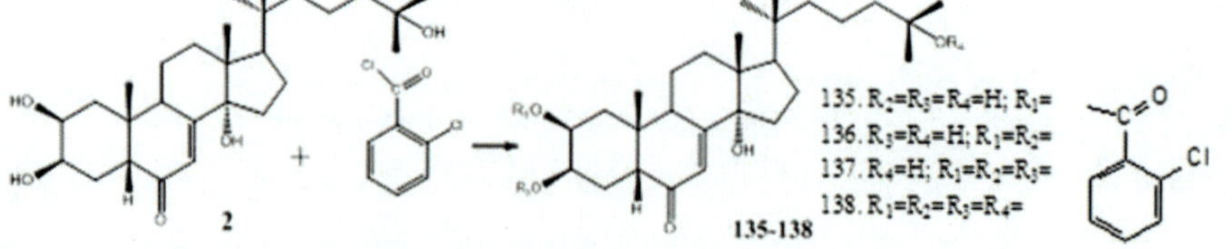

Fig. 5.12 Synthesis of O-chlorobenzoyl derivatives 20E

Table 5.7 Chemical shifts of ^{13}C NMR of compounds 2, 135,136, 137, and 138 (C_5D_5N, δ, ppm)

Atom C	DEPT	2	135	136	137	138
1	CH_2	38.49	34.10	35.36	35.25	35.20
2	CH	68.64	74.64	71.45	71.49	71.45
3	CH	68.55	65.59	70.12	70.17	70.09
4	CH_2	32.51	- > - > o JJ.ZJ	32.37	30.03	29.93
5	CH	51.90	51.66	52.36	52.39	52.33
6	C	203.97	203.11	201.54	201.53	201.42
7	CH	122.15	122.12	121.84	122.01	122.12
8	C	166.61	166.72	167.01	166.75	166.68
9	CH	34.93	34.84	34.10	34.88	34.78
10	C	39.16	39.41	39.28	39.33	39.35
11	CH_2	21.62	21.66	21.65	21.64	21.57
12	CH_2	32.28	33.47	32.18	32.41	32.36
13	C	48.61	48.49	48.49	48.65	48.59
14	C	84.67	84.64	84.57	84.54	84.45
15	CH_2	32.96	32.21	32.58	32.16	32.22
16	CH_2	21.99	21.93	21.91	22.02	22.06
17	CH	50.60	50.51	50.50	50.96	50.94
18	CH_3	18.40	18.34	18.33	18.33	18.34
19	CH_3	24.98	24.73	24.53	24.53	24.45
20	C	77.34	77.28	77.27	76.88	76.88
21	CH_3	22.21	22.15	22.17	22.94	22.88
22	CH	78.04	77.98	77.98	82.46	81.77
23	CH_2	27.98	27.98	27.99	26.72	26.21
24	CH_2	43.15	43.15	43.15	42.35	39.15
25	C	70.05	70.00	70.00	69.71	84.75
26	CH_3	30.52	30.45	30.07	30.13	26.44
27	CH_3	30.63	30.70	30.71	30.75	26.64
1’	C		131.82	131.68; 131.78	130.78; 131.14; 132.53	131–134
2’	C		133.93	134.01; 133.93	133.92; 134.09; 134.36	131–134
3’	CH		131.64	131.86; 131.98	132.25; 132.56; 132.69	131–134

(continued)

Table 5.7 (continued)

Atom C	DEPT	2	135	136	137	138
4'	CH		133.32	133.80; 133.76	133.11; 133.81; 133.84	131–134
5'	CH		127.53	127.64; 127.83	127.55; 127.68;	127.5–127.5
					127.86	
6'	CH		132.16	132.54; 132.66	131.72; 131.90; 132.01	131–134
7'	C		167.77	165.26; 165.47	165.31; 165.50 166.65	165.23; 165.47; 165.90; 166.63
1	CH_2					1.95, m; 2.41, m
2	CH	4.07, m	5.54, dt, (J = 1 1.9; 3.6)	5.73, dt, (J = 1Z5; 3.7)	5.59, dd, (J = 12.1; 3.5	5.68, dt, (J = 12.4; 3.8)
3	CH	4.11,m	4.45, s	5.99, s	5.83, s	5.99, s
4	CH_2					1.97, m, 2.16, m
5	CH	2.90, m	3.14, dd, (J = 13.5; 3.7)	2.90, dd, (J = 13.6; 4.2)	2.74, dd, (J = 13.4; 4.0)	2.89, dd, (J = 13.6; 4.2)
7	CH	6.14, d, (1 = 2.4)	6.31, d, (J = 2.2)	6.30, d, (1 = 2.3)	6.13, d, (J = 2.2)	6.28, d, (J = 2.4)
9	CH	3.47, t (J = 9.4)	3.76, t (J = 7.6)	3.74,m	3.59, t (J = 8.6)	3.76, t (J = 8.6)
11	CH_2					1.76, m; 2.09, m
12	CH_2					2.59, dd, (0.1 = 13.3; 3.8); 1.99, m
15	CH_2					1.95, m; 2.21, m
16	CH_2					2.24, m; 2.66, m
17	CH	2.88, m	3.05, t, (J = 8.9)	3.05, t. (J = 9.0)	2.93, t, (J = 8.6)	3.05, t. (J = 9.1)
18	CH_3	1.10,s	1.26, s	1.25, s	1.09, s	1.24, s
19	CH_3	0.95, s	1.17,s	1.15,s	0.99, s	1.14, s
21	CH_3	1.47, s	1.40, s	1.64, s	1.63, s	1.79, s
22	CH	3.76, d (J = 9.4)	3.92, d, (J = 9.6)	3.92, d, (J = 9.5)	5.68, dd, (J = 10.8; 1.9	5.82, d, (J = 10.3)

(continued)

Table 5.7 (continued)

Atom C	DEPT	2	135	136	137	138
23	CH_2					2.15, m; 2.37, m
24	CH_2					2.35, m; 2.40, m
26	CH_3	1.25,c	1.43, s	1.43,s	1.27, s	1.66, s
27	CH_3	1.25, s	1.43, s	1.43, s	1.27, s	1.66, s
3'	CH		7.47, dd, (1 = 7.9; 1.0)	7.24–7.51, m	7.90, dd, (J = 7.5; 1.9) 7.92, dd, (J = 7.5; 1.9) 7.94, dd, (J = 7.5; 1.9)	
4'	CH		7.36 td, (J = 7.7; 1.7)	7.24–7.51, m	7.16–7.26, m	
5'	CH		7.26, m	7.24–7.51, m	7.03–7.16,m	
6'	CH		7.94, dd, (J = 7.8; 1.7)	8.04–8.11,m	7.29, d, (J = 8.1) 7.31, d, (J = 8.1) 7.34, d, (J = 8.1)	

spectrum of compound 136, a paramagnetic shift of the signals of the carbon atoms C-2 (71.45 ppm) and C-3 (70.12 ppm) is observed. In addition, in the region of 127–134 ppm, the spectrum contains signals of 12 carbon atoms of the aromatic ring of the o-chlorobenzoyl group. In a weaker field at 165.26 and 165.47 ppm, signals of carbon atoms of two carbonyl groups appear. These data indicate that the ester groups in compound 136 are at positions C-2 and C-3. Therefore, compound 136 is 2,3-di-0-(o-chlorobenzoyl)-20-hydroxyecdysone.

To assign the signals of protons and carbon nuclei in the ^{1}H and ^{13}C NMR spectra of 137, the results of HetCor experiments were used. The signals of the H-2, H-3, and H-22 protons appear significantly in a weak field relative to those of 2 at 8 5.59, 5.83, and 5.68 ppm, respectively, which indicates the location of three o-chlorobenzoyl groups at C-2, C-3, and C-22. An additional three sets of signals of aromatic protons of three o-chlorobenzoyl functional groups appear in the weak-field part of the spectrum. Analysis of ^{13}C NMR and DEPT spectra of compound 137 showed the presence of 48 carbon atom signals. In the spectrum, the signals of carbon atoms C-2, C-3, and C-22 are shifted, as one would expect, to a weaker field (71.49 ppm, 70.17 ppm, 82.46 ppm, respectively). In addition, in the region of 127–134 ppm signals of 18 aromatic carbon atoms are observed. In a weaker field in the region of 165.31–166.65 ppm, signals of three carbon atoms of the carbonyl group of the o-chlorobenzoyl group are manifested. All of the above results indicate that the ester groups in compound 137 are in positions C-2, C-3, and C-22 of the steroid molecule, and it has the structure of 2,3,22-tri-O-(o-chlorobenzoyl)-20-hydroxyecdysone.

It should be noted that approximately the same chemical shifts are observed for H-2α, H-3 α, and H-22 signals (5.68 ppm, 5.99 ppm, 5.82 ppm) and in the ^{1}H NMR spectrum of the tetraester 138. This allows us to establish the addition of three of the four residues of o-chlorobenzoyl acid at C-2, C-3, and C-22. To determine the point of attachment of the fourth ester group in the molecule of compound 138, it is useful to compare the chemical shifts of the signals of 26- and 27-methyl groups in its ^{1}H NMR spectrum with similar characteristics of the spectrum of triester 138. In this case, it is indicative that the signals of carbon atoms 26- and the 27-methyl groups in the spectrum of tetraester 138 are significantly shifted to a strong field compared with the position of the signals of the same groups in the spectrum of the initial phytoecdysteroid 2. Analysis of ^{13}C NMR spectra of compound 138 shows a strong shift downfield signal carbon atom C-25 (84.75 ppm), together with the carbon atoms C-2, C-3, and C-22. From this, we can conclude that in the molecule of compound 138, the fourth ester group is in position C-25. Thus, compound 138 has the structure of 2,3,22,25-tetra-O-(o-chlorobenzoyl)-20-hydroxyecdysone (Silva et al. 2005).

5.4.5 *Synthesis of O-Methoxybenzoyl Derivatives of 20-Hydroxyecdysone*

In the synthesis of o-methoxybenzoyl derivatives of 20-hydroxyecdysone, the acylation reaction of 20-hydroxyecdysone (2) was carried out under the action of o-methoxybenzoic acid chloride in pyridine (Zatsny et al. 1973).

Column chromatography separated the reaction mixture into individual components. At the same time, new derivatives were obtained: 2-O-(o-methoxybenzoyl)-20-hydroxyecdysone (139), 2.22-O-di-(o-methoxybenzoyl)-20-hydroxyecdysone (140) 2,3,22-O-tri-(o-methoxybenzoyl)-20-hydroxy-ecdysone (141), 2,3,22,25-O-tetra-(o-methoxybenzoyl)-20-hydroxyecdysone (142) (Fig. 5.13).

The structure of o-methoxybenzoyl derivatives of 20-hydroxyecdysone was studied using the ^{1}H, ^{13}C, NMR spectra and the DEPT, HetCor experiments (Table 5.8).

The structure of 2,3,22-O-tri-(o-methoxybenzoyl)-20-hydroxyecdysone (141) was confirmed by ^{1}H NMR data. The proton signal H-2 manifests itself in the form of a doublet-triplet (J = 12.4; 3.4) at 5.71 ppm, and the proton signal H-3 is observed

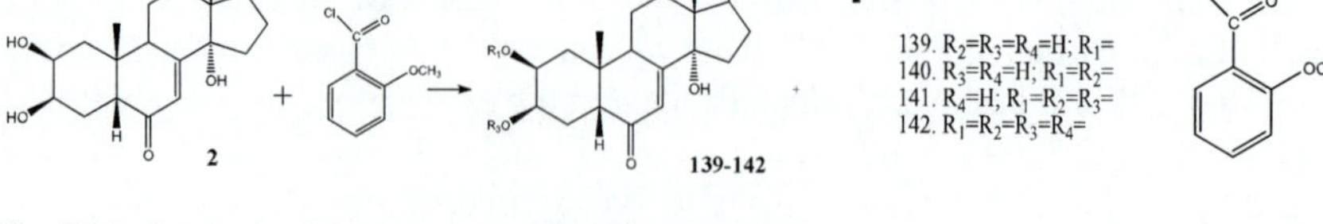

Fig. 5.13 Synthesis of O-methoxybenzoyl derivatives 20E

Table 5.8 Chemical shifts of ^1H and ^{13}C NMR of compound 141 (C_5D_5N (150.35 ppm), δ, ppm, J/Hz, 400 MHz) and HMBC experiment data

Atom C	δ_C	δ_H	HMBC (C)
1	35.41	2.00, m; 2.40, dd (13.2; 4.2)	19, 9, 10, 2; 10, 2, 3, 5
2	70.66	5.71, dt (12.4; 3.4)	
3	69.19	5.94, s	
4	30.29	1.90, m; 2.09, m	
5	52.43	2.92, dd(13.3; 3.8)	4, 9, 10
6	201.98		
7	122.00	6.27, d (2.1)	5,9, 14
8	166.69		
9	34.92	3.74, m	
10	39.25		
11	21.66	1.76, m; 2.10, m	
12	32.44	2.00, m; 2.57, m	14, 11, 18, 13
13	48.63		
14	84.53		
15	32.18	1.92, m; 2.21, m	16
16	22.05	2.22, m; 2.62, m	
17	50.76	3.07, t. (9.0)	16, 18, 13
18	18.37	1.22, s	12, 13, 17, 14
19	24.69	1.16,s	1, 10, 5
20	76.92		
21	23.07	1.73, s	17,20,22
22	81.81	5.78, dd, (J = 10.6; 2.0)	21,23,24, 20
23	26.77	2.15, m; 2.47, dd (12.6; 3.8)	24
24	42.30	2.01, m; 2.12, m	23, 26, 27, 25; 23, 26, 27, 25
25	69.69		
26	30.73	1.39, s	27, 24, 25
27	30.16	1.39, s	26, 24, 25
OH		6.51, s	
1'	121.32;21.53; 122.99		
2'	159.27;160.00 160.14		
3'	113.12; 113.20 113.28	6.88–7.04, m; 6.88- 7.04, m; 6.88–7.04, m	1', 5'; 1', 5';1,'5'
4'	133.75;134.38 134.45	7.37–7.48, m; 7.37- 7.48, m; 7.37–7.48, m	6', 2'; 6,'2'; 6',2'

(continued)

Table 5.8 (continued)

Atom C	δ C	δ H	HMBC (C)
5'	120.83;20.97; 121.09; 132.12	6.88–7.04, m; 6.88- 7.04, m; 6.88–7.04, m; dd, (7.8; 3.4); dd, (8.0; 3.5)	1'; 1'; 1';4',2',7';4',2', 7'
6'	132.49 (C2)	8.08, dd, (8.0; 1.7)	4',2', 7'
7'	166.02; 166.40; 167.86		22
OCH3	56.26, 56.34(C2)	3.68, s; 3.75, s; 3.80, s	2';2';2'

at 5.94 ppm (r.s.), the chemical shift has an H-22 signal at 5.78 ppm (dd, J = 10.6; 2.0).

Analysis of the ^{13}C NMR data of compound 141 showed a strong shift in the weak field of the signal of the carbon atom C-22 (81.81 ppm) as well as along with the carbon atoms C-2 (70.66 ppm) and C-3 (69.19 m. d.).

5.4.6 *Synthesis of O-Methoxybenzoyl-Substituted Turkesterone*

In the synthesis of o-methoxybenzoyl derivatives of turkesterone; turkesterone was acylated (143) under the influence of methoxybenzoic acid chloride (Zatsny et al. 1973).

The reaction was carried out in an absolutely dry pyridine solvent. The yield of o-methoxybenzoyl-turkesterone (144–147) was about 47–69%. By using a column chromatography, the mixture was separated into individual components. At the same time, new derivatives of 2-O-(o-methoxybenzoyl) -turkesterone (144), 2.11-O-di-(o-methoxybenzoyl)-turkesterone (145) 2,3,11-O-tri-(o-methoxybenzoyl)-turkesterone (147), 2,3,11,22-O-tetra-(o-methoxybenzoyl)-turkesterone (186) (Fig. 5.14).

In the NMR ^{1}H and ^{13}C spectra of ecdysteroids 144–147 (Table 5.9), the chemical shifts of the carbon atoms and hydrogen of the steroid ring are coinciding. In the ^{1}H NMR spectra of compounds 144–147, the signals of the methoxybenzoyl proton are H-11 in the region of 6.04–6.12 ppm (Table 5.9). In the ^{13}C NMR spectra of

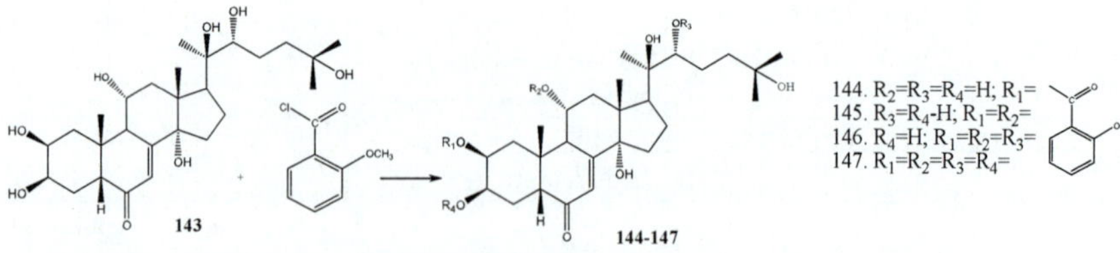

Fig. 5.14 Synthesis of o-methoxybenzoyl-substituted turkesterone

Table 5.9 Chemical shifts ^1H and ^{13}C NMR of compound 147 (C_5D_5N (150.35 ppm), δ, ppm, J/Hz, 400 MHz)

Atom C	δ C	δ H (J/Hz)
1	37.27	2.17, m; 2.49, m
2	70.40	5.99, dt (12.8; 3.5)
3	68.64	6.12, s
4	30.33	1.84, m; 2.04, m
5	53.14	2.92, dd (12.9; 4.5)
6	201.49	
7	123.52	6.37, d (2.5)
8	162.82	
9	39.93	4.29, dd (8.8; 1.9)
10	40.04	
11	72.84	6.12, s
12	39.43	2.95,m
13	48.65	
14	84.29	
15	32.18	1.98, m; 2.26, m
16	22.09	2.27, m; 2.66, m
17	50.65	3.15, dd (9.6; 8.8)
18	19.11	1.39, s
19	25.06	1.38, s
20	76.66	
21	22.85	1.65,s
22	81.69	5.76, d, (J = 9.3)
23	26.73	2.08, m; 2.42, m
24	42.26	2.03,m
25	69.70	
26	30.10	1.38, s
27	30.83	1.38, s
1’	121.28; 121.33; 122.86; 123.46	
2’	159.19; 159.98; 160.21; 160.53	
3’	113.09; 113.13; 113.16; 113.25	6.80–7.00; 6.80–7.00; 6.80–7.00; 6.80–7.00, m
4’	134.20; 134.33; 134.47; 134.63	7.34–7.42; 7.34–7.42; 7.34–7.42; 7.25, m
5’	120.53; 120.91; 121.02; 121.13	6.80–7.00; 6.80–7.00; 6.80–7.00; 6.80–7.00, m
6’	132.14	7.99, dd, (8.0; 1.8)
	132.28	7.90, dd, (8.0; 1.8)
	132.56	7.96, dd, (8,0; 1.8)

(continued)

Table 5.9 (continued)

Atom C	δ_C	δ_H (J/Hz)
	132.81	7.76, dd, (7.6; 1.7)
7'	165.59; 165.78; 166.40; 167.92	
OCH_3	56.19; 56.31; 56.36; 56.51	3.57, s; 3.73, s; 3.75, s; 3.84,s

compounds 144–147, the signals are characteristic, respectively, at C-11 (72.84 ppm) (Table 5.9).

The IR spectra of all synthesized derivatives of o-methoxybenzoyl-turkesterone (144–147) contain absorption bands in the regions of 1730–1797 cm^{-1}, characterizing stretching vibrations of the C = O group (complex ef.) And two bands in the region of 1647–1735 cm^{-1}, relating to bond vibrations C = 0 (keto).

Thus, for the first time, new derivatives of o-methoxy-benzoyl-turkesterone compounds 144–147 were synthesized.

5.4.7 *Synthesis of 20-Hydroxyecdysone Picolinate Derivatives*

In the reaction of 20-hydroxyecdysone (2) with picolinic acid chloride in a 1:4 ratio, respectively, picolinoyl derivatives of 20-hydroxyecdysone 148–151: 2-O-picolinoyl-20-hydroxyecdysone (148), 2,22-di-O- (picolinoyl) -20-hydroxyecdysone (149), 2,3,22-tri-O- (picolinoyl) -20-hydroxyecdysone (150), 2,3,22,25-tetra-O- (picolinoyl)—20- hydroxyecdysone (151) (Fig. .5.15).

The structure of the mono-picolinoyl-substituted 20-hydroxyecdysone product (2) as 2-O- (picolinoyl) (151) was confirmed by the ^{1}H NMR spectrum. In the spectrum, signals of methane protons N-3α and H-2α are present at 4.30 and 5.27 ppm, respectively. The chemical shift of compound 151 of the methane proton H-22 at 5.166 ppm the broadening of the doublet (J 10.5 Hz), in the spectrum of 20-hydroxyecdysone (2) H-22 at (5 4.03 ppm), the signal of which is shifted to a weak field with respect to the starting compound 2.

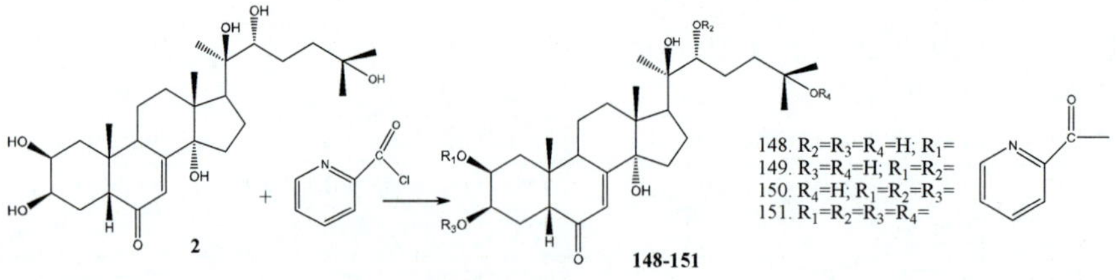

Fig. 5.15 Synthesis of 20-hydroxyecdysone picolinates

5.4.8 *Synthesis of 6-Chloronicotinate Derivatives of 20-Hydroxyecdysone*

The acylation reaction of 20-hydroxyecdysone (2) is carried out under the influence of 6-chloronicotinoyl chloride in pyridine. It was found that under these conditions a mixture of four 6-chloronicotinates is formed.

Column chromatography managed to separate this mixture into individual components. In this case, mono-, di-, tri-, and tetra-6-chloronicotinates were obtained (155–158).

Their structure was established as a result of the analysis of ^{1}H and ^{13}C NMR spectra. In this case, first of all, it is noteworthy that, judging by the spectral data, in all products of the acylation reaction the labile 14a-hydroxy Δ7-6-keto group characteristic of the initial 20-hydroxyecdysone (2) is retained. From the ^{1}H NMR spectra of the acylation products, the number of 6-chloronicotinic acid residues attached to the 20-hydroxyecdysone molecule can be determined quite easily from the number of proton signals of the pyridine ring (Fig. 5.16) (Bobaev et al. 2009).

The structure of the monoacylation product of 20-hydroxyecdysone (2) as 2-O-(b-chloronicotinate) (152) was proved according to the ^{1}H NMR spectrum (Table 5.10). In the spectrum, the signals of the methane protons H-3α and H-22 are present at 4.523 and 3.921 ppm, respectively, and the proton signal H-2α is blocked by an intense signal of water protons. The chemical shifts of the signal of the methane proton H-22 in the spectra of mono-6-chloro nicotinate (152) and 20-hydroxyecdysone (2) (5 3.908 ppm) practically coincide, which allows us to conclude that the structure contains the first free 22-hydroxy group.

In the spectrum of monoester 152, the signal of the equatorial proton H-3α is shifted to a weak field by about 0.32 ppm compared with the position of a similar proton in the spectrum of 20-hydroxyecdysone (6 4.20 ppm).

The structure of diester (153), as 2.22-di-O-(6-chloronicotinate), is reliably proved by the data of ^{1}H NMR spectrum. In it, the H-2 signals α and H-22 appear in a weak field at 5 5.279 and 5.166 ppm, respectively, which indicates their geminal location to the 6-chloronicotinate groups. At the same time, the value of the chemical shift of the H-3α proton (5 4.393 ppm) in the ^{1}H NMR spectrum indicates the presence in the molecule of compound 153 of a free 3β—hydroxy group.

The data of the ^{1}H NMR spectrum unambiguously make it possible to establish the structure of the triether as 2,3,22-tri-O-(6-chloronicotinate) 154. Moreover, to prove the structure, the presence of methane proton signals H-2α, H- in the ^{1}H NMR

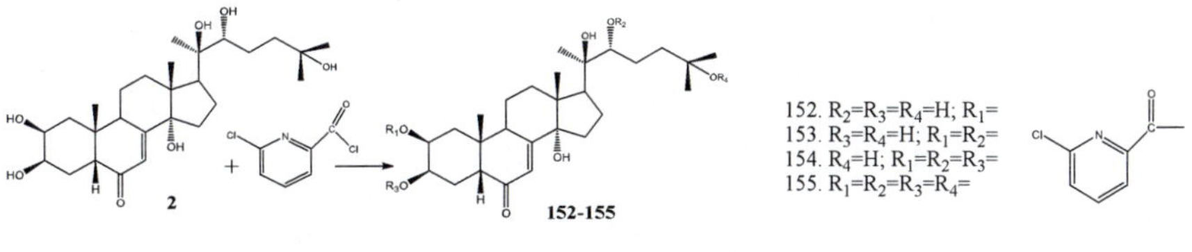

Fig. 5.16 Synthesis of derivatives of 6-chloronicotinate 20E

Table 5.10 Chemical shifts of carbon atoms (8, ppm) in ^{13}C 20- NMR spectra hydroxyecdysone (2) and its derivatives (152–155).

Atom	2	152	153	154
C-1	37.8	32.919	34.420	34.461
C-2	68.0	72.990	68.661	68.621
C-3	68.0	71.917	70.568	70.338
C-4	32.2	31.449	29.247	29.201
C-5	51.2	49.837	51.217	51.326
C-6	203.4	203.284	201.167	201.292
C-1	121.5	121.877	121.708	121.332
C-8	166.0	164.703	164.392	164.517
C-9	34.3	33.555	33.676	33.736
C-10	38.5	38.598	38.617	38.489
C-11	20.4	20.544	20.562	20.860
C-12	31.6	31.931	31.729	31.268
C-13	48.0	47.561	47.606	47.725
C-14	84.1	84.595	84.570	84.156
C-15	31.6	31.149	31.060	31.225
C-16	21.3	29.706	29.702	29.707
C-17	50.0	49.494	49.487	49.872
C-18	17.6	17.469	17.494	17.335
C-19	24.3	23.733	24.010	24.031
C-20	76.8	75.146	77.238	77.277
C-21	21.0	21.478	21.524	21.499
C-22	77.4	81.069	80.977	81.255
C-23	29.8	24.765	24.798	24.474
C-24	42.4	40.283	40.261	40.151
C-25	69.6	70.767	70.300	83.936
C-26	27.3	28.448	28.758	24.196
C-27	29.8	30.408	30.287	27.006
6-chloronicotinate		163.956; 165.382; 151.137; 151.256; 155.903; 155.952; 139.760; 139.833; 124.339; 124.385; 171.549	163.555; 163.715; 165.415; 151.061; 151.138; 155.945; 156.048; 156.337; 139.473; 139.600; 139.767; 124.335; 124.384; 124.604	162.953; 163.512; 163.595; 165.296; 150.512; 151.057; 151.143; 155.057; 156.007; 156.052; 156.324; 139.476; 139.608; 139.800; 140.075; 124;297; 124.404; 124.580
Solvent	C_5D_5N	$CDCl_3$	$CDCl_3$	$CDCl_3$

spectrum of this substance is significant 3α and H-22 in a weak field at 8, 5.480, 5.771, and 5.176 ppm, respectively. This clearly indicates that the ester groups in compound 154 are at positions 2, 3, and 22 of the steroid molecule.

It should be noted that the signals H-2α, H-3 α, and H-22 have approximately the same chemical shifts in the ^{1}H NMR spectrum of the tetraester 155. This allows us to establish the attachment of three of the four residues of 6-chloronicotinic acid at C-2, C-3, and C-22. To determine the point of attachment of the fourth ester moiety in the molecule of compound 155, it is useful to compare the chemical shifts of the signals of 26- and 27-methyl groups in its ^{1}H NMR spectrum with the similar characteristics of the spectrum of triester 154. It is significant that the signals of 26- and 27-methyl groups in the spectrum of tetraether 155 are significantly shifted to a weak field compared to the position of the signals of the same groups in the spectrum of triester 154. From this, we can conclude that in the molecule of compound 155 at position 25 there is a complex ester rnaya group, and thus, this substance has a structure 20- hydroxyecdysone 2,3,22,25-tetra-O-(6-chloronicotinate).

5.4.9 Synthesis of 20-Hydroxyecdysone Hydrazones

Hydrazones are widely used in the synthesis of new compounds and exhibit diverse biological activity. Hydrazones of ecdysteroids and their chemistry are poorly understood. We first synthesized 6-M-semicarbazone-20-hydroxyecdysone (156), which is converted into the stable compound -6-H-hydrazono-20-hydroxyecdysone (157) and 6-M-hydrazone stachysterone B (158). Direct action of 2 with semicarbazide hydrochloride in methanol led to the formation of semicarbazone (156), hydrazone 157, and the dehydration of compound 158, respectively (Fig. 5.17).

At the end of the reaction, the resulting mixture was partitioned into individual compounds by column chromatography. In this case, hydrazone 157 and dehydrated compound 158 were obtained (Fig. 5.17). The structure of the reaction product as 6-N-hydrazone-20-hydroxyecdysone (199) and impurities as 6-N-hydrazone-stachysterone B (158) were determined by analyzing ^{1}H and ^{13}C NMR spectra (CD_3OD (49.00), δ, ppm, J/Hz, 400 MHz) and experimental data from DEPT, HMBC. δC 100 MHz.

Chemical shifts of ^{13}C NMR compounds 20-hydroxyecdysone (2) C-6 at 203.97 ppm, N-hydrazone-20-hydroxyecdysone (157) C-6 at 152.11 ppm, 6-N-hydrazone stachysterone B (158) C-6 at 151.47 ppm. The characteristic peaks in ESI–MS are m/z = 462.0 $[M + H_20]^+$, Mm = 480.16, $C_{27}H_{44}O_6$ 20-hydroxyecdysone (2). The characteristic peaks in ESI–MS are m/z = 495.46 $[M + H]^+$, Mm = 494.67, $C_{27}H_{46}N_20_6$. 6-N-hydrazono-20-hydroxyecdysone (157). The characteristic peaks in ESI–MS are m/z = 477.45 $[M + H]^+$, Mm = 476.66, $C_{27}H_{46}N_2O_6$. 6-N-hydrazone-20-hydroxyecdysone (158).

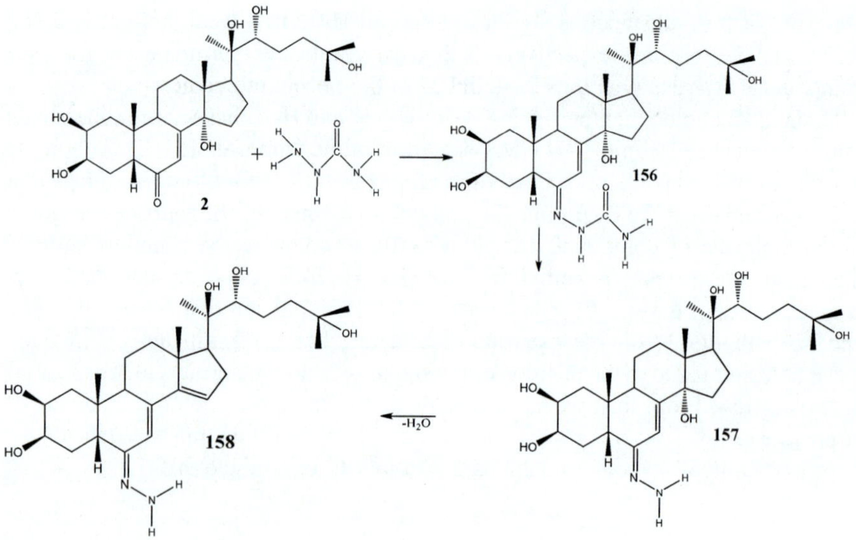

Fig. 5.17 Synthesis of hydrazones of 20-hydroxyecdysone

5.5 Alkaline Hydrolysis of Ecdysteroids

5.5.1 Alkaline Hydrolysis of Tomentesterone A

To a solution of 60 mg of tomentesterone A (46) in 10 ml of methanol was added 72 mg of potassium bicarbonate in 5 ml of water. The reaction mixture was left for 48 h in a thermostat at 39 °C. Then it was diluted with water, neutralized, methanol was distilled off in vacuo. The products were extracted with ethyl acetate. The solvent was distilled off to dryness, and the residue was chromatographed on a silica gel column. Elution with a system (chloroform–methanol (15:1)) gave 17.1 mg of 5α-2-deoxy-α-ecdysone-22-acetate, $C_{29}H_{46}O_6$, m.p. 158–160 °C (methanol–water), $[\alpha]_D + 51.4 \pm 2°$(c 0.20; methanol). IR spectrum (KBr, ν, cm^{-1}): 3440 (OH), 1735, 1260 (ester group), 1660 (7-en-6-keto grouping), M^+ 490 (Ramazonov et al. 1995; Ramazanov et al. 2007).

Subsequent elution with the same solvent mixture gave 12 mg of 5α-2-deoxy-α-ecdysone (22), $C_{27}H_{44}O_5$, m.p. 225–227 °C (methanol–water) (Saatov et al. 1987a).

Benzoic acid was isolated from the aqueous residue after acidification with dilute hydrochloric acid (1:1) and extraction with ethyl acetate.

5.5.2 Alkaline Hydrolysis of Tomentesterone B

3 ml of a 0.5% aqueous potassium bicarbonate solution were added to 5 mg of tomentesterone B (47) in 5 ml of methanol. The reaction mixture was placed in a thermostat for 10 days at 37 °C. Processing was performed as described above. An ethyl acetate extract was evaporated to dryness. The residue was recrystallized from aqueous methanol. Received 2 mg of ecdysteroid with m.p. 225–227 °C, according to the IR spectrum (KBr; v; cm^{-1}): 3440 (OH), 1645 (7-en-6-keto grouping) identical to 5α-2-deoxy-ecdysone (Saatov et al. 1987b).

Benzoic acid was isolated from the aqueous residue after acidification with dilute hydrochloric acid (1: 1) and extraction with ethyl acetate.

5.5.3 Alkaline Hydrolysis of 3-O-Benzoate-2-Deoxyecdysterone

To a solution of 50 mg of 3-O-benzoate-2-deoxyecdysterone (49) in 5 ml of methanol was added 30 ml of potassium bicarbonate in 3 ml of water. The reaction mixture was left in a thermostat for three days at 37–38 °C, then diluted with water, neutralized, and extracted with ethyl acetate (A). From the aqueous solution, after acidification with dilute (1:1) hydrochloric acid and extraction with ethyl acetate, 7 mg of benzoic acid was isolated with m.p. 122 °C. Solvent (A) was distilled off to dryness and the residue was chromatographed on a silica gel column. Washing with a system (chloroform–methanol (10: 1)) gave 22 mg of 5α-2-deoxyecdysterone (22), $C_{27}H_{44}O_6$, m.p. 235–237 °C (from methanol–water mixture). IR spectrum (KBr, ν, cm^{-1}): 3347 (OH), 1670 (Δ7-6-keto grouping).

Mass spectrum, m/z: 428 (M + -2H_2O), 429, 410, 395, 330, 329, 312, 286, 285, 99, 81, 69.

Further washing of the column with the same solvent system resulted in 24 mg of 2-deoxyecdysterone (23), m.p. 254–255 °C (from aqueous ethanol) identified with the known sample (Mamadalieva et al. 2000; Ramazanov et al. 1997), using TLC and the melting point of the mixing sample.

5.5.4 Alkaline Hydrolysis of 22-O-Benzoate-Polypodine B

22-O-Benzoate-polypodine B (42) (10 mg) was dissolved in 3 ml of a 0.5% methanolic sodium hydroxide solution. The reaction mixture was left at room temperature for a day, then diluted with water (10 ml) and neutralized. The methanol was evaporated, extracted with ethyl acetate. The solvent was distilled off to dryness, the residue was chromatographed on a silica gel column, eluting with a chloroform–methanol system (9: 1). Received 6 mg of substance with m.p. 252–254 °C (acetone), identified with the notorious sample of polypodine B (41) (Matsuoka et al. 1969). After acidification with dilute (1:1) hydrochloric acid and extraction with ethyl acetate, benzoic acid was isolated with a melting point of 122 °C.

5.6 Acylation Reactions

5.6.1 Obtaining 3,22-O-Diacetate-Tomentesterone B

20 mg of tomentesterone B (47) dissolved in 3 ml of pyridine was acetylated with 3 ml of acetic anhydride at room temperature for 24 h. After distillation of the solvent *in vacuo*, the residue was chromatographed on a silica gel column. Elution of the column with system 3 yielded 10 mg of 3.22-O-diacetate-tomentesterone B, $C_{38}H_{52}O_8$.

IR spectrum (KBr; v; cm^{-1}): 3420 (OH), 1650 (Δ7-6 keto grouping) 1720, 1730, 1290, 1250, 1225 (ester group), 1610, 1590, 715 (benzene ring).

Mass spectrum (EU, 70 eV), m/z (I_{oTH},,%): 618 (M + -H_2O + ; 0.2), 514 (8), 496 (22), 486 (33), 454 (66), 430 (66), 426 (36), 374 (27), 373 (60), 327 (72), 326 (60), 313 (72), 276 (60), 275 (71), 216 (66), 215 (97), 122 (100), 105 (98), 81 (78), 77 (96).

5.6.2 Obtaining of Tomentesterone A from 3,22-O-Diacetate-Tomentesterone B

3,22-O-diacetate-tomentesterone B (5 mg) was dissolved in 5 ml of aqueous methanol 0.5% $KHCO_3$ solution. The reaction mixture was left at room temperature for 4 h, after which the solution was diluted with water, neutralized with acetic acid, and the reaction product was extracted with ethyl acetate. The solvent was distilled, the residue was chromatographed on a silica gel column. Elution of the column, as described in the previous paragraph, yielded 3 mg of tomentesterone A (46) with m.p. 139–140 °C (methanol–water) (Ramazanov et al. 1995).

5.6.3 Obtaining 2,3-O-Diacetate-22-O-Benzoyl-Polypodine B

A solution of 10 mg of 22-O-benzoyl-polypodine B (42) in 2 ml of pyridine was acetylated with 2 ml of acetic anhydride at room temperature for one day. Then the reaction mixture was evaporated. The dry residue was chromatographed on a silica gel column. Elution with a chloroform–methanol system (30: 1) gave 9 mg of 2,3-O-diacetate-22-O-benzoyl-polypodine B, $C_{38}H_{52}O_{11}$, m.p. 160–162 °C (from methanol–water).

IR spectrum (KBr.ν, cm^{-1}): 3446 (OH), 1722, 1744, 1695, 1252, 1218 (ester groups), 1603, 716 (benzene ring), 1669 (Δ7-6-keto grouping).

Mass spectrum (EU, 70 eV), m/z (I_{oTH}, %): 532 (M + -C_6H_5COOH-2CH_3; 23), 528 (23), 509 (74), 495 (53), 492 (27), 463 (50), 445 (39), 350 (44), 122 (45), 105 (100), 81 (69), 71 (64), 69 (64).

5.7 Oxidation Reactions

5.7.1 Oxidation of 3-Keto-22-O-Acetate-2-Deoxy-A-Ecdysone

To 500 mg of 22-O-acetate-2-deoxy-α-ecdysone in 30 ml of acetone was added 2 drops of water and 275 mg of chromic anhydride. The mixture was stirred and left at room temperature for 40 min. The reaction mixture was poured into water and the product was recovered with ethyl acetate. The ethyl acetate fraction was chromatographed on a silica gel column, and eluting with a chloroform–methanol system (50: 1), 120 mg of 3-keto-22-acetate-2-deoxy-α-ecdysone, $C_{29}H_{44}O_6$, m.p. was isolated, with characteristics—210–211 °C (from methanol-hexane mixture), $[\alpha]_D$ 22 + 43.5 + 2° (c 0.16; methanol). IR spectrum (KBr, ν, cm^{-1}): 3480 (OH), 1735, 1250 (ester group), 1710 (C = O at C-3), 1670 (7-en-6-keto grouping).

Mass spectrum, m/z (%): 488 (M+; 1), 470 (7), 458 (5), 452 (4), 442 (5), 428 (0.5), 410 (16), 395 (35), 392 (20), 377 (13), 330 (2), 329 (4), 312 (10), 311 (5), 300 (15), 283 (45), 282 (57), 281 (30), 233 (100), 232 (89), 99 (25), 81 (20).

5.7.2 Oxidation of 3-Keto-2-Deoxy-A-Ecdysone

80 mg of 3-keto-22-O-monoacetonide-2-deoxy-α-ecdysone was dissolved in 2.5 ml of a 0.5% methanol solution of KOH. The reaction mixture was left for a day. After neutralization, it was diluted with water and extracted with ethyl acetate. The solvent was distilled off to dryness and the residue was chromatographed on a silica gel column. Elution with a chloroform–methanol system (50: 1) gave 35 mg of 3-keto-2-deoxy-α-ecdysone, $C_{27}H_{42}O_5$, m.p. 114–115 °C (chloroform-benzene). $[\alpha]_D^{25}$ + 86.5 + 2° (c 0.35; methanol). IR spectrum (KBr, ν, cm $^{-1}$): 3425 (OH), 1710 (C = O), 1666 (7-en-6-keto grouping).

Mass spectrum m/z (%): 446 (M + 9), 428 (98), 413 (40), 410 (50), 395 (98), 377 (8), 359 (40), 341 (50), 330 (95), 312 (96), 297 (50) 283 (30), 232 (35), 99 (100), 81 (98.).

References

Afonkina CP, Shafikov PB, Savchenko RG, Galyautdinov IV, Odinokov V (2006) Obtaining 7,8-dihydro analogs in the interaction of derivatives of 20-hydroxyecdysone with lithium aluminum hydride. Russ J Org Chem 42(8):1252–1253

Akhrem AA, Levina IS, Titov YA (1973) Ecdysones are insect steroid hormones 232

Baltaev UA (2000) Phytoecdysteroids: structure, sources and pathways of biosynthesis in plants. Russ J Bio Chem 26:892–925. https://doi.org/10.1023/a:1026662505403

Bobaev ID, Kovganko NV, Chernov YG et al (2009) Synthesis of 20-hydroxyecdysone 6-chloronicotinates. Chem Nat Compd 45:385–388. https://doi.org/10.1007/s10600-009-9338-6

Chong YK, Galbraith MN, Horn DHS (1970) Isolation of deoxycrustecdysone, deoxycrustecdysone and α-ecdysone from the Fern Blechnum minus. Chem Soc Chem Communs 1217–1218. https://doi.org/10.1039/C29700001217

Dzhukharova MK, Saatov Z, Gorovits MB et al (1991) Phytoecdysteroids of plants of the genus *Silene*. XVIII. 2-Deoxyecdysterone 20,22-monoacetonide from *Silene brahuica*. Chem Nat Compd 27:207–209. https://doi.org/10.1007/BF00629760

Galbraith MN, Horn DHS, Middleton EJ et al (1973) Structure of podecdysone C, a steroid with moulting hormone activity from the bark of *Podocarpus elatus* R. Br. Experientia 29:782. https://doi.org/10.1007/BF01946280

Hardman R (1987) Recent developments in our knowledge of steroids. Planta Med 53(3):233–238. https://doi.org/10.1055/s-2006-962690

Hikino H, Hikino Y, Nomoto K, Takemoto T (1968) Cyasterone, an insect metamorphosing substance from *Cyathulacapitata*: structure. Tetrahedron 24:4895–4906. https://doi.org/10.1016/S0040-4020(01)98686-7

Karlson P, Koolman J (1973) Metabolites fate of ecdysone and 3-dehydroecdysone in Calliphora vicina. Insect Biochem 3:409–417. https://doi.org/10.1016/0020-1790(73)90074-7

Karlson P, Hoffmeister H, Hoummel H et al (1965) Reaktionen des Ecdysonmolekus. Chem Ber 98(7):2394–2402. https://doi.org/10.1002/cber.19650980743

Karlson P, Bugany H, Döpp H, Hoyer GA (1972) 3-Dehydroecdyson, ein Stoffwechselprodukt des Ecdysons bei der Schmeibfliege Calliphora erythrocephala Meigen 353(2):1610–1614. https://doi.org/10.1515/bchm2.1972.353.2.1610

Karlson P (1994) On the use of ecdysteroid nomenclature. XI Ecdysone Workshop: Program and Abstracts. Ćeske Budèjovice 7–8

Levina TS, Kulikova LE, Kamernitzky AV, Kachala VV, Pokrovskaya EV, Smirnov AN (2008) 3-and 19-Oximes of 16α,17α-cyclohexanoprogesterone derivatives: synthesis and interactions with progesterone receptor and other proteins. Steroids 73:815–827. https://doi.org/10.1016/j.steroids.2008.03.003

Mamadalieva NZ, Ramazanov NS, Dinan LN et al (2000) Phytoecdysteroids of plants of the *Silene* Genus. 2-Dehydroxyecdysterone-3-O-benzoate from *Silene wallichiana*. Chem Nat Compd 36:513–515. https://doi.org/10.1023/A:1002847725472

Matsuoka T, Imai S, Sakai M, Kamada M (1969) Studies on phytoecdysones—a review of our works. Ann Rep Takeda Res Lab 28:221–271

Odinokov VN, Savchenko RG, Shafikov RV et al (2005) Stereochemistry of hydride reduction of 20-hydroxyecdysone derivatives. Russ J Org Chem 41:1296–1305. https://doi.org/10.1007/s11178-005-0338-0

Ramazanov NS, Maksimov ES, Saatov Z et al (1995) Phytoecdysteroids of plants of the genus *Silene*. XVII. Tomentesterone from *Silene tomentella*. Chem Nat Compd 31:600–603. https://doi.org/10.1007/BF01164888

Ramazanov NS, Maksimov ES, Saatov Z et al (1997) Phytoecdysteroids of plants of the genus *Rhaponticum* I. Carthamosterone A from *Rh. carthamoides*. Chem Nat Compd 33:301–302. https://doi.org/10.1007/BF02234880

Ramazonov NS, Mamadalieva NZ, Bobaev ID (2007) Phytoecdysteroids from five species of the genus *Silene*. Chem Nat Compd 43:117–118. https://doi.org/10.1007/s10600-007-0049-6

Ramazonov NS, Syrov VN (2006) Synthesis of cyasteronylthiocarbamate derivatives. Chem Nat Compd 42:558–561. https://doi.org/10.1007/s10600-006-0213-4

Saatov Z, Usmanov BZ, Abubakirov NK (1979) Phytoecdysones of *Silene praemixta* I. Silenosterone. Chem Nat Compd 15:700–703. https://doi.org/10.1007/BF00565565

Saatov Z, Abdullaev ND, Gorovits MB et al (1986) Phytoecdysteroids of plants of the genus *Silene* X. Sileneoside E—2-Deoxy-α-ecdysone 3-O-β-D-glucopyranoside—from *Silene brahuica*. Chem Nat Compd 22:297–300. https://doi.org/10.1007/BF00598300

Saatov Z, Gorovits MB, Abubakirov NK (1987a) Phytoecdysteroids of plants of the genus *Silene* XIII. Ecdysterone 20,22-monoacetonide from *Silene scabrifolia*. Chem Nat Compd 23:643. https://doi.org/10.1007/BF00598705

Saatov Z, Gorovits MB, Abubakirov NK (1987b) Phytoecdysteroids of plants of the genus *Silene*. XV. 2-Deoxy-α-ecdysone 22-O-benzoate from *Silene wallichiana*. Chem Nat Compd 23:708–711. https://doi.org/10.1007/BF00596648

Saatov Z, Gorovits MB, Abubakirov NK (1987c) Phytoecdysteroids of plants of the genus *Silene* XIII. Ecdysterone 20,22-monoacetonide from *Silene scabrifolia*. Chem Nat Compd 23:643. https://doi.org/10.1007/BF00598705

Saatov Z, Gorovits MB, Abubakirov NK (1988a) Phytoecdysteroids of plants of the genus *Silene*. XVI. Viticosterone E 22-O-benzoate from *Silene wallichiana*. Chem Nat Compd 24:463–465. https://doi.org/10.1007/BF00598532

Shafikov RV, Afonkina SR, Savchenko RG, Galyautdinov IV, Odinokov VN (2007) Stereoselectivity of the hydrogenolysis of the 14α-hydroxyl group in the diacetonides of 6-dihydro-5α-20-hydroxyecdysones. J Organic Chem 43(2):310–311

Silva MC, Riva S, Melo ML (2005) Regioselective enzymatic acylation of vicinal diols of steroids. Tetrahedron 61:3065–3073. https://doi.org/10.1016/j.tet.2005.01.104

Sing SB, Thakur RS (1982) Structure and stereochemistry of paristerone, a novel phytoecdysone from the tubers of *Paris polyphylla*. Tetrahedron 38:2189–2194. https://doi.org/10.1016/0040-4020(82)85166-1

Usmanov BZ, Saatov Z, Abubakirov NK (1977) Phytoecdysones of *Ajuga turkestanica*. V. Chem Nat Compd 13:595. https://doi.org/10.1007/BF00569612

Ves'kina NA, Odinokov VN (2012) Transformations of ecdysteroids in the synthesis of their structural analogs and minor phytoecdysteroids. Russ J Org Chem 48:1137–1161. https://doi.org/10.1134/S1070428012090011

Zatsny IL, Gorovits MB, Abubakirov NK (1973) Phytoecdysones of *Serratula* II. Viticosterone E from *Serratula sogdiana* and its partial synthesis. Chem Nat Compd 9:170–173. https://doi.org/10.1007/BF00563337

Zatsny IL, Gorovits MB, Rashkes YV et al (1975) Phytoecdysones of *Serratula* III. Mass spectrometric study of the acetates and acetonides of ecdysterone and viticosterone E. Chem Nat Compd 11:170–172. https://doi.org/10.1007/BF00570661

Chapter 6
Biological Activities of Phytoecdisteroids

Abstract Phytoecdysteroids are allelochemical toxins and antioxidants for non-adapted phytophage insect species. For the manifestation of hormonal activity, the structural organization of the phytoecdysteroid molecule is important. The most essential elements are the cis-junction of rings A and B, the presence of a D 7–6-conjugated keto group, and a complete sterol side chain. The presence of 14α-OH, 2β-OH, 3β-OH, and 20R-OH increases the biological activity. A very significant indicator of the manifestation of the protein-anabolic action of the compounds. However, the majority of phytoecdysteroids that have been isolated to date have a significantly lower ability to activate protein biosynthesis in mammals than steranobol. Of the other biological effects of phytoecdysteroids associated with their effect on metabolic processes in animals. No increase in insulin secretion was observed in animal experiments. Thus, the improvement in the activity of the central nervous system. Phytoecdysteroids did not show any toxic effects to the higher animals. In experiments on mice, it was found that LD50 for ecdysterone, when administered intravenously, was 6.4 g/kg, and for inocosterone –7.8 g/kg. Ecdysten which is based on ecdysteroids mental and physical performance improved, attention concentration increased, and reactive and personal anxiety normalized. In this chapter, postulates that phytoecdysteroids as metabolically active substances with the ability to increase the general nonspecific resistance of the body to various destabilizing environmental factors are of significant interest both in terms of further theoretical research and in terms of their practical use.

Keywords Phytoecdysteroids · Pharmacological activity · Biological activity · Anabolic · Protein biosynthesis

According to one of the most substantiated hypotheses, phytoecdysteroids are allelochemical toxins and antifidants for non-adapted phytophage insect species. This point of view corresponds to the general concept of the role of substances of secondary metabolism of plants as protective factors.

The antifungal, radioprotective, hemorheological, antitumor activity of extracts and isolated fractions of *Lychnis chalcedonica* and other plants was revealed. It has been established that ecdysterone exhibits hemorheological and radioprotective, and polypodine B antifungal action (Zibareva 2003). Phytoecdysteroids are considered

U. Y. Yusupova et al., *Phytoecdysteroids*,
https://doi.org/10.1007/978-981-16-6711-4_6

as part of a multicomponent chemical defense strategy that has been developed in the process of co-evolution of plants and herbivorous invertebrates (Dinan 1998). Phytoecdysteroids, however, are not universal protective factors of plants, they are effective only against non-adapted species of insects and are often present in the plant together with other toxic allelochemicals (for example, ayugarins, vitanolides, alkaloids, saponins, flavonoids), possibly interacting with them in synergism (Dinan 1994; Abubakirov 1981; Bergamasco and Horn 1983).

6.1 Hormonal and Growth-Stimulating Activity of Phytoecdysteroids

The effect of exogenous ecdysteroids on the growth and development of insects is studied using various biotests: feeding phytoecdysteroids with food, injection into the body, and by placing larvae in solutions containing ecdysteroids. The effect of ecdysteroids is evaluated according to a number of indicators: in larvae, survival, growth characteristics, changes in the processes of molting and pupation are determined, in adult forms, the survival, fecundity and viability of the offspring are evaluated. The antifidant activity of phytoecdysteroids is also determined. It turned out that the effect of exogenous phytoecdysteroids significantly depends on the type of insect and the stage of its development, the applied concentration of solutions, the method of introduction of the insect into the body, as well as the chemical structure of the compound. For example, some insect species (scoop—*Heliothis* spp., Migratory locust—*Locusta migratoria*) are resistant to very high concentrations of phytoecdysteroids in the diet (400–1000 ppm of 20-hydroxyecdysone), while others (cabbage—*Pieris brassicae* and cabbage scoop—*Mamestra brassicae*) reject food containing only 5–60 ppm 20-hydroxyecdysone (Smagghe et al. 1998).

For the manifestation of hormonal activity, the structural organization of the phytoecdysteroid molecule is important. The most essential elements are the cis-junction of rings A and B, the presence of a D 7–6-conjugated keto group, and a complete sterol side chain. The presence of 14α-OH, 2β-OH, 3β-OH, and 20R-OH increases the biological activity (Sláma 1993; Carlisle et al. 1963). Thus, the results of numerous experiments give reason to consider phytoecdysteroids as allelochemical substances produced by certain types of plants to protect against insect species susceptible to them.

In the first experiments devoted to elucidating the possible role of phytoecdysteroids as regulators of plant growth and development, it was reported that ecdysone extract from the locust thoracic glands stimulated the growth of internodes of dwarf peas. The stimulation effect was 10% of the effect of gibberellic acid (Matsuoka et al. 1969). The presence of gibberellin-like activity was recorded in inocosterone, which stimulated root and leaf growth in rice seedlings, while 20-hydroxyecdysone and ponasterone A did not have such activity (Hendrix et al. 1972). At the same time, a negative result was obtained when studying the activity of 20-hydroxyecdysone in four gibberellin tests (Jacobs and Suthers 1971). Based on these data, it was suggested that in the experiment of (Karlson et al. 1976), the extract could contain a substance of diterpenic nature, similar in structure to gibberellic acid, which caused the observed effect.

20-hydroxyecdysone did not affect shoot growth, the flowering process in cultivated apical buds of the waterlily of Pencilwan *Xanthium pennsylvanicum* (Lafont et al. 2011). 20-Hydroxyecdysone, ponasterone A, and ponasterone C isolated from the culture fluid of cultured gametophytes of the common bracken fern, *Pteridium aquilinum*, did not exhibit biological activity in the biotest by the induction of anteridia in gametophytes of some species of ferns from the genera *Pteridium*, *Onocleaeem*, and *Anemia*. However, 20-hydroxyecdysone inhibited the formation of female flowers and stimulated the formation of male flowers in squash—*Cucurbita pepo* (Hardman 1987).

A study of the effect of 20-hydroxyecdysone on the rate of photosynthesis and respiration in isolated *Vicea faba* leaves showed that it does not affect the respiration rate and only at low concentrations (30 ppm) stimulates the accumulation of photosynthesis products (Nakanishi et al. 1966).

6.2 Pharmacological Activity of Phytoecdysteroids

The discovery of large amounts of ecdysteroids in plants opened up an accessible source for obtaining the required amounts of these compounds for large-scale biological studies on their effect on those organisms that are not capable of endogenous production of steroids of this type (Bruno et al. 2002; Sláma et al. 1996; Buniam et al. 2020). The greatest interest, of course, was the study of the effect of phytoecdysteroids on the body of higher animals and humans. Of particular interest in this problem arose after it was established that many plants producing ecdysteroids have long been famous for valuable feed qualities, were widely known in the folk medicine of different countries for their tonic and adaptogenic properties, and were used to increase mental and physical performance, improve metabolism substances (various species of *Achyrantes, Cyathulla, Rhaponticum, Serratula, Silene, Ajuga,* etc.), however, the chemical composition of these plants, known before the discovery of ecdysteroids in them, did not satisfy nogo explanation described effects (Syrov 1996; Lafont 1997). Apparently, the listed factors, along with the available data on the ability of phytoecdysteroids in lower animals to cause various metabolic shifts (Akhrem et al. 1973), not only contributed to the intensive pharmacological study of these compounds, but also determined the initial approach to assessing the biological activity of the studied steroids in the mammalian organism also from the position of the metabolic orientation of their action. One of the first works in this direction was the study of Japanese scientists who showed that the administration of phytoecdysteroids to mice leads to a noticeable increase in protein-synthesizing processes in their bodies (Okui et al. 1968; Otaka et al. 1968; Otaka et al. 1969). Thanks to this discovery, the next few years passed under the sign of a comprehensive study of the revealed anabolic activity of these compounds. In a large number of publications, it was reported that with repeated administration of phytoecdysteroids to various animals (mice, rats) at doses of 5–10 mg/kg, an increase in body weight, as well as the mass of their internal organs and skeletal muscles (Syrov 1996; Syrov and Kurmukov 1976; Syrov et al. 1981a). In the work of Masuoka et al. (1970), an increase in their brain mass in animals was also noted under their influence. These changes were unambiguously associated with an increase in protein biosynthesis under the influence of phytoecdysteroids and, correspondingly, an increase in its total content in the liver, heart, kidneys, anterior tibial, and other muscles (Syrov 1996; Syrov and Kurmukov 1976; Syrov et al. 1981b). This process under the action of phytoecdysteroids, as a rule, proceeded equally in all organs and tissues, as evidenced by the similarity of the effect of the incorporation of 14C-amino acids into their protein structures and approximately the same percentage increase in their mass compared to the corresponding control values (Syrov 1996). A very significant indicator of the manifestation of the protein-anabolic action of the compounds in animals was the increase in their total protein in the blood under their influence, mainly due to albumin, an increase in the number of red blood cells in the peripheral blood, and an increase in their hemoglobin (Syrov 1996), stimulation of immune processes in the body (Shakhmurova et al. 2012), as well as an increase in protein synthesis by human skin fibroblasts and mammalian liver cells in in vitro experiments (Burdette and Soda 1963). Within the framework of the same effect (as a special case of activation of protein-synthesizing processes), an increase in glutamate decarboxylase activity in various organs and tissues can also be observed under the influence of phytoecdysteroids (Chaudhary et al. 1969), including acetylcholinesterase (Catalan et al. 1984). Na+, K+ -ATPases, succinate dehydrogenase, lactate dehydrogenase (Dzik and Kaczor 2019), alkaline phosphatase (Golovatskaya 2004), and many other enzyme systems of the body (Tashmukhamedova et al. 1986a). With the anabolic effect of phytoecdysteroids, it was noticed by Masuoka et al. (1970) that there is a decrease in the serum urea content under their influence. A noticeable anabolic activity of phytoecdysteroids was detected not only in relation to mammals. In a number of works (Golovatskaya 2004), it was also shown that these compounds can stimulate an increase in the body weight of chickens, while the effect was manifested not only in intact, but also in chickens that are undergoing rickets (Golovatskaya 2004). Long-term supplementation of phytoecdysteroids in the diet stimulated the synthesis of duodenal mucosa proteins in them (Golovatskaya, 2004). In a series of experimental works, it was found that in their anabolizing effect on animals, phytoecdysteroids are in many ways similar to steroidal anabolic drugs—synthetic analogs of male sex hormones, and in some cases the effect of some of them (Turkesterone, Ponasterone A) was even at the level of such well-known drugs of this series, like methandrostenolone (nerobol) and 4-chlortestosterone (Syrov 1996; Okui et al. 1968). However, the majority of phytoecdysteroids that have been isolated to date have a significantly lower ability to activate protein biosynthesis in mammals than steranobol (Syrov 2000). Significant differences were revealed in the nature of the anabolic effect they caused. So, phytoecdysteroids, in contrast to sterenobols, had a markedly weakened anabolic effect on animals with a disturbed hormonal background of the body (castration, hypophysectomy) (Syrov 1996). With their introduction, as a rule, no selective anabolic-organotropic action was found that was characteristic of derivatives of male sex hormones (Syrov 1996). Also, not found in phytoecdysteroids and any specific hormonal effects inherent in steranabol (androgenic, antigonadotropic, uterotropic, thymolytic) (Syrov 2000; Syrov 1996). The anabolic effect of phytoecdysteroids in experiments on female rats was characterized, in contrast to the steranobol (Lazarev et al. 2020), by the same patterns as in males. All the revealed differences in the effect on the body as a whole in representatives of the two classes of compounds mentioned are explained by the ambiguous mechanism of their regulation of protein-synthesizing processes in higher animals. Thus, in experiments on mice, it was found that stimulation of protein biosynthesis under the action of phytoecdysteroids is not related to their effect on the synthesis of i-RNA and is only a reflection of the acceleration of translation processes (Syrov et al. 2012; Syrov 1984). As a result, the synthesis of the proteins characteristic of a given organism is activated and only against the background of their genetically determined induction (Syrov 1984). The anabolic effect of sterenabols, as well as male sex hormones (Baulieu 1975), is primarily associated with effects on transcription processes with subsequent generalized stimulation of protein biosynthesis at the translation level due to an increase in the number and activity of ribosomes (Syrov 1985). Therefore, the effect of phytoecdysteroids in a holistic organism is characterized by a "harmonious" course of anabolic reactions (Syrov 1996). The anabolizing effect of testosterone derivatives, varying quantitatively, is nevertheless qualitatively characterized by the same specificity of action as that of male sex hormones (Syrov 1996). This can be confirmed by pharmacokinetic studies, which showed that 3H-ecdysterone does not have an affinity for the nuclei of mammalian cells (Dzhukharova et al. 1987), in contrast to male sex hormones and their analogs (Baulieu 1975), and found in significant quantities only in the hyaloplasm (Dzhukharova et al. 1987). It was also not found characteristic of male sex hormones and their analogs (Baulieu 1975), selective accumulation of 3H-ecdysterone in androgen-dependent target organs (Djukharova et al. 1987). All these data indicate that phytoecdysteroids may be of some interest as potential agents for stimulating protein-synthesizing processes in the body with a general weakening and only in the absence of serious endocrine disorders. Under these conditions, even in spite of a weaker anabolic effect, they can have certain advantages over steranobol because of the absence of side effects that limit the use of the latter, especially when prescribed to women and children (Milhorn 2018). Recently, quite a lot of work has also been devoted to the study of the effects of phytoecdysteroids on other types of metabolism closely associated with protein. Their effect on carbohydrate metabolism has been studied in most detail. Thus, in a series of experimental studies, the hypoglycemic effect of these compounds was noted, especially under experimental hyperglycemia (Tashmukhamedova et al. 1986b; Syrov et al. 2012). It has been suggested that the hypoglycemic effect of phytoecdysteroids is not associated with an increase in insulin secretion, but is due to their activating effect on the biosynthesis of receptor proteins involved in the implementation of the insulin stimulus and increased glucose uptake by tissues

(Kosovsky et al. 1989). The latter finds confirmation in the work of Tashmukhamedova et al. (1986a), which describes pronounced positive dynamics under the influence of phytoecdysteroids in the content of lactic and pyruvic acid in the organs and tissues of animals, reflecting the improvement in them of aerobic oxidation processes. In studies relating to the study of long-term use of phytoecdysteroids, an increase in the glycogen content in organs in normal animals was also noted (Syrov et al. 1975), and very importantly, the expressed glycogen-saving effect of these substances in many pathological conditions is described, accompanied by a decrease in glycogen content (Syrov et al. 1986; Syrov et al. 1988). Phytoecdysteroids have a beneficial effect on energy reactions in the body by enhancing the synthesis and increasing the activity of a number of enzymes that play a prominent role in them (glutamate, succinate, NAD-H-dehydrogenases, succinate-NAD-N-oxidases. As a result, an increase in the intensity of oxidative phosphorylation and an increase in the production of macroergic phosphorus compounds are observed (Almatov et al. 1985).

It has been shown that phytoecdysteroids lower blood cholesterol and triglycerides in intact rats and inhibit the development of severe hypercholesterolemia and hypertriglyceridemia after injection of newt WR-1339 (Mironova et al. 1982; Syrov et al. 1983). In rabbits with advanced atherosclerosis, phytoecdysteroids also reduce serum cholesterol and triglycerides, lower the aortic atherosclerotic lesion index, and improve lipid metabolism in the liver (Syrov et al. 1983; Matsuda et al. 1974). One of the factors affecting the reduction of cholesterol and triglycerides in animals after the introduction of phytoecdysteroids may be an increase in the lipolytic activity of triglyceride lipase in the body.

Under the action of phytoecdysteroids, significant changes were also revealed both in the qualitative and quantitative content of phospholipids in various organs and skeletal muscles. The introduction of ecdysteroids to young intact animals under conditions of persistent stimulation of anabolic processes (Syrov 1996) leads to an increase in the metabolic activity of phospholipids in the cardiac and skeletal muscles (an increase in the inclusion of labeled precursors in the composition of phospholipids occurs against a background of a decrease in the quantitative content of phospholipids in the studied organs), aimed at covering the growing needs of cells for energy production substrates. Under conditions of pathological violations of the fractional composition of phospholipids, identified in particular in the membranes of the mitochondria of the liver during experimental insulin deficiency and consisting primarily in a decrease in the content of phosphatidylcholine, phosphatidylserine, phosphatidylethanolamine, and sphingomyelin against the background of an increase in the level of lysophospholipids and phosphatidic acids, turcidene normal and rather quickly contributed to the restoration of the phospholipid spectrum of mitochondrial membranes to of expressions characteristic of intact animals (Mirtalipov 1990). It is also important to note the ability of some phytoecdysteroids to inhibit lipid peroxidation (Kim et al. 2009). When introduced into the body, phytoecdysteroids are characterized by a potassium-fixing effect, which was revealed in experiments on mammals, birds (Golovatskaya 2004), as well as on mollusks for which these compounds are also not endogenous hormones (Golovatskaya 2004). So, when studying the processes of ion regulation in the neurons of the cochlea Helix pomatia, it was found that ecdysterone increased the intracellular concentration of K^+. It was shown that incubation of the nerve ganglia in the presence of ecdysterone led not only to a dose-dependent increase in the concentration of intracellular K^+, but also to a slight increase in the concentration of Ca^{2+}. The Na^+ concentration did not change (Golovatskaya 2004). It has been suggested that ecdysterone enhances the transport of K^+ and, possibly, H^+ against their electrochemical gradients (Golovatskaya 2004). In rats, in the blood serum after administration of ecdysterone, a decrease in the content of calcium and phosphorus was revealed (Syrov et al. 1986).

Of the other biological effects of phytoecdysteroids associated with their effect on metabolic processes in animals, a pronounced fibrinolytic effect, as well as increased levels of prostaglandin F_2 in liver tissue, heart, and blood plasma should be mentioned. An increase in the intracellular content of c-AMP in the myocardium, and the adrenopositive effect are found in experiments on the spleen in vitro (Syrov et al. 1989).

Many researchers noticed that under certain conditions ecdysteroids exhibit a membrane-stabilizing effect. In this regard, the data presented in (Golovatskaya 2004) are particularly interesting, where it was shown that incubation of rabbit erythrocytes with ecdysterone, viticosterone E, polypodine B, α-ecdysone, and cyasterone led to the fact that these compounds increased their resistance index calculated according to acid erythrograms. It is possible that the stabilizing effect of ecdysteroids on membranes is the basis of their anti-inflammatory effect in some forms of experimental inflammation.

A small number of publications are devoted to the study of the effect of phytoecdysteroids on the hormonal background of higher animals. There is only sporadic evidence that, when ecdysterone is administered to rats, the level of somatotropic hormone, testosterone in the blood slightly increases with a sufficiently pronounced increase in corticosterone. No increase in insulin secretion was observed in animal experiments. However, an increase in the content of this hormone with respect to the initial level was revealed with prolonged intake of ecdysterone in the human body. Moreover, an increase in the content of free thyroxine (T4) and thyroid-stimulating hormone was also recorded in blood serum (Syrov et al. 2008). Favorable changes found in metabolic processes in various organs and tissues under the influence of phytoecdysteroids were also accompanied by well-defined functional effects. First of all, the general tonic effect of these compounds on the body was quite pronounced. Thus, the improvement in the activity of the central nervous system when animals were injected with ecdysterone could be judged by the activating effect revealed by him on EEG rabbits (Yusupova et al. 2020). The drug contributed to an increase in potentials, a decrease in their amplitude in the cortex, and ordering of potentials in subcortical formations. Against the background of the influence of ecdysterone, the assimilation of the rhythm of light flickering by the cerebral cortex increased at a frequency of 5 and 20 per second. In addition, ecdysterone showed a noticeable antagonism with respect to the action of chloral hydrate and hexenal in mice and gave an awakening effect during the action of thiopental sodium and chloralose in experiments on rabbits (Syrov and Kurmukov 1977). Similar results were obtained earlier, in the study of individual substances isolated from safflower levzea and later identified with phytoecdysteroids—inocosterone and ecdysterone. Under the influence of phytoecdysteroids, the production of conditioned defense reflexes accelerated (Syrov and Kurmukov 1977), the endurance of animals in relation to various physical activities increased (Syrov et al. 1988; Syrov and Kurmukov 1977). Their sexual behavior intensified. An increase in the resistance of animals after the introduction of phytoecdysteroids to stressful environmental factors was noted (Volodin et al. 2012), spontaneous infection with giardiasis (Islamova et al. 2012). Some attention should be paid to the data on the ability to resist various helminth infections using ecdysteroids (Lafont 1997). Ecdysterone prevented the development of aconitine, calcium chloride arrhythmias, and the death of animals caused by arrhythmogens. Improvement under the influence of ecdysterone and cyasterone of bile secretion function of the liver is described (Syrov et al. 1986).

Phytoecdysteroids as metabolically active substances most clearly showed their effect in a wide variety of experimental pathological conditions of the body, which were based on the weakening or even serious violation of metabolic processes. So, in a number of publications (Syrov et al. 1988, 2010), not only the stimulating effect of phytoecdysteroids on physical performance was noted, but also a significant effect on the prevention of fatigue syndrome, which occurred due to favorable biochemical and functional changes aimed at maintaining tissue metabolism homeostasis and energy production. In this case, the reaction of intracellular myocyte hypertrophy with a predominant increase in the polyribosomal apparatus was morphologically revealed.

Phytoecdysteroids had a pronounced therapeutic effect in experimental myocarditis due to the normalization of the content of creatine phosphate and glycogen in the heart muscle. Much earlier than in the control, under their influence, the electrolyte balance and the redox potential of cardiomyocytes were restored. It is known that the most optimizing effect of phytoecdysteroids, in this case, was when they were combined with cardiac glycosides. The introduction of ecdysterone into animals after reproducing an experimental myocardial infarction in them could achieve a significant decrease in the volume of necrosis, and subsequently, cause stimulation of reparative processes in the necrotic zone. The drug in conditions of experimental myocardial infarction also contributed to a significant increase in the level of systolic pressure in the left ventricle of the heart and the contractility index, sharply reduced with this pathology (Kurmukov and Ermishina 1991).

In experimental hepatitis caused by heliotrin or carbon tetrachloride, ecdysteroids prevented the development of dystrophic necrotic changes in the liver and helped restore the structure and function of hepatocytes. Moreover, a favorable rearrangement of the activity of the polyenzymatic systems of mitochondrial membranes, an increase in their stability, and resistance to the action of exogenous factors contributing to the degradation of mitochondria was also noted in the liver. These changes led to the normalization of the entire respiratory chain and the external path of electron transfer in the liver cells. In hepatitis (compared with intact animals), the stimulating effect of phytoecdysteroids on the secretion of bile was more pronounced. Moreover, their positive effect on the exchange of bilirubin, the synthesis of bile acids, and the secretion of cholesterol was revealed. The ability of phytoecdysteroids to reduce the number of destructive injuries of the stomach caused by various ulcerogens due to the normalization of plastic processes in its tissues was also experimentally proved. The introduction of phytoecdysteroids to laboratory animals with diabetes caused by alloxan led to a decrease in blood sugar, an improvement in carbohydrate metabolism, and normalization of the processes of oxidative phosphorylation and Ca^{2+} transporting function of the liver mitochondria. Ecdysterone accelerated the healing of bone fractures in the experiment (Syrov et al. 1981b, 1986).

Of the general pharmacological effects of phytoecdysteroids, the increase under their influence of blood pressure in acute animal experiments is noteworthy, accompanied by a slight increase in the frequency of respiratory movements. In most cases, this increase was short-term and rather unstable (Syrov 1996). Phytoecdysteroids did not show any toxic effects in relation to higher animals. In experiments on mice, it was found that LD_{50} for ecdysterone when administered intravenously was 6.4 g/kg, and for inocosterone −7.8 g/kg. When administered orally in both cases, the LD_{50} exceeded 9 g/kg (Chong et al. 1970). No toxic manifestations were noted with prolonged administration of phytoecdysteroids into the body of various experimental animals (Ashrafova and Syrov 1979).

Thus, biological activity studies of this genus are crucial for phytochemical investigations because of their therapeutic features. Various biological activity studies have been applied on pure natural compounds and extracts from different *Dianthus* species (Yusupova et al. 2020) such as: proliferative (Tong et al 2012), antioxidant, cytotoxic, antimicrobial [Casiglia et al. 2014], analgesic and antihepatotoxic (Gilbert and Rewitz 2009), immunomodulatory properties (Mutlu et al. 2018). Phytoecdysteroids isolated from *Dianthus (fam. Caryophyllaceae)* and their nootropic, tremorogenic and anticonvulsive activity were studied for the first time (Yusupova et al. 2020).

Thus, phytoecdysteroids as a new class of biologically active natural compounds that are of great interest not only in the theoretical aspect, as a tool for elucidating many issues related to the regulation of metabolic processes in living beings that are far apart from each other in evolutionary terms, but also in practical terms as means for the correction of impaired metabolism in the body of higher animals and humans in a variety of pathological conditions. This opened up a broad perspective for their study in clinical practice. One of the first successes in this direction was the creation on the basis of a plant steroid of this class—ecdysterone isolated from *Safflower levzea* (Abubakirov et al. 1988), a drug undergoing a test called "ratibol" (ecdysten).

Ecdysten has proven itself to be quite effective in patients with manifestations of stress syndrome due to the optimizing effect on the metabolism and functions of the central nervous system and, in the first place, higher nervous activity. This was quite clearly manifested by the general tonic effect in asthenic conditions and weakness in the period of convalescence after severe somatic and infectious diseases, surgical interventions, and prolonged exposure to ecologically unfavorable conditions. Under the influence of ecdysten, mental and physical performance improved, attention concentration increased, and reactive and personal anxiety normalized (Syrov et al. 2011a; Sadykova et al. 2011).

No less significant results were obtained with the use of ecdysten in cardiology practice. In the work (Abdullaev and Nagaeva 2007), it was shown that in patients with myocarditis under the influence of ecdysten there was an improvement in indicators characterizing the state of metabolic processes in the heart muscle (according to biochemical analysis and ECG results). There was an increase in exercise tolerance, and hemodynamic processes improved. In the treatment of ecdystenic myocardial infarction, a decrease in the severity of post-infarction angina, an increase in static endurance and physical performance, and an improvement in the hemodynamic support of static and dynamic loads were revealed. Apparently, the use of ecdysten (as an anabolic active agent) in the treatment of patients with chronic cardiovascular insufficiency together with cardiac glycosides: celanide, erisimoside (isolated from cultivated in Uzbekistan jaundice—*Erysimum diffusum* Ehrh. and strophanthusid bisramnoside (apparently) will also be of certain interest, obtained by partial synthesis on the basis of agrofon strofantidina). In a series of experiments, it was shown that in this case the hemodynamics and function parameters improved the Ionic state of the heart, increased cardiac output, and the rate of contraction of the circular fibers of the left ventricle of the myocardium. Along with this, there are favorable shifts towards normalization of impaired lipid, energy, and, most importantly, electrolyte metabolism of the myocardium. The latter largely eliminates the negative effect of cardenolides (especially with their long-term use) (Syrov and Khushbaktova 2004).

Also noteworthy are the results of the study, which showed pronounced hepatoprotective activity of ecdysten, since Uzbekistan belongs to the regions of highly endemic for viral hepatitis. In experimental conditions, on a model of heliotrin liver damage in rats, ecdysten contributed to the normalization of the enzymatic activity of blood serum, stimulation of the protein and glycogen synthesizing functions of the liver, inhibition of lipid peroxidation, and induction of all microsomal oxidation units in hepatocytes. The use of ecdysten in the complex therapy of patients with chronically viral hepatitis B much earlier than in patients whose complex treatment this drug was not included in, led to a decrease in intoxication, the disappearance of hyperfermemia, hyperbilirubinemia, and an increase in the protein content in blood serum. A positive effect on humoral (IgA, IgM, IgG) and cellular (CD^{3+}, CD^{4+}, CD^{8+}) immune factors, non-specific resistance factors (increased phagocytic activity of neutrophils), as well as a corrective effect on autoimmune processes (Syrov et al. 2004). It had an ecdysten and pharmacotherapeutic effect in the treatment of patients with gastric ulcer and duodenal ulcer. In this case, its effect was manifested by positive dynamics of blood biochemical parameters, a tendency to normalize gastric secretion, a high percentage of faster scarring of ulcers and an increase in the duration of disease remission (Syrov et al. 2008). With the addition of ecdysten to the basic treatment of some nephrological patients, along with an improvement in well-being, a decrease in anemia, proteinuria, a decrease in the content of urea and creatinine in the blood serum, and restoration of its protein composition was revealed. The administration of ecdysten to patients with chronic glomerulonephritis also led to the normalization of morphometric indicators of microcirculation of the bulboconjunctiva, a decrease in the frequency of vascular, intravascular, and perivascular changes (Saatov et al. 1999).

A favorable result has been described in the treatment of ecdysthenic diabetes mellitus. It was found that the sugar-lowering effect of ecdysten is a reflection of its activating effect on the synthesis of enzyme systems of carbohydrate metabolism, which contribute to more intensive consumption of blood glucose by organs and tissues of the body for the energy supply of protein-anabolic processes stimulated by it (Syrov et al. 2012).

The restorative effect of ecdysten when prescribing to patients of medical institutions of a different profile, in addition to the described effects, most favorably affected other vital functions of the body. The use of ecdysten in men suffering from infertility associated with impaired spermatogenesis, as a complication of urological diseases, causes an improvement in sperm quality indicators. At the same time, due to the general tonic effect, restoration of copulative function is noted. A similar effect was observed when prescribing ecdysten in the recovery period after myocardial infarction (Syrov et al. 2000).

Ecdysten has a therapeutic effect in pregnant women with iron deficiency anemia, suffering from euthyroid goiter. During treatment, most women noted an improvement in general condition, a decrease in weakness, hypotension, and increased appetite. They observed a dynamic increase in red blood counts, an increase in the total protein in the blood serum mainly due to albumin, and the body absorbed essential microelements (due to an increase in the synthesis of transport proteins involved in these processes) (Amonov and Syrov 2003).

An interesting experience is the use of ecdysten in individuals with giardiasis infection. Under its influence, the elimination of astheno-vegetative manifestations was observed, the immune status was normalized, the state of the gastrointestinal tract, hepatobiliary system improved, and in most cases, the complete elimination of parasites was detected. As with other painful conditions, the pharmacotherapeutic effect in giardiasis, apparently, is primarily determined in many respects by an increase in the adaptive and protective capabilities of the macroorganism itself (Islamova 2014).

Ecdysten promotes the rapid regeneration of damaged epidermal cells; therefore, it can be used as an adjuvant to accelerate the healing of wounds and burns. Of practical importance is also the fact of the regulatory influence of ecdysten detected in the body of old animals on compensatory-adaptive processes, including stimulation of normal apoptosis, apparently directed in this case to the elimination of damaged cellular elements carrying genomic disorders whose frequency increases markedly with age, which may be of interest in geriatric practice (Syrov et al. 2007).

Ecdysten as a metabolically active agent with a predominant effect on the processes of protein synthesis (in contrast to sterenabols having specific hormonal effects) also proved to be very good in sports medicine (Syrov et al. 2011b). It is in this case more clearly than elsewhere. It revealed its general strengthening and protein-anabolic effect on the body as a whole. It was shown that the use of ecdysten by athletes accelerated recovery and adaptation to physical activity in the training and competitive periods. According to the self-assessment (questionnaire method), it was found that the vast majority of athletes from the experimental groups noted less fatigue during training, better tolerance of loads, a decrease in apathy and irritation when doing work to the limit, a fairly quick relief of the phenomena of fatigue. The effect, as a rule, was detected at 4–5 days from the start of the drug.

Objectively, in comparison to control, the application of ecdysten showed a positive dynamics of weight indicators; during anthropometric studies of individuals who used ecdysten many times, an increase in muscle mass was observed with a significant decrease in fat. The biochemically anabolic effect of ecdysten was confirmed by nitrogen retention in the body, faster normalization of urea and lactate levels in the blood after physical exercise, and also an increase in hemoglobin and total serum protein. Moreover, despite the pronounced anabolic activity, no dangerous side effects of the drug, characteristic of anabolic steroids, such as shifts in the content of endogenous hormones and negative effects on the function of the adrenal cortex, were found. The use of ecdysten in sports practice did not contradict the rules of anti-doping control (Syrov et al. 2011b). The experience of combining ecdysten with certain cytamines has been accumulated; in particular, it has been shown that its administration with heptamine gives a more significant effect in terms of correcting performance than taking ecdysten or heptamine alone (Emirova et al. 2004). It is also shown that the use of ecdysten in cyclic sports during intense physical activity of a large volume in combination with protein preparations contributes to a more pronounced maintenance of muscle mass and does not prevent the use of lipid energy depots in the body.

Speaking about ecdysten, one can't get around another side of its emerging use as an anabolic type of drug without side hormonal effects—this is a pediatric practice. Under experimental conditions, the rationale for this approach to the use of ecdysten was the work that showed the benefits of this drug in children over synthetic steroidal anabolic agents (Utemuratova et al. 2006), as well as its ability to delay the development of hypoxic and traumatic hypotrophy and restore the weight and size of the fetus. In the clinic, the effect of ecdysten in newborns with intrauterine malnutrition is realized in shorter periods of restoration of body weight, normalization of tissue turgor and skin elasticity, normalization of its color, stabilization of muscle tone (Pak 2001). The use of ecdysten contributes to a more rapid relief of clinical symptoms of intrauterine growth retardation, regardless of its stage (Alimov et al. 2009). It was also shown that the use of ecdysten (especially together with immunomodulin) in the treatment of perinatal sepsis of mixed (bacterial-fungal) etiology in premature infants leads to a more rapid improvement in the general condition of patients, reduction of intoxication, increase of motor activity, improvement of reflex activity, normalization of immunological parameters. Ecdysten was found to be highly effective (as part of complex therapy) in the treatment of chronic viral hepatitis delta.

In recent years, based on ecdysterone and other phytoecdysteroids isolated from various plant sources, effective biological active additives have also been created that optimize the state of metabolism in various organs and tissues, which can significantly improve their functional state (Syrov et al. 2007).

Thus, the analysis of the presented material on the use of ecdysten and biologically active additives containing phytoecdysteroids in clinical and sports medicine opens up the prospect of their widespread use as effective pharmacocorrectors of one or another pre-pathological and pathological condition, as well as adaptogenic agents, due to the ability significantly expand the boundaries of the body's resistance to the effects of various extreme factors.

References

Abubakirov NK (1981) Ecdysteroids of flowering plants (Angiospemae). Chem Nat Compd 40:685–702. https://doi.org/10.1007/BF00574363

Abdullaev TA, Nagaeva GA (2007) The use of the domestic drug "Ecdysten" in the treatment of patients with myocarditis. J Pharm 1:71–74

Abubakirov NK, Sultanov MB, Syrov VN (1988) Tonic. 1312774 USSR. Discoveries 17

Akhrem AA, Levina IS, Titov YA (1973) Ecdysones are insect steroid hormones 232

Alimov AV, Pak EA, Amizyan NM et al (2009) Newborns with intrauterine growth retardation. Methodical recommendations 16

Almatov KT, Tashmukhamedova MA, Khushbaktova ZA et al (1985) The effect of ecdysterone, turkesterone and nerobol on respiration and oxidative phosphorylation of rat liver mitochondria in toxic hepatitis. Med J Uzb 9:69–71

Amonov II, Syrov VN (2003) The use of ecdystene in the complex treatment of iron deficiency anemia of pregnant women suffering from euthyroid goiter. Pathology 1:32–35

Ashrafova RA, Syrov VN (1979) The general effect, toxicity and changes in the microstructure of internal organs and tissues of experimental animals with prolonged administration of ecdysterone. Pharm Nat Comp 147−153

Baulieu EE (1975) Some aspects of the mechanism of action of steroid hormones. Mol Cell Biochem 7:157–174. https://doi.org/10.1007/BF01731406

Bergamasco R, Horn DHS (1983) Distribution and role of insect hormones in plants. In: Applied organic chemistry, editor/s. Endocrinology of Insects. Alan R. Liss, Inc., New York 627–654. http://hdl.handle.net/102.100.100/285692

Bruno LB, Jeon-philippe A, Fabrice M, Francois A (2002) Ecdysteroids: one potential new anabolic family in breeding animals. Analytica Chimica Acta 473:1–2:89–97.https://doi.org/10.1016/S0003-2670(02)00874-7

Buniam J, Chukijrungroat N, Rattanavichit Y et al (2020) 20-Hydroxyecdysone ameliorates metabolic and cardiovascular dysfunction in high-fat-high-fructose-fed ovariectomized rats. BMC Complement Med Ther 20:140.https://doi.org/10.1186/s12906-020-02936-1

Burdette WJ, Soda RL (1963) Effect of ecdysone on incorporation of 14C-leucine into hepatic protein in vitro. Proc Soc Exptl Biol Med 112:216−217.https://doi.org/10.3181/00379727-112-27996

Carlisle D, Osborne D, Ellis P et al (1963) Reciprocal effects of insect and plant-growth substances. Nature 200:1230. https://doi.org/10.1038/2001230a0

Casiglia S, Bruno M, Senatore F (2014) Volatile constituents of *Dianthus rupicola* Biv. from Sicily: activity against microorganisms affecting cellulosic objects. Nat Prod Res 28:1739–1746. https://doi.org/10.1080/14786419.2014.945087

Catalan RE, Aragones MD, Godoy JE, Martinez AM (1984) Ecdysterone induces acetyl-cholinesterase in mammalian brain. Comparative biochemistry and physiology. C, comparative pharmacology and toxicology 78(1):193–195. https://doi.org/10.1016/0742-8413(84)90068-9

Chaudhary KD, Lupien PJ, Hinse C (1969) Effect of ecdysone on glutamic decarboxylase in rat brain. Experientia 25:250–251. https://doi.org/10.1007/BF02034373

Chong YK, Galbraith MN, Horn DHS (1970) Isolation of deoxycrustecdysone, deoxycrustecdysone and α-ecdysone from the Fern Blechnum minus. Chem Soc Chem Communs 1217–1218. https://doi.org/10.1039/C29700001217

Dinan L (1994) Phytoecdysteroids in Kochia scoparia (burning buch). J Chromatograph 658:69–76. https://doi.org/10.1016/0021-9673(94)85207-3

Dinan L (1998) A strategy towards the elucidation of the contribution made by phytoecdysteroids to the deterrence of invertebrate predators on plants. Rus J Plant Physiol 45(3):347–359

Dzhukharova MK, Sakhibov AD, Kasymov B et al (1987) Pharmacokinetics of ecdysterone in experiments. Pharm Chem J 21:689–692. https://doi.org/10.1007/BF00758127

Dzik KP, Kaczor JJ (2019) Mechanisms of vitamin D on skeletal muscle function: oxidative stress, energy metabolism and anabolic state. Eur J Appl Physiol 119:825–839. https://doi.org/10.1007/s00421-019-04104-x

Emirova LR, Rozhkova EA, Panyushkin VV et al (2004) The effect of cytamines and their combination with ecdysten, apilak, vitomax and essentials on the performance of athletes. Experim and Clinic Pharmacol 67(3):66–68

Gilbert LI, Rewitz KF (2009) The function and evolution of the Halloween genes: the pathway to the arthropod molting hormone. In: Smagghe G (eds) Ecdysone: structures and functions. Springer, Dordrecht. https://doi.org/10.1007/978-1-4020-9112-4_9

Golovatskaya IF (2004) Effect of ecdysterone on morphological and physiological processes in plants. Russ J Plant Physiol 51:407–413. https://doi.org/10.1023/B:RUPP.0000028689.97402.d5

Hardman R (1987) Recent developments in our knowledge of steroids. Planta Med 53(3):233–238. https://doi.org/10.1055/s-2006-962690

Hendrix SD, Jones RL (1972) The activity of β-ecdysone in four gibberellin bioassays. Plant Physiol 50(1):199–200. https://doi.org/10.1104/pp.50.1.199

Islamova ZhI, Davis NA, Syrov VN, Osipova SO (2012) Prospects for the use of drugs created on the basis of phytoecdysteroids in the treatment of giardiasis. Theor Appl Ecol 1:57–61

Islamova ZI (2014) An experimental-clinical study of ecdisten as an effective tool in the treatment of giardiasis. Collection of articles of the VII international scientific conference: "Priority Directions in the Field of Science and Technology in the 21st Century". Publishing House CHINOR ENK, Tashkent 1:154–158

Jacobs WP, Suthers HB (1971) The culture of apical buds of Xanthium and their use as a bioassay for flowering activity of ecdysterone. Am J Bot 58:836–843. https://doi.org/10.1002/j.1537-2197.1971.tb10037.x

Karlson P, Koolman J (1976) Ecdysone: mode of action, inactivation and elimination. Actualites sur les hormones d'invertebres. Ed M.M. Durchon, Paris 403–412

Kim KY, Cho SE, Yu BS (2009) Intragastrically applicated CCl_4-thiopental sodium enhanced lipid peroxidation and liver fibrosis (Cirrhosis) in rat: malonedialdehyde as a parameter of lipid peroxidation correlated with hydroxyproline as a parameter of collagen synthesis (Deposition). Toxicol Res 25:71–78. https://doi.org/10.5487/TR.2009.25.2.071

Kosovsky MI, Syrov VN, Mirakhmedov MM et al (1989) The effect of nerobol and ecdysterone on some associated insulin-dependent processes is normal and with insulin resistance. Probl Endocrinol 5:77–81

Kurmukov AG, Ermishina OA (1991) The effect of ecdysterone on experimental arrhythmias, changes in hemodynamics and myocardial contractility caused by occlusion of the coronary artery. Farmacol Toxicol 1:27–29

Lafont R (1997) Ecdysteroids and related molecules in animals and plants. Arch Insect Biochem Physiol 35(1–2):3–20. https://doi.org/10.1002/(SICI)1520-6327(1997)35:1/2

Lafont R, Ho R, Raharivelomanana P, Dinan L (2011) Ecdysteroids in ferns: distribution, diversity, biosynthesis, and functions. In: Kumar A, Fernández H, Revilla M (eds) Working with Ferns. Springer, New York. https://doi.org/10.1007/978-1-4419-7162-3_22

Lazarev SI, Kovalev SV, Khorokhorina IV et al (2020) Design and technological calculation of a combined electric baromembrane apparatus for separating solutions in chemical production processes. Chem Petrol Eng 56:475–481. https://doi.org/10.1007/s10556-020-00797-6

Matsuda H, Kawaba T, Yamamoto Y, Ogawa S (1974) Effect of ecdysterone on experimental atherosclerosis in rabbits. Folia Pharmacol Jap 70(3):325–329. https://doi.org/10.1254/fpj.70.325

Matsuoka T, Imai S, Sakai M, Kamada M (1969) Studies on phytoecdysones—a review of our works. Ann Rep Takeda Res Lab 28:221–271

Milhorn HT (2018) Anabolic steroid dependence. In: Substance use disorders. Springer. https://doi.org/10.1007/978-3-319-63040-3_13

Mironova VN, Kholodova YD, Skachkova TF et al (1982) Hypocholesterolemic effect of phytoecdysones in experimental hypercholesterolemia in rats. Chemistry. Tashkent 3:101 − 105

Mirtalipov DT (1990) Phospholipids of liver mitochondrial membranes in alloxan diabetes and the role of endogenous lipolytic enzymes in their modification: Dis. cand. biol. sciences. Tashkent

Mutlu K, Sarikahya N, Nalbantsoy A, Kirmizigul S (2018) Chemical constituents and biological activities of *Dianthus elegans* var. *elegans*. Nat Prod Res 32(11):1245–1253. https://doi.org/10.1080/14786419.2017.1333988

Nakanishi K, Koreeda M, Sasaki S et al (1966) Insect hormones. The structure of ponasterone A, insect-moulting hormone from the leaves of *Podocarpus nakaii* Hay. Chem Commun 24:915–917. https://doi.org/10.1039/C19660000915

Okui S, Otaka T, Uchiyama M et al (1968) Stimulation of protein synthesis in mouse liver by insect-molding steroids. Chem Pharm Bull 16(2):384–387. https://doi.org/10.1248/cpb.16.384

Otaka T, Uchiyama M, Okui S et al (1968) Stimulatory effect of insect-methamorfosing steroids from *Achyranthes* and *Cyathula* on protein synthesis in mouse liver. Chem Pharm Bull 16(12):2426–2429. https://doi.org/10.1248/cpb.16.2426

Otaka T, Okui S, Uchiyama M (1969) Stimulation of protein synthesis in mouse liver by ecdysterone. Chem Pharm Bull 17:75–81. https://doi.org/10.1248/cpb.17.75

Pak EA (2001) The rationale and effectiveness of ecdystene in the complex treatment of newborns with intrauterine malnutrition. New Medicine 11:21–25

Saatov Z, Agzamkhodzhaeva DA, Syrov VN (1999) Distribution of phytoecdysteroids in plants of Uzbekistan and the possibility of using drugs based on them in neurological practice. Chem Nat Compd 35:186–191. https://doi.org/10.1007/BF02234932

Sadykova GA, Alyavi AL, Abdullaev AK et al (2011) The study of the Ecdysten preparation in the complex treatment of patients with chronic obstructive pulmonary disease according to psychological tests. Ther J Uzbekistan 2–3:39

Shakhmurova GA, Mamadalieva NZ, Zhanibekov AA et al (2012) Effect of total ecdysteroid preparation from *Silene viridiflora* on the immune state of experimental animals under normal and secondary immunodeficiency conditions. Pharm Chem J 46:222–224. https://doi.org/10.1007/s11094-012-0766-4

Sláma K (1993) Ecdysteroids: insect hormones, plant defensive factors, or human medicine. Phytoparasitica 21:3–8. https://doi.org/10.1007/BF02980867

Sláma K, Koudela K, Tenora J et al (1996) Insect hormones in vertebrates: anabolic effects of 20-hydroxyecdysone in Japanese quail. Experientia 52:702–706. https://doi.org/10.1007/BF01925578

Smagghe G, Degheele D (1998) Ecdysone agonists: mechanism and biological activity. In: Ishaaya I., Degheele D (eds) Insecticides with novel modes of action. applied agriculture. Springer, Berlin, Heidelberg. https://doi.org/10.1007/978-3-662-03565-8_2

Syrov VN (2000) A comparative study of the anabolic activity of phytoecdysteroids and sterenabols in the experiment. Chem Farm J 4:31–34

Syrov VN, Kurmukov AG (1976) About the anabolic activity of phytoecdysone—ecdysterone isolated from *Rhaponthicum carthamoides* (Willd) Iljin. Farmacol Toxicol 6:690–693 (in Russ)

Syrov VN, Kurmukov AG (1977) On the tonic properties of ecdysterone isolated from safflower levzea. Report Acad Sci Uzbek SSR 12:27–30 (in Russ)

Syrov VN, Melnikova EV, Sultanov MB (1981) The effect of phytoecdysone ecdysterone on the course of toxic hepatitis in rats caused by heliotrin. Report Acad Sci Uzbek SSR 5:36–38 (in Russ)

Syrov VN, Kurmukov AG, Sultanov MB (1981) Evaluation of the anabolic effect of phytoecdysones and their 6-keto analogs in experiments on female rats. Report Acad Sci Uzbek SSR 3:31–33 (in Russ)

Syrov VN, Nabiev AN, Sultanov MB (1986) The effect of phytoecdysteroids on the bile secretion function of the liver is normal and in experimental hepatitis. Farmacol Toxicol 3:100–103

Syrov VN, Khushbaktova ZA, Khalikov TR et al (1988) The influence of ecdysterone and saparal on functional, biochemical and morphological indicators of working capacity. J Uzb Biol 3:61–65 (in Russ)

Syrov VN, Khushbaktova ZA, Dzhakhangirova MA, Sharipov AK (2011) Ecdysteroid-containing drugs and dosed physical activity in the training of athletes. Leader Press, Tashkent, p 250

Syrov VN, Shakhmurova GA, Khushbaktova ZA et al (2012) A comparative study of the regulatory effect of ecdysterone and retabolil on protein synthesizing processes in higher animals. Theor Appl Ecol 1:13–17 (in Russ)

Syrov VN, Khushbaktova ZA (2004) Experimental evaluation of the integrated use of celanide and ecdysten in conditions of cardiovascular pathology. Bulletin of Tinbo. Tashkent 1:56–63

Syrov VN, Aizikov MI, Kurmukov AG (1975) The effect of ecdysterone on protein, glycogen, and fat in the liver, heart, and muscle of white rats. Report Academy of Sciences of the Uzbek SSR. Tashkent 8:37−38 (in Russ)

Syrov VN, Khushbaktova ZA, Abzalova MK, Sultanov MB (1983) On the lipid-lowering and antiatherosclerotic properties of phytoecdysteroids. Report Academy of Sciences of the Uzbek SSR 9:44−45 (in Russ)

Syrov VN, Khushbaktova ZA, Mirzaev YR, Baltaev UA (1989) About antiulcer activity of phytoecdysteroids. J Chem Farm 4:441–445 (in Russ)

Syrov VN, Mirzaev YR, Khrushchev SA, Iskanderova SD (2000) The effect of ecdysten on indicators of sexual function in an experiment and in clinical conditions. Exp Clin Pharmacol 63(4):35–37 (in Russ)

Syrov VN, Gildieva MS, Khushbaktova ZA (2007) The study of proliferation and apoptosis in tissues of various genesis under the influence of phytoecdysteroids, cycloartan glycosides, flavonoids and lactones in the body of old rats. Abstracts of the XIV Russian National Congress "Man and Medicine". Moscow 781 (in Russ)

Syrov VN, Khushbaktova ZA, Yun VL, Dzhakhangirova MA (2008) Ecdysteroid-containing preparations-representatives of a new generation of actoprotective agents. Bulletin of TINBO. Tashkent 2:92–101

Syrov VN, Islamova ZhI, Khushbaktova ZA et al (2010) Possibility of pharmacocorrection with ecdysteroid-containing drugs of pathological changes in adverse environmental conditions. Vestnik Tinbo, Tashkent 2:77–83 (in Russ)

Syrov VN, Khushbaktova ZA, Kerimov FA, Sharipov AK (2011b) The use of the Ekdysten preparation in the training of young athletes. Materials of the republican scientific and practical conference "Scientific, Pedagogical and Biomedical Problems of Ensuring the Sport of Highest Achievements". Tashkent 254–256

Syrov VN (1984) To the mechanism of anabolic action of phytoecdysteroids in mammals. Scientific. doc. higher Schools. Biol. Sciences 11:16−20

Syrov VN (1985) To the mechanism of action of sterenabols. Report. Acad Sci Uzbek SSR 1:48−50

Syrov VN (1996) Pharmacological study of phytoecdysteroids of plants of the genus *Rhaponthicum*, *Silene* and *Ajuga*: Dis. Doctor of Medical Science. Tashkent 287

Tashmukhamedova MA, Almatov KT, Khushbaktova ZA et al (1986a) Effect of phytoecdysteroids and sterenabols on the activity and stability of membrane-bound enzymes of liver mitochondria in experimental hepatitis. Med Chemistry. Tashkent 1:81–84 (in Russ)

Tashmukhamedova MA, Almatov KT, Syrov VN et al (1986b) A comparative study of the effects of ecdysterone, turkesterone, and nerobol on the function of rat liver mitochondria in experimental diabetes. Med Chemistry. Tashkent 5:24–28 (in Russ)

Tong Y, Luo JG, Wang R, Wang XB, Kong LY (2012) New cyclic peptides with obsteoblastic proliferative activity from *Dianthus superbus*. Bioorg Med Chem Lett 22:1908–1911. https://doi.org/10.1016/j.bmcl.2012.01.058

Utemuratova DS, Syrov VN, Khushbaktova ZA (2006) The results of a comparative study of ecdysten and nerobol in the age aspect. Abstracts of the XIII Russian National Congress "Man and Medicine". Moscow 598

Volodin VV, Syrov VN, Khushbaktova ZA, Volodina SO (2012) Stress-protective effect of ecdysteroid-containing substance Serpisten. Theor Appl Ecol 1:18–24

Yusupova U, Usmanov D, Azamatov A, Ramazonov N, Rejepov J (2020) Phytochemical constituents and biological activities of *Dianthus helenae* Vved., growing in Uzbekistan, Nat Prod Res. https://doi.org/10.1080/14786419.2020.1862834

Zibareva LN (2003) Phytoecdysteroids of plants of the family *Caryophyllaceae*. Abstract. dis. doctor Chem. Sciences. Novosibirsk 31

Chapter 7
Technologies for Producing Drugs and Biologically Active Additives Based on Ecdysteroids from Plants *Rhaponticum Carthamoides*, *Silene Praemixta*, and *Ajuga Turkestanica*

Abstract One of the ways of medicinal plant materials' rational use is the development of integrated low-waste technologies for the production of herbal medicines and biologically active additives. This allows us to solve the raw material problem in pharmaceutical production, reduce the cost of production by allocating costs to several products (preparations), and make production more profitable. The effectiveness of complex technologies increases significantly if modern methods of processing plant materials are used, such as extraction methods while preserving the complex of native biologically active substances, fractionating extracts using ultrafiltration, purification of target products by adsorption and ion-exchange chromatography, and, in addition, by introducing good manufacturing practice and quality systems at all stages of pharmaceutical production. The selection of biologically active substances (BAS) from plant materials is important and the process of separation of a multicomponent system is essential here, and in order to achieve this separation it is necessary to use a combination of known methods of processing raw materials in a certain sequence. For many years at the Institute of Plant Chemistry named after Academician S. Yu., the Yunusov Academy of Sciences of the Republic of Uzbekistan also conducts multifaceted comprehensive studies of compounds of this class isolated from plants of Uzbekistan and neighboring republics. This monograph is devoted to the generalization of the data obtained.

Keywords Technology · *Silene praemixta* · *Rhaponticum integrifolium* · BAA · Ecdysten

7.1 Technology Overview

The main technological methods used to obtain ecdysteroid preparations are the same as in the case of other herbal medicines, and they include (a) grinding of raw materials; (b) extraction; (c) separation of solid and liquid phases; (d) evaporation of the extract; (e) isolation and purification of the target product.

The extraction of plant materials is an important stage of its processing, the effectiveness of which is determined by the properties of plant tissue, the physicochemical

U. Y. Yusupova et al., *Phytoecdysteroids*,
https://doi.org/10.1007/978-981-16-6711-4_7

affinity of the solvent and the recoverable substances, and the general laws of mass transfer.

There is no unified theory of the extraction of plant compounds, which depends on engineering calculation of the technological process and the selection of optimal extraction conditions. Although the hypothesized mechanism of the process is known (Saatov et al. 1993), the process itself is very complex due to its multicomponent nature of the mathematical description. The main objective is the selective extraction of the main product from plant materials, and therefore, the selection of the most selective extractant. The latter depends on the type of raw materials, on the properties of the extracted substance and is solved only experimentally.

Plant raw materials. This is one of the main components involved in the technological process. Plant material is a complex system consisting of a large number of different groups of organic compounds that are interconnected, and, therefore, the nature of the raw material is not stable in composition (Chueshov 2002a, b).

As shown in the works of academician Yunusov S. Yu. and several other authors, the qualitative and quantitative composition of BAS in plants depends on soil, climatic, geographical, and a number of other environmental conditions. In addition to a change in the composition of the main substances, the composition and content of other related compounds also fluctuate. The quality requirements for emitted biologically active substances, which are the basis of medical or veterinary drugs, are very stringent. Therefore, the development of a reproducible, stable in all its qualitative and quantitative indicators of the technological process of production of biologically active substances is a difficult task.

Methods of extraction of plant materials. Existing methods for the extraction of compounds from plant materials are classified into static and dynamic. Among these extraction methods, periodic ones are distinguished when the supply of the initial materials (extractant and/or plant material) to the extractors is carried out periodically and the extraction of one or more portions of the raw material is carried out for a certain time.

Static periodic methods include single-stage—maceration, and multi-stage—remaceration, circulation with periodic discharge (these are multi-stage direct-flow), as well as multi-stage countercurrent—re-colation with periodic discharge.

Dynamic periodic methods include single-stage—percolation and multi-stage—re-colation. Among the dynamic methods, continuous are distinguished—direct-flow and counter-current.

Currently, new forms of extraction are being sought and implemented with the maximum dynamization of all types of diffusion. Examples of such modifications of maceration are:

- vortex extraction—turbo extraction;
- extraction using ultrasound (acoustic);
- electric pulse and other methods of pulse processing of raw materials;
- centrifugal extraction;
- fractional maceration and others.

Among the many factors of the intensification of extraction, the leading ones are rapid changes in temperature and pressure. Changes in these parameters are achieved using various physical and electrophysical methods of a pulsed magnetic field, thermal and mechanical action of a laser beam on a mixture of extractant with raw materials, acoustic radiation and pulsed electric discharge (Chueshov 2002a). Thus, there is a method for the rapid extraction of valuable components from dried plant materials by extraction with water at a temperature of 40–60 °C and the creation of ultrasonic vibrations in a mixture with a frequency in the range of 100–150 kHz, which provides an extract with a high degree of transparency without loss of flavoring compounds and without extracting components with an unpleasant bitter or astringent taste. At the same time, due to ultrasonic waves activity, the extract is sterilized (Vorobiev 2008).

In the production of pharmaceuticals, various extraction methods are used, such as percolation, repercolation, maceration with stirring, continuous extraction. All these methods allow getting a good result, thanks to the hydrodynamic conditions. Under the conditions of these methods, the extractant moves relative to the feed. In the work (Abdukadirov 2007), the authors, using the battery extraction method from 5 diffuser extractors, achieved a fourfold reduction in the hydraulic module of the process and a fivefold extraction time compared to the extraction method.

Improving the hydrodynamic conditions is also possible when using mixing, low- and high-frequency vibration, circulation of the extractant, creating oscillations, the effects of centrifugal forces, the effects of powerful vortices, etc. Currently, there are a number of works using various factors. For example, in the course of studies it was found that mixing during the extraction of Rosa canina fruits positively affected the yield of water-soluble substances, and also allowed to extract of 84.6% of flavonoids (Sagdullaev 2007; Sagdullaev et al. 2001, 2002).

Conducting extraction in a vortex flow made it possible to extract water-soluble polysaccharides from the marshmallow roots by 4.1% more in a shorter time (30 min) than in the case of extraction with mechanical stirring (150 min) (Shakhidoyatov 2004).

Extraction by liquefied gases is one of the promising methods of material extraction. When using liquefied butane, butanopropane, nitrogen, ammonia, carbon dioxide, freons, argon, etc., having a boiling point below room temperature, as an extractant, oxidation, decomposition, and loss of substances and changes in their properties during evaporation do not occur. These extractants volatilize at room temperature.

In addition, the use of liquefied gases simplifies the technological scheme of the extraction process by eliminating the stage of the regeneration of the extractant. The extraction process is several times faster than when using other extractants, which is economically more profitable, and is often characterized by an almost complete absence of water-soluble ballast substances. Among liquefied gases, liquefied CO_2 has found the greatest application at present (Petrenko et al. 2021).

As it is known, a biologically active complex recoverable by liquefied gases differs from that recovered by classical solvents and is characterized by an increased content of fat-soluble and lower content of water-soluble substances. Therefore, liquefied gases are more often used to extract lipophilic complexes from plant materials (Chueshov 2002a; Ikonnikov et al. 2010).

To obtain the extraction of a complex composition that includes all biologically active substances present in the feedstock, the developers propose to use a mixture of solvents based on liquefied gases or to extract plant materials in an extraction unit with a carbon dioxide regeneration and ultrasonic treatment system. Literature sources describe extraction methods using immiscible solvent systems of different polarity, i.e., two-phase systems of extractants (Ivanova et al. 2003; Kaukhova et al. 2004). The most important feature of biphasic extraction, which distinguishes it from other extraction methods, is that two extractants come into contact with the feed simultaneously, each of which individually is capable of extracting either hydrophilic or lipophilic compounds. This technology, as shown by the research results, allows quick and with high efficiency to carry out complex processing of raw materials and get two products (extracts) with a high content of biologically active substances in one technological stage (Kaukhova 2006).

Thus, the effectiveness of the extraction process depends on many technological factors.

7.2 Methods for the Isolation of Ecdysteroids

A wide range of positive effects of ecdysteroids, in particular ecdysterone, allow us to consider compounds of this class as very promising for the development of effective drugs based on them with improved properties.

According to their chemical structure, ecdysteroids are polyoxysteroids containing characteristic structural elements that allow them to be combined into one group of natural compounds. These include the 14α-hydroxy-Δ^7-6-ketone group in cycle B, the diol group in cycle A at positions 2, 3, and the sterol side chain with one or more hydroxy groups.

Usually, ecdysteroids are divided into six groups according to the number of carbon atoms in the steroid skeleton: C-19, C-21, C-24, C-27, C-28, and C-29 ecdysteroids (Akhrem et al. 1973).

Ecdysteroid molting hormones isolated from animal organisms mainly belong to the third and fourth groups of compounds, while all six groups of ecdysteroids are found in plant materials.

For ecdysteroids, including ecdysterone, there is no universal way of isolation. In each case, the most suitable method or combination of methods is used, taking into account the properties of the target substances and the characteristics of plant materials. The literature describes some of the methods for the isolation and purification of ecdysteroids.

To isolate 20-hydroxyecdysone and 2-deoxy-20-hydroxyecdzone from *Lagochillus platycalyx* (Boymatov et al. 2010), the dried and ground raw materials were extracted with methanol. The extract was concentrated to dryness, diluted with water, and the formed precipitate was removed by filtration. The aqueous portion was sequentially extracted with chloroform, ethyl acetate, and then with butanol. After evaporation of the solvents *in vacuo*, ethyl acetate and butanol fractions were obtained. The butanol fraction was separated on a silica gel column using a chloroform-methanol solvent system. As a result, 20-hydroxyecdysone was isolated, and 2-deoxy-20-hydroxyecdysone was detected in the purified fractions.

The authors (Mamadalieva et al. 2002), in their work isolated from the aerial part of *Silene linicola* 7 ecdysteroids, including ecdysterone and turkesterone, according to the following technology. The air-dry ground material was extracted with ethanol. The extract was then concentrated, diluted with water, filtered. After further distillation of ethanol *in vacuo*, the aqueous residue was treated with chloroform. Ecdysteroids from the aqueous residue were exhaustively extracted with ethyl acetate, then with butanol. After evaporation of the solvents in vacuo, ethyl acetate and butanol fractions were obtained. The ethyl acetate fraction was chromatographed on an alumina column, then on a silica gel column. In both cases, the elution was carried out with a solvent system of chloroform-methanol, as a result of which Viticosterone E (yield 0.0008% by weight of the feed) and 2-deoxy-ecdysterone (yield 0.017%) were obtained. After washing the column with a chloroform-methanol system, 0.001% α-ecdysone and 0.073% polypodine B were obtained.

The butanol amount of ecdysteroids was separated on a silica gel column, eluting with a chloroform-methanol solvent system. 0.3% Ecdysterone and 0.00083% Turkesterone are isolated. Using the chloroform-methanol-water solvent system further, 0.0015% integristerone was obtained.

Phytoecdysteroids were isolated from *Leuzea carthamoides* seeds by methanol extraction (Borovikova et al. 1999; Baltaev 1991, 1992). Phytoecdysteroids were obtained by treating the condensed methanol extract with hexane, followed by extraction with butanol. Next, rapisterone B, rapisterone C, stesterone, rapisterone D-20-acetate, and makisterone A were isolated by phytoecdysteroid adsorption chromatography.

In the work (Bobaev et al. 2009), the dried and grounded aerial part of *Scutellaria guttata* was used to extract a composition of compounds with methanol. The extract was sequentially treated with chloroform, ethyl acetate, then with butanol-1. Fractions were isolated by chromatographic separation on a silica gel column from butanol extraction of methanol extract, from which ecdysterone was isolated by rechromatography, eluting with chloroform-methanol solvent systems, and α-ecdysone and a mixture of low-polar ecdysteroids were found in the purified fractions.

Phytoecdysteroids were extracted from ground plant material *Serratula coronata* L. with 40% ethanol, the extract was treated with zinc sulfate solution, and the sum of ecdysteroids was separated on a column filled with a Diasorb-130-C16T sorbent. After separation of the total extract, α-ecdysone, 20-hydroxyecdysone and inocosterone were obtained (Patov 2006).

From the plant *Silene italica* ssp., nemoralis ecdysteroids were isolated according to the following scheme (Fig. 7.1) (Bathori and Pongracz 2005).

As it can be seen from the above examples, the methods for isolation of ecdysteroids from plant materials differ a little bit from each other, and, analyzing these methods in stages, the following can be noted. In the extraction of ecdysteroids from

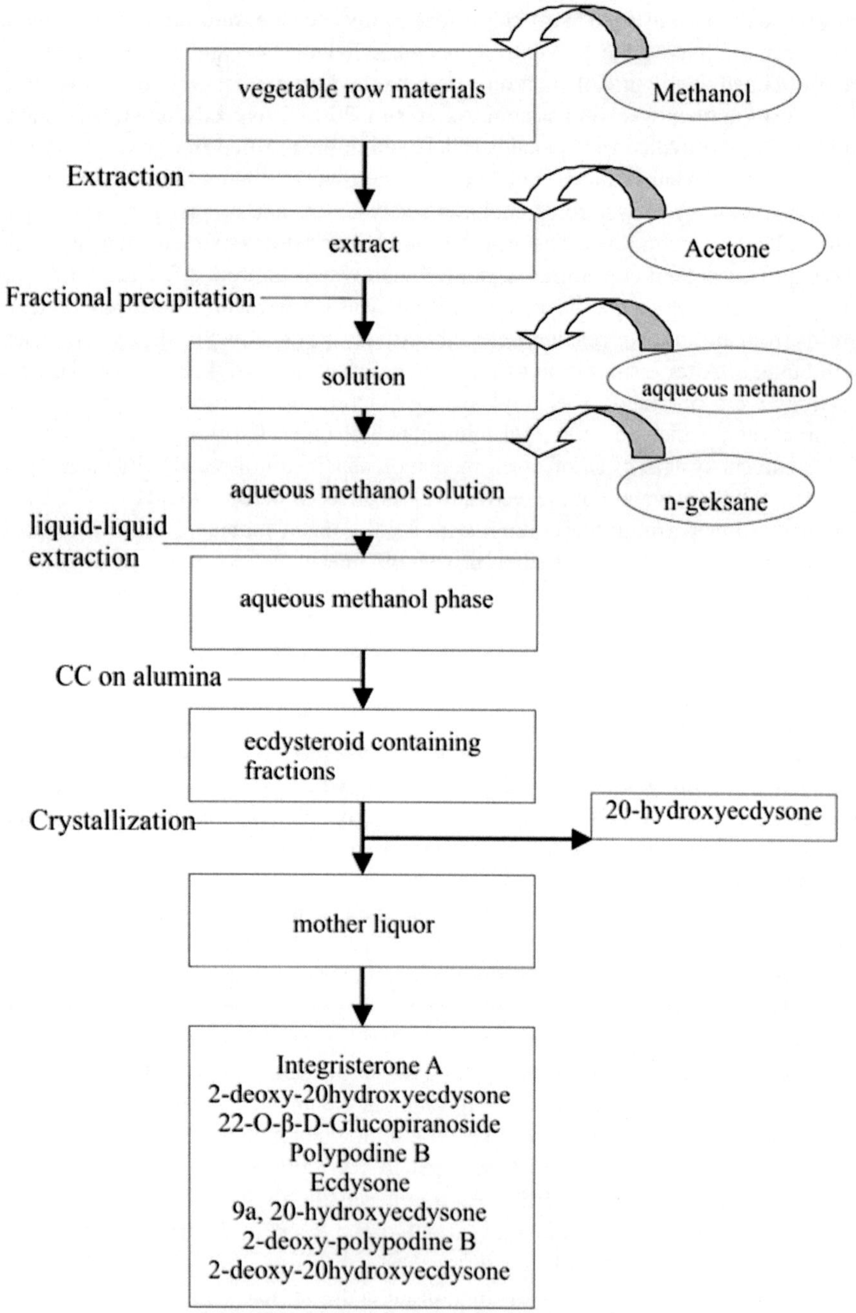

Fig. 7.1 Isolation of ecdysteroids from *Silene italica sp. nemoralis*

raw materials, methanol and ethanol are used in many cases (both at room temperature and when heated). The process of removing hydrophobic impurities from condensed extracts is carried out mainly by treating them with hexane, chloroform, diethyl, and petroleum ethers. In some methods for producing ecdysteroids, the removal of hydrophobic impurities is decided differently: they are precipitated, increasing the polarity of the solvents.

In most cases, n-butanol, ethyl acetate, and in some cases, solvent mixtures, for example, ethyl acetate-alcohol, are used to extract from purified aqueous solutions.

For the purification of ecdysteroids, as a rule, adsorption chromatography of their sum on alumina and distribution chromatography on silica gel are used, and at the final stage, crystallization of ecdysteroids, mixtures of various solvents are used depending on their physicochemical properties.

In the considered methods, the yield of ecdysteroids is hundredths and thousandths of a percent of the mass of raw materials, and they are designed for research purposes, without taking into account the production's economical effectiveness and technological process overall.

It is worth noting that industrial technologies for the production of ecdysteroids from representatives of the genera *Rhaponticum*, *Silene*, and *Ajuga* have been developed at the Institute of Chemical Engineering of the Republic of Uzbekistan.

7.3 The Technology to Produce Ecdysterone from Roots with Rhizomes and Seeds of *Rhaponticum Carthamoides*

7.3.1 Characteristics of Raw Materials

Usually, the plant material contains 0.11–0.16% ecdysteroids, as well as inulin, calcium oxalate, phosphoric acid salts, ascorbic acid, gums, resins, essential oil, essential oil glycosides, alkaloids, catechins, flavonoids (Chikov et al. 1976), lipids, including fatty acids with a predominance of omega 6 linoleic acid, sterols, glycolipids and other components (Khomova et al. 1995).

According to our data, the ecdysterone content in the organs of different plant species of leuzea, depending on the place of growth, ranges from 0.11 to 0.49%. The richest raw materials are flower baskets containing ecdysterone in the range of 0.3–0.5% by weight of raw materials (Table 7.1). Consequently, not only roots with rhizomes, but also the aerial part of the leuzea, irrespective of the place of its growth, can be used as raw materials to produce ecdysterone.

The data in Table 7.1 show that the raw material contains a complex composition of extractable substances in an amount exceeding more than 20% of its mass, therefore, the development of a technology for its complex processing with the release of ecdysteroids and lipid concentrate presented certain difficulties.

A large series of experiments were conducted on the selection of optimal extractants for obtaining a crude extract of ecdysteroids and purification of the extract by

Table 7.1 The content of ecdysterone in plants of the genus *Rhaponticum* from various places of growth, in %

Plant supplier	Plant organ	The ecdysteroids content in raw materials, in %	Isolate ecdysterone, in %	
			From content in raw materials	From the mass of raw materials
Rh. Carthamoides				
Krasnogorsk factory	The roots	0.11	45.4	0.05
State farm «Hakkaskiy»	Roots	0.14	54.5	0.06
	Leaves, stalks	0.19	26.4	0.05
	Inflorescence	0.30	53.3	0.16
Irtysh branch «Lekrasprom»	Roots	0.12	50.0	0.06
	Leaves, stalks	0.18	22.2	0.04
	Inflorescence	0.32	50.0	0.16
Kuznetsk State farm	Roots	0.16	51.2	0.082
	Leaves, stalks	0.18	33.3	0.06
	Inflorescence	0.44	45.4	0.2
Rh. Integrifolium				
Lab. med.-techn. Plants. ICPS AS RUz	Roots	0.22	45.4	0.1
	Leaves, stalks	0.12	33.3	0.04
	Inflorescence	0.49	44.8	0.22

liquid–liquid extraction, the optimal conditions for purification of target products by membrane filtration, chromatography, and ion exchange purification methods, as well as the crystallization conditions of the final product were determined.

7.3.2 *Selection of Selective Extractant*

An important stage in the isolation of biologically active substances from plants is the correct selection of solvent polarity at various stages of processing raw materials, extracts, and sediments, because the degree of solubility of substances in solvents depends on the strength of the molecular interaction between them.

The solubility of ecdysterone in various organic solvents and their mixtures is shown in Table 7.2 (Mamatkhanov et al. 1980).

From the data in Table 7.2, it can be seen that methanol (81.4%), as well as 70–80% ethyl alcohol (80.5%, 81.1%) extract the substances in approximately the same way and surpass the other studied extractants in extraction capacity (Mamatkhanov et al. 1980). These solvents do not extract such groups of compounds like proteins, pectins, polysaccharides, and the solubility of ecdysteroids in lower alcohols is maximum.

Table 7.2 Solubility of ecdysterone at 20 °C

Solvent, concentration %	Solubility, g/100 ml
Acetone 100	0.193
– //– 95	0.575
– //– 90	0.737
– //– 85	1.130
– //– 80	1.180
Ethanol 95	4.80
– //– 90	5.24
– //– 80	5.91
– //– 70	6.10
Methyl alcohol	7.50
Water	0.19
Chloroform–methanol 4:1	1.0
– //– 2:1	3.21
n-buthyl alcohol, saturated with water	3.21
Ethylacetate, saturated with water	4.10

When using 80% ethyl alcohol, the extraction of ecdysterone amounted to 95.9% of its content in raw materials.

The influence of the degree of grinding of raw materials on the extraction of ecdysterone. During the extraction of unmilled and coarsely ground raw materials, the contact surface of the particles with the solvent is small, the process is slow, and a large consumption of solvent is required. With very fine grinding of raw materials less than 1 mm, the process accelerates, however, due to poor filterability of the raw material layer, difficulties arise in the implementation of the process. For existing extractors, the optimal degree of grinding of the roots with rhizomes of the leuzea is 1–5 mm.

The effect of temperature on the extraction of ecdysterone. During extraction at different temperature conditions, and if all other conditions are being the same, it was found that an increase in temperature does not significantly affect the yield of ecdysterone. With increasing temperature, the yield of extractive substances increases, but the amount of ecdysterone remains almost unchanged. In addition, an increase in temperature negatively affects the purity of the extract, because it increases the recovery of ballast substances, which complicates the subsequent purification of the extract. Isolation of ecdysterone from such an extract is difficult, therefore, extraction at room temperature is most optimal, as it provides a high yield of the final product and does not require extra energy.

The use of hot solvents for extraction, as suggested in the literature, is impractical because of the significant loss of solvent and the complexity of the equipment. In addition, at a high temperature, significantly more associated substances pass into the extract.

7.3.3 Optimization of the Ecdysterone Extraction Process

For forecasting and optimization of technological processes, mathematical planning of the experiment is successfully applied, which has firmly entered into technological science and practice. This method allows one to obtain mathematical models linking the optimization parameter with factors influencing it and makes it possible to identify their optimal technological modes without a long process.

Having received a priori information that affects the process of extracting ecdysterone from raw materials, a further search for the optimal process conditions was carried out by the method of mathematical planning of the Box-Wilson experiment. At the same time, the following factors varied: X_1—extraction time; X_2—alcohol concentration in the extractant; X_3—the degree of grinding of raw materials; X_4—extraction temperature; X_5—l/d (ratio of height to diameter of the extractor).

When planning, j replicas from a full factorial experiment of type 25 were used. Factor levels and intervals of their variation were selected.

The optimization parameter (Y) selected the ecdysterone output at the first phase contact. For the experiments, crushed rhizomes with the roots of safflower leuzea and with the ecdysterone content of 0.12% by weight of the raw material were used. The experiments were conducted under static conditions in extractors with a capacity of about 7 l, where 1 kg of raw material was loaded. The matrix of experimental design and the results are given in Table 7.3.

The experimental results are presented in the form of a regression equation:

$$Y = b_0 + b_1x_1 + b_2x_2 + b_3x_3 + b_4x_4 + b_5x_5; \quad (7.1)$$

where, b_0, b_1, b_2, b_3, b_4, b_5—regression coefficients that are calculated by the formula:

$$b_i = \frac{\sum_{i=1}^{n} (X_{ij} \times Y_i)}{N}; \quad (7.2)$$

Table 7.3 Matrix of experiment planning and their results

No. experience	Factor						Response function			
	X_0	X_1	X_2	X_3	X_4	X_5	Y_1	Y_2	Y_{exp}	Y_{calc}
1	+	–	–	+	+	–	30.2	30.8	30.5	30.58
2	+	–	+	–	+	+	25.6	26.1	25.85	25.6
3	+	+	–	–	+	–	38.4	37.6	38.0	37.94
4	+	+	+	+	+	+	27.1	27.5	27.3	27.56
5	+	–	–	–	–	+	25.3	24.6	24.95	25.2
6	+	–	+	+	–	–	35.8	36.4	36.1	36.02
7	+	+	–	+	–	+	27.2	27.6	27.4	27.16
8	+	+	+	–	–	–	43.0	43.6	43.3	43.38

where, i is experiment number (1, 2, …, 8); j factor number (1, 2, 3, 4, 5);
X_{ij}-coded the value of factors;
n-the number of experiments performed under independent conditions.

$$b_0 = 31.68;\ b_1 = 2.33;\ b_2 = 1.46;\ b_3 = -1.35;\ b_4 = -1.26;\ b_5 = -5.3.$$

Substituting the calculated values of the coefficients in Eq. 7.1, it is obtained:

$$Y = 31.68 + 2.33X_1 + 1.46X_2 - 1.35X_3 - 1.26X_4 - 5.3X_5$$

The calculation of the variance of reproducibility is presented in Table 7.4.
The adequacy of dispersions was verified by the Cochren criterion:

$$G_{\text{exp}} = \frac{S^2_{\max}}{\sum_{i=1}^{N} S_i^2} \le G_{cr}; \tag{7.3}$$

Degree of freedom $f_1 = (n-1) = 2\text{–}1 = 1; f_2 = m(n-1) = 8 \cdot (2\text{–}1) = 8$.
n—number of parallel experiments; m—number of experiments.
$G_{\text{tabl.}} = 0.6798$ [373]; The dispersions are homogeneous, since $G_{\text{calc.}} < G_{\text{tabl}}$.
The reproducibility dispersion calculated by the formula:

$$S_y^2 = \frac{2\sum_{i=1}^{N}(Y_{iq} - Y)^2}{N} = \frac{\sum_{i=1}^{N} S_i^2}{N} = 1.392/8 = 0.174 \tag{7.4}$$

Determine the confidence interval: $\Delta b_i = \pm t\, S_{bi}$;
where $S_{bi} = \sqrt{S^2}_{bi}$;
t—table value of student criterion.

$$t = 2.306 \text{ in } f = 8.$$

Table 7.4 Statistical analysis

No. experience	Response function			ΔY_i	$\Delta Y_i^{\ 2}$	S_i^2	$\Delta Y_i'$	$(\Delta Y_i')^2$
	Y_1	Y_2	Y_{av}					
1	30.2	30.8	30.5	0.3	0.09	0.10	−0.08	0.006
2	25.6	26.1	25.85	0.25	0.063	0.126	0.25	0.06
3	38.4	37.6	38.0	−0.4	0.16	0.32	0.06	0.004
4	27.1	27.5	27.3	0.2	0.04	0.08	−0.26	0.07
5	25.3	24.6	24.95	−0.35	0.123	0.246	−0.25	0.06
6	35.8	36.4	36.1	0.3	0.09	0.18	0.08	0.006
7	27.2	27.6	27.4	0.2	0.04	0.08	0.24	0.058
8	43.0	43.6	43.3	0.3	0.09	0.18	−0.08	0.006
Sum			253.4			1.312		0.27

$$\Delta b_i = \pm 2.306\,(0.021 = \pm 0.334);$$

The coefficients are significant when $b > \Delta b$.

The adequacy of the model was checked by the Fisher criterion:

$$0.27/2 = 0.135$$

where n—the number of parallel experiments in the ith row of the matrix;

$$f = n - (K + 1) = 8 - (5 + 1) = 2$$

K—the number of model coefficients.

$$F_{\text{exp}} = \frac{S_{\text{ad}}^2}{S_{\text{y}}^2} = \frac{46{,}3322}{10{,}305} = 4{,}496$$

$$F_{\text{tabl}} = 4.5 \text{ for } f_1 = 2,\ f_2 = 8$$

In this case, $F_{\text{exp}} < F_{\text{tabl}}$; 4.496 < 4.5; therefore, the model adequately describes a linear approximation to the optimum.

As a result, it was found that the main influence on the process of extraction of ecdysterone from rhizomes with safflower-root roots is exerted by factors X_1—extraction time and X_2—alcohol concentration in the extractant.

A sharp ascent was not carried out, since the experimental yield of 43.3%, which is quite acceptable at the first contact of the phases.

Thus, the method of mathematical planning of the experiment revealed the optimal conditions for the extraction of ecdysterone from raw materials at the first contact of the phases, namely, extraction with 80% alcohol at a temperature of 20 °C, the degree of grinding of raw materials—70% of the passage through a sieve with a hole diameter of 3 mm, at l/d extractor = 1 and an extraction time of 6 h.

Kinetics of Ecdysterone Extraction. When extracting any substance from plant material, where it is accompanied by a large number of other components, it is advisable to study the kinetics of extraction of the excreted substance. Insufficient contact time of the raw material with the solvent reduces the yield of the product, and an excessive increase in the contact time leads to extracts with a high content of related substances, which complicates further technological operations. To determine the duration of extraction, the kinetics of the ecdysterone extraction process was studied by analyzing the change in ecdysterone concentration over time.

The concentration of ecdysterone in individual extracts was determined by the chromato-spectrophotometric method developed at the Institute of Chemical Physics of the Republic of Uzbekistan (Yakubova et al. 1983). It was found that the extraction curves are typical isotherms tending to equilibrium. Moreover, the necessary infusion time for the first contact of the phases was 5 h, for the second contact of the phases—3 h, for the third—2 h, for the fourth and fifth—for 1 h, which characterizes the change in the rate of extraction of ecdysterone depending on the degree of depletion of raw materials.

Purification of the condensed extract from impurities. Carried out using the method of liquid–liquid extraction.

Removal of hydrophobic impurities from the extract. Plant tissues contain organic compounds of a lipid nature, which physically and chemically differ sharply from ecdysteroids and can be relatively easily separated by selective extraction with non-polar solvents, for example, hexane.

The use of hexane, petroleum, and diethyl ethers to remove hydrophobic substances from extracts under industrial conditions has certain disadvantages, because these extractants are volatile and explosive, in addition, the solvent capacity of petroleum ether with respect to plant hydrophobic substances is insufficient. According to the results of our research, in the production of ecdysterone, the removal of lipids and chlorophyll can be carried out by treating the extract with solvents ranging from hexane to ethyl acetate. However, when using ethyl acetate, the loss of the final product due to the partial dissolution of ecdysterone occurs. Solvents in the range from hexane to ether have less dissolving power than lipophilic compounds. Therefore, the most suitable solvent for removing hydrophobic substances from extracts obtained from the processing of safflower leuzea, *Silene praemixta* M. Pop. (Smolevka mixed) and *Ajuga turkestanica* Regel (tenacity Turkestan), both in terms of its dissolving ability and ease of use turned out to be chloroform.

Removing hydrophilic impurities from the extract. Of all the groups of organic compounds that are extracted together with ecdysteroids, the most difficult is purification from hydrophilic impurities. The right choice of solvent for this purpose largely determines the efficiency of the process. To extract ecdysteroids from a purified aqueous solution, an effective solvent mixture was selected consisting of 1:1 chloroform and isopropyl alcohol, which combines a higher dissolving ability of the isocompound compared to its normal isomer with a low solubility of the mixture in water.

Sorption Purification of the Amount of Ecdysteroids

When ecdysterone is extracted from an aqueous solution with a mixture of chloroform and isopropyl alcohol, inulin and other sugars remain in the aqueous phase, and some of the accompanying ecdysteroids, as well as tannins, catechins, flavonoids, and other compounds, are extracted into the organic phase. The most rational purification of ecdysterone from these substances is adsorption chromatography. Chromatographic alumina was used as a sorbent; mixtures of organic solvents: methanol, ethanol, chloroform, propanol-2, ethyl acetate, and acetone served as eluents.

It was found that the best result is obtained by eluting ecdysterone with a mixture of chloroform and methanol.

On alumina, more polar compounds (ecdysteroids) than ecdysterone are sorbed, while less polar are extracted by chloroform together with hydrophobic substances.

Getting the Final Product

Double recrystallization of ecdysterone from a mixture of methanol and ethyl acetate allowed us to obtain a pharmaceutical product that meets the requirements of VFS, with a yield of more than 89% of the content in the technical product.

Thus, the preparation of ecdysterone includes the following process steps:

- extraction of roots with rhizomes of leuzea with 80% ethyl alcohol;
- evaporation of the extract to a bottom residue;
- treatment of the aqueous residue with chloroform to separate low-polar impurities (resins, etc.);
- extraction of ecdysteroids from the aqueous residue with a mixture of chloroform with isopropanol;
- sorption purification of ecdysteroids on alumina with the separation of ecdysterone;
- double recrystallization of ecdysterone from a mixture of methanol with ethyl acetate;
- separation of crystals;
- drying the finished product;
- packing, marking, and packaging.

The results of the studies made it possible to develop a rational technology for the production of ecdysterone from the roots with rhizomes of safflower leuzea (Egamberdieva and Jabborova 2018), which was tested at the Institute's facilities and introduced at the Chimkent Chemical Pharmaceutical Plant (Kazakhstan) and the Manturovsky Plant of Russian Medicines (Russia).

Schematic diagram of the production of ecdysterone from the roots with rhizomes of safflower leuzea consists of the following stages (Fig. 7.2).

Air-dry roots with rhizomes of safflower-headed leuzea with a moisture content of 6% are ground in a hammer mill of the DDK type (1). The degree of grinding of raw materials—80% of the passage through a sieve with a hole diameter of 6 mm Ground raw materials are bagged, weighed (2) and transferred to the stage to extract the sum of ecdysteroids from it.

Extraction of ecdysteroids from raw materials is carried out with methanol or 80% ethyl alcohol in an extractor (3) by infusion at room temperature. Filtered diagonal is placed on the grid of the loaded extractor, ground raw material is poured, and 80% ethanol from the dipstick is poured into it (4). The extracts are filtered through a druk filter (5) and collected in a collector (6, 7).

After extraction, the meal is discharged from the extractor and loaded into a vacuum drum dryer (8) for alcohol recovery. The distillation of alcohol from the meal is collected in a collection box (9) and sent to a measuring device (4) for reuse. Meal with an ecdysterone content of not more than 0.048% for dry weight is sent to the dump.

The extract is served in portions for thickening in a vacuum circulation apparatus (10). The alcohol-based distillation is poured into the collection tank (11) and transferred to the measuring device (4) for reuse. The thickened extract is poured

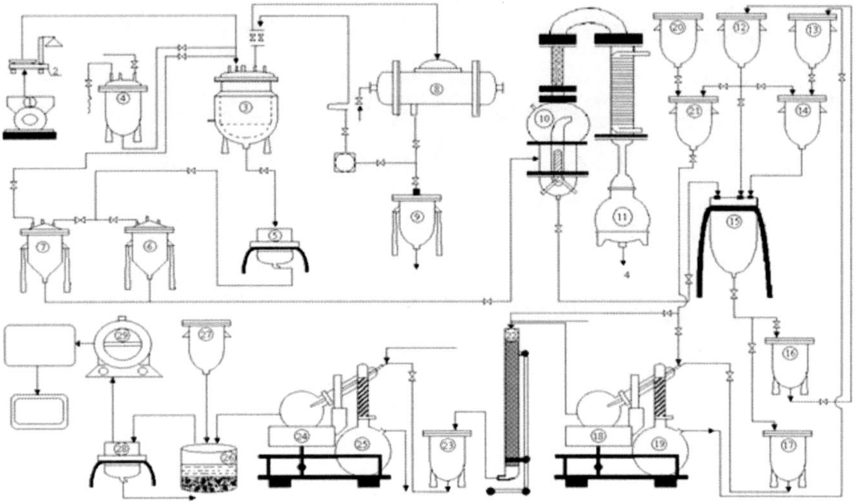

Fig. 7.2 Hardware technological scheme for the production of ecdysterone from *Rh. carthamoides*

into the reactor (15) and the hydrophobic impurities are removed by treatment with chloroform, which is fed from the mernik (12). Chloroform extraction is collected in a collection tank (16), chloroform is regenerated and sent to a mernik (12) for reuse, and the residue is sent to obtain a bioadditive.

The solution treated with chloroform is extracted with a mixture of chloroform with isopropyl alcohol, which is fed from a mixer (14). The extracts are collected in a collection tank (17), from where they are sent in portions to a rotary evaporator (18), where they are evaporated until the solvent is completely removed. The distillate is collected in a collection (19). The distillation of a mixture of chloroform with isopropyl alcohol is sent to a mixer (14), where it is adjusted by specific gravity to their ratio (1:1). The aqueous mother liquor is sent to the dump.

The residue obtained after removing the mixture of chloroform with isopropyl alcohol is dissolved in a mixture of methanol with chloroform (1:2), which is fed from the mixer (21) and passed through a column (22) loaded with aluminum oxide. The eluate is collected in a collection (23); the treated alumina is sent for regeneration.

The eluate is fed to a rotary evaporator (24), where it is evaporated to dryness. The distillation of the mixture of methanol with chloroform is poured into the collection tank (25), sent to the mixer (21), where it is brought to the initial concentration by specific gravity. The dry residue is dissolved in methanol, which is transferred from the measuring unit (20) and poured into the crystallizer (26), where ethyl acetate is added from the measuring unit (27). After a day, the crystals are separated by filtration (28) to obtain a technical ecdysten of 73% purity. The mother liquor is poured into the collection. Technical ecdysten with a basic substance content of 73% is dissolved in methanol (26) in methanol, ethyl acetate is poured therefrom (27) and

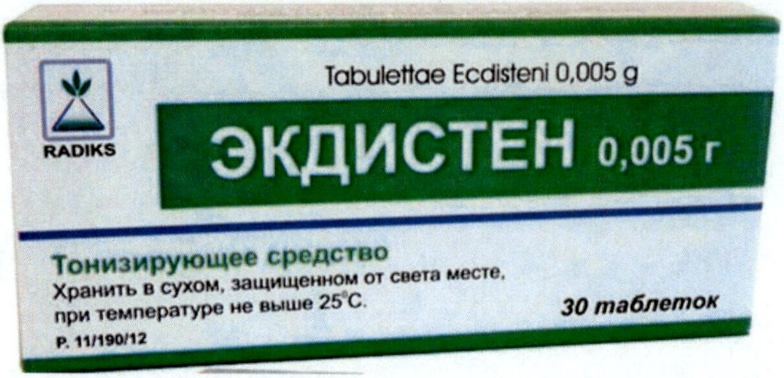

Fig. 7.3 Tonic–Ecdysten

left to recrystallize. After a day, the precipitated crystals are separated by filtration (28) and dried in an oven at a temperature of 105 °C to constant weight.

Obtaining the ecdysterone with a basic substance content of 98% (Fig. 7.3).

To control production, a chromatographic mass spectrophotometric method for determining the amount of ecdysterone by technological stages was developed (Yakubova et al. 1983). The analysis is based on the quantitative determination of ecdysterone in plant materials by chromatographic separation of the amount of extractives in a thin layer of a mixture of aluminum oxide and silica gel in a solvent system of chloroform-methanol-acetone (6:2:1), elution of zones containing ecdysterone and spectrophotometric determination of it in eluate at a wavelength of 241 nm.

7.4 Supplements for Household Chemicals Based on Lipid Waste Ecdysteroid Production

The chloroform extract obtained by removing hydrophobic impurities from condensed ethanol extraction of ecdysteroids contains biologically active substances of lipid and phenolic nature. According to the results of chemical analysis, biologically active lipids and lipophilic compounds make up a third of the hydrophobic waste, while the sum of triterpenols, sterols, and their esters in this waste reaches almost 8% (Khomova et al. 1995). Toxicological and hygienic studies have shown that lipid-containing waste products of ecdysterone production are classified as low-toxic compounds (hazard class IV) and by the decision of the Ministry of Health of the Uzbek SSR of 04.03.1987, it is approved for use as a dietary supplement in household chemical products and cosmetics. According to biological tests, a lipid-containing waste improves energy and plastic processes in the skin (carbohydrate, protein metabolism), increases its redox potential and the intensity of aerobic processes that awaken dormant hair follicles.

These data substantiated the feasibility of comprehensive processing of raw materials by developing a technology for the production of dietary supplements for household chemicals based on lipid-containing waste (Khomova et al. 1994). Supplements were produced by the experimental production of ICPS AS RUz along with ecdystene. It is part of the Limonchiki children's assortment of foaming agents, which was produced by the Almalyk plant of household chemicals in the period 1991–1994.

7.5 Isolation of Ecdysterone from *Rhaponticum Carthamoides* Seeds

Seeds of *Rh. carthamoides* contain more than 1% ecdysterone by weight of raw material. A rational technology was developed for the production of ecdysterone from *leuzea* seeds (Mamatkhanov et al. 1980).

The seeds crushed in a roller mill are loaded into the extractor and poured with chloroform. The chloroform extract was evaporated, the oily distillation residue was treated with hexane. The precipitate formed is separated off, loaded onto an alumina column, and eluted with chloroform-methanol, the eluate is evaporated, the residue is crystallized from methanol-ethyl acetate. The yield of ecdysterone is 1.15% by weight of raw materials with a basic substance content of 98%.

7.6 The Technological Process of Ecdysterone Production from Inflorescences of *Rhaponticum Integrifolium*

Another species of large-headed *Rhaponticum integrifolium* (large-headed deciduous), fam. *Asteraceae*, under natural conditions, is found on gravelly slopes in the middle zone of the mountains of Central Asia, in particular, in the Surkhandarya region, on the Ferghana Range and the Pamir-Alai.

The maximum amount of ecdysterone in the above-ground organs is *Rh. integrifolium*, introduced at the Institute's experimental site (Sagdullaev et al. 1999), which is contained in the budding period; ecdysterone is accumulated in seeds in the greatest amount *Rh. integrifolium*.

Ecdysterone from flower baskets is isolated as follows (Sagdullaev et al. 1999):

Flower baskets with an ecdysterone content of 0.46% by weight of the raw material are crushed in a roller mill and extracted with methanol in an extractor by the infusion method. The extract is concentrated and diluted with water. The methanol-water solution is treated with chloroform and the ecdysteroids are also extracted from the purified aqueous solution with a mixture of chloroform with isopropyl alcohol. The combined chloroform-isopropanol extraction is chromatographed on an alumina column. The resulting eluate was evaporated to dryness, the residue was crystallized from a solvent mixture of methanol-ethyl acetate. After a day, precipitated crystals with an ecdysterone content of 89% were separated and recrystallized from the same mixture, the crystals were separated by filtration, washed with ethyl acetate, and dried at a temperature of 100 °C. Obtained about 0.15% by weight of the raw ecdysterone with a basic substance content of 96%.

7.7 Technological Process for the Production of Ecdysterone from *Silene Praemixta* M. Pop

A promising source of raw materials for ecdysterone production is plants of the genus *Silene* (Smolevka, family *Caryophyllaceae*). This genus includes annual, perennial herbaceous or shrub plants (Bondarenko 1971). In Central Asia, 84 species of Smolevka grow (Bondarenko 1971).

Silene praemixta M. Pop (mixed resin) is a perennial plant growing in the Western Tien Shan. The aerial part (aerial part) of the mixed tar contains ecdysteroids, such as Viticosterone E, 2-deoxy-α-ecdysone, 2-deoxyecdysterone, 2-deoxyecdysterone-3-acetate, primixisterone, silenosterone, α-ecdysone, ecdisterone.

In ICPS, a technology was developed for the production of ecdysterone from *S. praemixta* aerial part (Mamathanov et al. 1993, 1999). The results of a study of the factors that influenced the stage of technological processes showed that the technology for the production of ecdysterone from rhizomes with safflower roots is also possible for aerial part mixed tar.

According to chemical analysis, the ecdysterone content in the *S. praemixta* aerial part is 1.1% by weight of raw materials.

Selection of selective extractant. To select a selective extractant, we studied the process of extraction of raw materials with a number of organic solvents and their aqueous mixtures. The raw material of the mixed tar was extracted with ethanol of various concentrations under the same conditions at room temperature.

The combined extracts are evaporated on a rotary evaporator at a temperature of 70–80 °C and vacuum (−0.6–0.4 kg/cm^2). The concentrated extracts were diluted with water and treated with chloroform. The chloroform-treated solution was extracted with a mixture of chloroform with isopropyl alcohol. The extracts were collected and evaporated until the solvent was completely removed. The residue obtained after removal of the mixture of chloroform with isopropyl alcohol was dissolved in a mixture of methanol with chloroform, and passed through a column of aluminum oxide.

The eluate was evaporated to dryness and crystallized in a solvent mixture of methanol-ethyl acetate.

The crystals were separated, recrystallized, and separated again. The resulting substance was dried in an oven at a temperature of 105 °C to constant weight.

The results showed that the optimal extractant for the extraction of ecdysteroids from aerial part tar is 70–80% ethyl alcohol and methanol, and the optimal ratio of raw materials and extractant was hydraulic module 1: (17 ÷ 20).

The influence of the degree of grinding of raw materials. When studying the influence of the degree of grinding of raw materials on the extraction of ecdysterone from *S. praemixta* aerial part, it was found that the process is slow during the extraction of the crushed raw materials. The extraction of crushed raw materials with a particle size of less than 1 mm extracts ecdysterone faster, but the extract is cloudy and difficult to filter. Thus, it is recommended to use ground vegetable raw materials with a particle size of 1–5 mm to isolate ecdysterone from n/h mixed tar.

The effect of temperature. For the extraction of ecdysteroids fro the aerial part *S. praemixta* was studied in the range of 20–70 °C. The maximum yield of ecdysterone was obtained by extraction of raw materials with 80% alcohol at a temperature of 50–60 °C.

Kinetics of the extraction of ecdysterone from aerial part S. praemixta. To develop a rational regime for extracting ecdysterone from raw materials, the kinetics of extraction of aerial part mixed tar is studied. It was 96.8%, which is quite acceptable for the extraction stage.

Purification of ecdysteroid extract from aerial part S. praemixta. A study of the purification process of the extract showed that a fivefold treatment with chloroform removes the bulk of the accompanying substances of lipophilic nature, and it is advisable to extract ecdysterone from the purified solution with a mixture of chloroform and isopropyl alcohol. Subsequent chromatographic purification of the sum of ecdysteroids on alumina with a mixture of chloroform and methanol makes it possible to isolate ecdysterone.

Recrystallization of ecdysterone from mixtures of methanol with ethyl acetate and ethanol with acetone made it possible to obtain the target product.

Description of the technological scheme for the production of ecdysterone from aerial part *S. praemixta.* We present a scheme for the preparation of ecdysterone from *S. praemixta* aerial part (Fig. 2.4), developed on the basis of the results obtained by (Mamatkhanov et al. 1993). Air-dry aerial part is crushed, weighed, and transferred to the stage to extract the sum of ecdysteroids from it. The extraction of raw materials is carried out with aqueous alcohol. The extracts are combined, filtered, and concentrated. Distillation of the alcohol is passed for reuse.

The Condensed Extract is Treated with Chloroform

Treated with chloroform, the solution was extracted with a mixture of chloroform and isopropyl alcohol and evaporated until the solvent was completely removed.

The residue obtained after removal of the mixture of chloroform with isopropyl alcohol was dissolved in a mixture of methanol with chloroform and passed through a column of aluminum oxide. Treated alumina is sent for regeneration.

The eluate is evaporated to dryness. The residue is crystallized from a mixture of methanol with ethyl acetate, and technical ecdysterone is obtained.

The distillation of the mixture of ethyl acetate and methanol was adjusted to the initial concentration by specific gravity and sent for reuse.

Technical ecdysterone is recrystallized, crystals are separated, and dried to constant weight. Ecdysterone is obtained with a basic substance content of 98%.

7.8 *Ajuga Turkestanica* Aerial Processing Technology

Endemic plant *Ajuga turkestanica* Regel. Brig. (tenacity Turkestan). *Lamiaceae*—a small-branched shrub with a narrow growing area. The plant is distributed in the Baysun district of the Surkhandarya region of Uzbekistan at an altitude of 800–1300 m above sea level.

The ecdysterone content in the aerial part of Turkestan survivability is 0.22–0.35% (Abdukadirov et al. 2005). In addition to ecdysterone, the aerial part contains a number of known phytoecdysteroids: α-ecdysone, cyasterone, ajugalactone, ajugasterone, 2,3-monoacetonide of ecdysterone, turkesterone, and 22-acetylcyasterone. From the aerial part of the plant two iridoids, garpagide and 8-O-acetylgarpagide, with a yield of 2.0 and 2.5%, respectively, were isolated. The plant is rich in biologically active components of a lipid nature (hydrocarbons, carotenoids, sterols and their esters, triterpenols, chlorophylls, polar lipids), which provides them with high activity (Talibaev et al. 1994).

Based on these biologically active substances, ICPS AS RUz created drugs and dietary supplements, such as ecdysterone (Mamathanov et al. 2008), turkesterone (Abdukadirov et al. 2005), ajustan (Abdukadirov et al. 2005). Dietary supplement jistenin, an eksumid for the preparation of which a specific technology was developed with the integrated processing of this unique raw material, combining the specific technology for producing these drugs.

7.8.1 *The Technological Process for Producing Ecdysterone from Aerial Part* Ajuga Turkestanica

A method for producing ecdysterone from A. *turkestanica* was developed at the Institute of Chemical Physics, Academy of Sciences of the Republic of Uzbekistan (Mamathanov et al. 2008).

Solvent selection for the extraction of ecdysterone (Abdukadirov et al. 2005). We studied the effect of methanol and ethanol on the degree of extraction of ecdysterone from aerial part extracts. To assess the yield, ecdysterone, turkesterone, and the sum of iridoids were chosen as marker compounds, since they are the main biologically active substances of aerial part *A. turkestanica.*

The best results were obtained during the extraction of ecdysterone, turkesterone, and iridoids from raw materials with aqueous alcohol and methanol. However, in this case, the yield of extractives increases, and the yield of biologically active substances changes insignificantly. This is because the degree of solubility of substances in solvents depends on the strength of the molecular interaction between them, i.e., dielectric constant. A chemical study of aerial part *A. turkestanica* (Abdukadirov et al. 2004) revealed the presence of alcohol-soluble sugars in the plant in the amount of 7.5%, water-soluble polysaccharides (VRPS)—4.9%, pectin substances (PV)—4.5%, lipophilic substances—6.75% and iridoid glycosides—5% (Kotenko et al.

1993). Therefore, during the extraction of aerial part *Ajuga turkestanica* with more polar solvents, accompanying substances pass into the extract, which impedes the further purification process. In addition, the use of methanol in large volumes as an extractant due to its toxicity is undesirable. Based on this, ethanol was used as an extractant to obtain an extract from aerial part survivors with a high content of biologically active substances and a lower amount of concomitant substances.

The optimal degree of grinding of the raw material turned out, as in the case of the leuzea and tar, 1–5 mm.

A study of the effect of temperature on the extraction of ecdysterone from this raw material within 20–60 °C showed that extraction at room temperature (20–40 °C) is optimal.

Optimization of the extraction of ecdysterone from raw materials. To assess the degree of influence of factors on the extraction of ecdysterone from the aerial part *A. turkestanica,* as well as to determine the conditions for the maximum yield of ecdysterone, the Box–Wilson experiment planning method was used. The optimization parameter was the ecdysterone yield at the first phase contact. Based on the results of one-factor experiments, factors and ranges of variation were selected: X_1—degree of grinding of raw materials, mm; X_2 is the concentration of ethyl alcohol,%; X_3—the duration of the process, h; X_4—extraction temperature, °C. Two levels of four factors are established, i.e. full factorial experiment of type 2^4. We used fractional replica 2, replicas of the full factorial experiment 2^4 using type 2^{4-1} planning with generating ratios $X_4 = X_1 * X_2$.

The calculations were performed using the created Microsoft Excel.

Based on the calculation results, we obtained the following equation:

$$Y = 32.2875 + 6.425X_1 + 6.4875X_2 + 8.5625X_3 + 4.85X_4$$

It was established that according to the quantitative contribution, the factors are arranged in the following order:

$$X_3 > X_2 > X_1 > X_4$$

After optimizing the extraction process, the yield increased by 8%.

Kinetics of Ecdysterone Extraction from Raw Materials. To determine the duration of extraction, the kinetics of ecdysterone extraction in time was studied. The experiments were carried out similarly to experiments in studying the kinetics of ecdysterone extraction from safflower leuzea. As extractant, 95% ethyl alcohol was used. The analysis was carried out by a chromato-spectrophotometric method, chromatographic separation of ecdysterone from related substances was carried out on plates with a fixed layer of silica gel of the L5/40μ grade in the solvent system chloroform-methanol-acetone. The optical density of the eluates was determined at a wavelength of 241 nm.

According to the results of the experiments, the necessary brewing time was established. The degree of extraction was 96.6%, which is quite acceptable for the extraction stage.

Removal of concomitant substances and extraction of ecdysterone from aqueous solution.

In addition to ecdysterone, the extract contains other ecdysteroids, iridoids (Kotenko et al. 1993), carbohydrates, polysaccharides, pectin substances (Abdukadirov et al. 2004), neutral lipids (Syrov et al. 1994).

The compounds were separated from hydrophobic substances by treating a condensed alcohol solution diluted with water and chloroform. It was found that during the extraction of an aqueous solution with chloroform, 3% (by weight of raw materials) of hydrophobic accompanying substances is recovered.

After the removal of hydrophobic substances in an aqueous solution, a large number of hydrophilic substances are left. When extracting ecdysterone from an aqueous solution by liquid-liquid extraction, a significant portion of these compounds remains in the aqueous solution.

Description of the technology for the production of ecdysterone from aerial part *A. turkestanica*.

Based on the studies, the technological scheme for the production of ecdysterone from Turkestan survivors (Mamathanov et al. 2008). Ground shoots of the plant are weighed and loaded into the extractor. Then the compounds were extracted with aqueous alcohol, the extract is evaporated, diluted with water. The aqueous bottoms are treated with chloroform. Ecdysteroids are extracted from the treated aqueous solution with butyl alcohol or with a mixture of chloroform—isopropyl alcohol. The amount of ecdysteroids is thickened and poured with acetone to precipitate the accompanying substances. The precipitate is separated and removed from the process cycle. The filtrate is concentrated and chromatographed on a column of aluminum oxide, eluents are a mixture of chloroform—isopropyl alcohol or chloroform—methyl alcohol.

The eluate was evaporated to dryness, the residue was crystallized from methanol-ethyl acetate. The precipitated crystals are separated and dried.

The obtained 0.1% by weight of ecdysterone had a purity of 97.6% (Mamathanov et al. 2008).

Quantification of Ecdysterone

For the quantitative determination of ecdysterone according to the stages of the technological process, conditions for its analysis by HPLC were developed (Abdukadirov et al. 2005). Ecdysterone was analyzed by HPLC on a Zorbax Eclipse—C 18, 3 × 150 mm, 3.5 μm column (Agilent Technologies) using an Agilent LC 1100 chromatograph with a four-element pump, a degasser and a variable wavelength UV detector. The chromatograph was controlled and processed by Agilent ChemStation liquid chromatography software. UV detection of the separated ecdysteroids was carried out at a working wavelength of 260 nm.

The quantitative determination of ecdysterone was carried out using the external standard method based on a two-point calibration with the calibration equation:

$$Y = mx + b$$

where *Y* is the peak area, *x* is the amount of substance,%.

The constants *m* and *b* are determined respectively for ecdysterone:

$$m = 275180.00;\ b = 5.68 * 10^{-12}.$$

For quantitative determination of ecdysterone, calibration solutions were prepared as 50% ethanol with a concentration of 0.004 and 0.1%, and the sample was introduced through a loop of 20 μl of Agilent hand injector (Abdukadirov et al. 2005).

The obtained chromatograms indicated the identity of the resulting ecdysterone from *A. turkestanica* and *Rh. carthamoides.*

7.8.2 *The Technological Process for Producing Turkesterone from Aerial Part* A. Turkestanica

The technological process includes the following stages (Mamatkhanov 2009):

- grinding n/h survivals of Turkestan;
- extraction of raw materials with 95% ethyl alcohol;
- thickening of alcoholic extract;
- removal of hydrophobic impurities with hexane or gasoline;
- extraction of Turkestone from an aqueous solution with butanol;
- thickening of butanol extraction;
- precipitation of iridoids and other impurities with acetone;
- chromatographic purification of Turkestone;
- thickening of the eluate;
- crystallization of Turkesterone from a mixture of methanol-ethyl acetate;
- filtration and drying of Turkestone.

7.8.3 *Technology for Isolating Ajustan Substance from* A. Turkestanica

A technology has been developed for the preparation of ayustan, a drug with lactostimulating and adaptogenic effects, from Abdukadirov et al. (2004, 2005) whose active principle is the sum of ecdysteroids and iridoids mixed with 10% pectin. The ecdysterone content in astana is at least 1.5%.

It was established by experiments that the maximum removal of associated hydrophobic substances is achieved by treating an aqueous solution of ayustan with gasoline, then by treatment with chloroform.

Removal of high molecular weight-related substances by ultrafiltration. The processes of separation of liquid systems play an important role in many sectors of the economy. For their implementation, methods such as distillation, rectification, extraction, and adsorption are used. However, the most universal method is separation using semipermeable membranes (membrane methods).

Among membrane methods for separating liquid systems, reverse osmosis, ultrafiltration, and microfiltration are widely used in engineering and laboratory practice.

Ultrafiltration is the process of membrane separation of solutions of high molecular weight and low molecular weight compounds, as well as fractionation and concentration of high molecular weight compounds. It flows due to the unique throughput properties of the membranes.

Conditional boundaries of the application of membrane methods for the separation of liquid systems:

- reverse osmosis 0.0001–0.001 μm (particle diameter);
- ultrafiltration 0.001–0.02 μm;
- microfiltration 0.02–10 μm.

Simplicity of design, speed of separation, low cost are the main advantages of the ultrafiltration method compared to conventional separation processes. The method is especially convenient when separating solutions containing traces of thermally and chemically unstable substances, which include food products, medicinal substances, and other biological objects. (Mamatkhanov et al. 1994).

To remove high molecular weight related substances from the aqueous solution of ayustan, the ultrafiltration method was used on an ALPMA-A-404-90 installation (Germany) with a Carbosep ceramic membrane (France) with a nominal pore size of 10 kDa. This membrane is chemically stable when using cleaning agents in the range of 20–85 °C and pH 1–14 and thermally during sterilization with hot steam up to 121 °C.

An aqueous solution purified from hydrophobic impurities (with a content of 20.5% solids) was passed through ultrafilters. Using literature data and obtaining a priori information affecting the process, we optimized the conditions for cleaning ayustan using a plan of the Latin square 3×3 type.

The optimization parameter was chosen as the content of dry matter in the filtrate. As a result, to remove high molecular weight related substances by ultrafiltration, the best result was achieved with a fluid circulation speed of 600 l/h, an ultrafiltration temperature of −60 °C, and a filtration rate through the membranes of −30 l/h.

Membrane filters with a throughput of 10 kDa allowed to reduce the amount of high molecular weight impurities in ayustan to 22% without loss of basic substance.

Drying the purified ajustan water extract. The purified extract (PE) contained 16% solids, including ecdysterone, turkesterone, and the amount of iridoids.

Iridoids mixed with glucose are highly hygroscopic and difficult to dry. To obtain a dry extract, we used an ShSV-45K vacuum drying oven, an IKS-2M infrared radiation dryer, and an Anhydro N_2 nozzle-type spray dryer.

Schematic diagram of the production of ajustan from aerial part A. turkestanica. As a result of the studies, a technology was developed to obtain the drug ayustan from the aerial part Turkestan plant parts (Mamatkhanov 2009) and the drug is approved for use in medicine.

The dried aerial part of the plant is crushed in a mill (1) to an average particle size of 1–3 mm. The crushed raw materials with the content of ecdysterone 0.34%, turkesterone 0.31%, the amount of iridoid 4.2% are weighed on a balance (2) and loaded into an extractor (4). Aqueous alcohol is poured from the measuring device (3) into the extractor (4). Alcoholic extract from the collection (5) is served in portions in a vacuum evaporation apparatus (6). The evaporation is carried out at a temperature of 60–70 °C and vacuum (0.6–0.8 kg/cm^2). Distillation of alcohol is collected and used for the extraction of raw materials.

Distilled water is added to the evaporated thick extract and the alcohol residue is removed by evaporation. The aqueous extract is poured into a glass column (7), where it is treated three times with gasoline and twice with chloroform. Used gasoline and chloroform are regenerated for subsequent use.

After treatment with solvents, the extract was evaporated (8) to remove residual solvents. The resulting aqueous concentrate is diluted with water in a collection tank (9) and passed through an ultrafiltration unit (10) with cermet membranes with pores of 10 kDa. The fluid circulation rate is 600 l/h, the ultrafiltration temperature is 60° C, and the filtration rate through the membranes is 30 l/h. The filtrate is collected in a collection (11). The result is an intermediate—"purified extract".

The purified solution after ultrafiltration contains 15–16% solids. From the collector (11), the solution is fed into the reactor (12) with a stirrer.

The solution is diluted with distilled water to a dry solids content of 10%, and pumped using a peristaltic pump (13) with a controlled flow of the solution into the spray dryer (14). To obtain a uniform powder during drying, the solution is mixed. During the drying process, a dusty air mixture forms, which is separated in a cyclone. From the cyclone, the powder is collected in a collector (15), after which it is sent to the packaging. The result is a dry powder of the drug ayustan (Fig. 7.4).

7.8.4 *The Technological Process for Producing Jistenin's BAA from Aerial Part* Ajuga Turkestanica

Jistenin is a biologically active additive (BAA) in cosmetics, which is obtained from aerial part *A. turkestanica*. The content of dry matter in jistenin is not more than 5%.

Extraction of dietary supplements jistenin. To obtain jistenin, a method for the selective extraction of ecdysteroids from raw materials was required, which ensured a minimum transition to the iridoid extract along with colored impurities. Preliminary studies have shown that ethyl acetate is the most suitable solvent for producing dietary supplements. Although ecdysterone and turkesterone in their pure form are poorly soluble in ethyl acetate, during the extraction process, ethyl acetate is saturated with soluble components of the feed, and in this solution the solubility of ecdysteroids increases, while the solubility of iridoids remains low. The extract obtained using

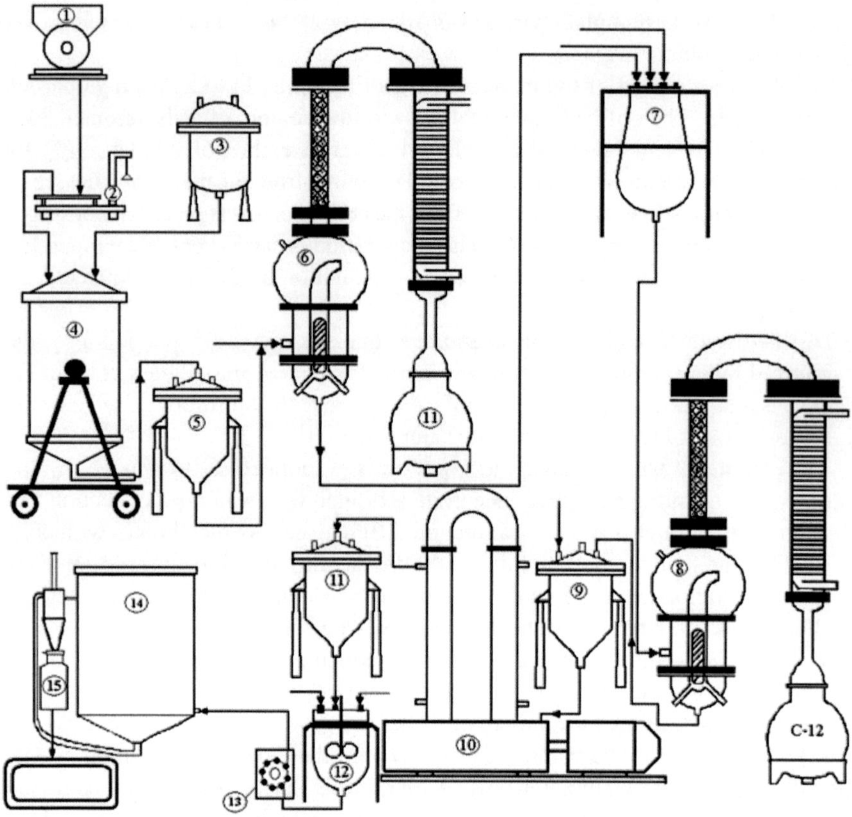

Fig. 7.4 Hardware scheme for obtaining *Ajustan* substance

ethyl acetate, practically does not require the inclusion of the stage of removal of lipid-like substances from the thick extract. Ecdysteroids are easily extracted from the concentrated extract with water, while the main non-polar components remain in ethyl acetate. The main factor in the process is the degree of concentration of the extract and the amount of water to be extracted. During the concentration of the extract, an abundant precipitate forms, which impedes the further process. The optimal concentration of the extract was established empirically and obtained the degree of concentration of ethyl acetate extract to the mass of the feedstock.

To extract ecdysteroids from concentrated ethyl acetate extract.

Purification of the aqueous extract. The main quantity by processing the aqueous extract with butanol is extracted turkesterone and ecdysterone. The resulting butanol extraction was evaporated until the solvent was completely removed, the residue was diluted with water and passed through ion exchangers. When the aqueous extract passes through the anion exchange resin, the solution acquires an alkaline reaction. A cation exchanger was used to neutralize the solution. The necessary purity of the product is ensured when the volume of anion exchange resin is 0.5–1.0 L.

Schematic diagram of the production of dietary supplements jistenin from aerial part A. turkestanica.

Air-dry aerial part plant parts are crushed in a mill (1) to a degree of grinding—80% of the passage through a sieve with a hole diameter of 6 mm. Ground raw material is packed in bags, weighed (2) and transferred to the stage of extraction of raw materials.

Ethyl acetate from the measuring unit (3) is poured into the extractor (4) and the raw material is extracted.

The ethyl acetate extract is collected in a collector (5), from where it is fed in portions into a vacuum circulation apparatus (6), where it is concentrated. The distillation of ethyl acetate is collected in a collection tank (7) and the measuring device (3) is sent for reuse. A meal is fed to a dryer to extract ethyl acetate, then taken to a dump.

The thickened ethyl acetate extract was poured into a glass reactor (8) and extracted with water. The aqueous extracts are evaporated in a vacuum evaporation apparatus (11) until the residual ethyl acetate is completely removed, concentrated, and poured into a reactor (9) to extract ecdysteroids with butanol. The ethyl acetate phase is directed to the regeneration of ethyl acetate.

The concentrated aqueous extract in the reactor (9) is extracted with butanol, which is fed from the measuring unit (10). The butanol extracts were evaporated in (11) until the butanol was completely removed, then water was added. The aqueous concentrate is poured into a collection tank (13) and diluted with water. The distillation of butanol is collected in a collection tank (12) and sent to a measuring device (10) for reuse.

An aqueous concentrate with a solids content is passed through ion-exchange columns (14, 15), connected in series, and filled with ion-exchange resins. When displaced from the columns, the solution purified on ion exchangers is diluted with water, then it is concentrated in (16) to a solids content of 11%. The concentrate is poured into a collection tank (18), where it is diluted with alcohol.

The water distillate is collected in the collector (12) and sent to the sewer.

The water-alcohol solution is poured into a glass reactor (19). Stirred for 1 h and filtered on a suction filter (20). Then its concentration is brought to 50% with alcohol and sent to the packaging. The yield of jistenin is 35% (Fig. 7.5).

According to this technology, the experimental production of jistenin dietary supplements was started by ICPS AS RUz, which was then sold to Christian Dior to produce various cosmetic creams (Fig. 7.6).

7.8.5 *Technological Process for Producing Eksumid from Aerial Part* Ajuga Turkestanica

The preparation eksumid is a purified extract of aerial part survivability with an equal content of ecdysterone and Turkesterone. It has a tonic effect.

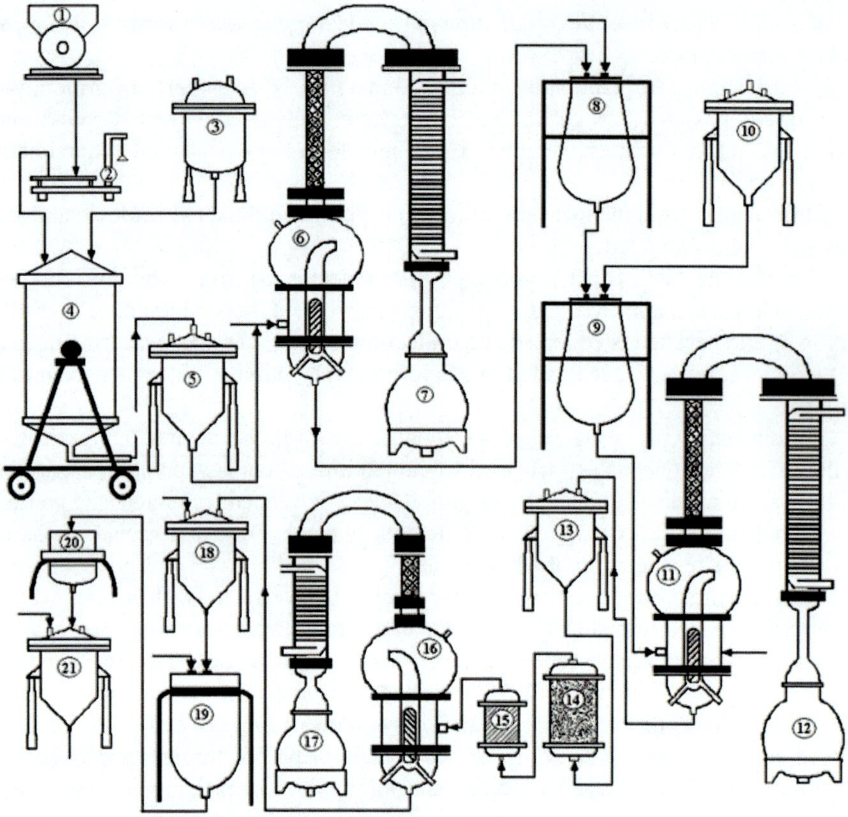

Fig. 7.5 Hardware diagram of *Jistenin* production

Fig. 7.6 Products of the company Christian Dior (France)

The eksumid substance differs from astana in a higher content of ecdysteroids and a relatively lower content of hygroscopic components such as iridoids and low molecular weight carbohydrates. The intermediate product of eksumid is an aqueous solution after purification of the extraction of ecdysteroids in an ultrafiltration unit.

Eksumid is obtained by the following technology. The aqueous solution of ecdysteroids purified on an ultrafiltration unit (see Fig. 7.4) is fed into a vacuum evaporation apparatus (VEA) (11) using vacuum, concentrated to a dry solids content of 45–50% and drained into a collection tank (12). The concentrate was extracted with butanol in a reactor (13). All extracts are combined and evaporated in VEA (14) to remove the extractant. Water is added to the concentrate, poured into a collection tank (15), diluted with water, and passed through ion-exchange columns (16). The purified solution is concentrated in VEA (17) to a solids content of 11%. Then the concentrate is poured into the reactor (18), where it is diluted with ethyl alcohol and treated with activated carbon with stirring. The solution is filtered on a suction filter (19) and evaporated in VEA (20). The resulting 25% solution is pumped (22) to a spray dryer (23). The dried product is collected in a collection (24). A powdered product is obtained which is stable during storage (Fig. 7.7).

A technology has also been developed to produce Turkestan tenacious lipid concentrate, which improves skin metabolic processes (Talibaev et al. 1994), which is recommended as an active additive in the manufacture of cosmetics and household chemical products.

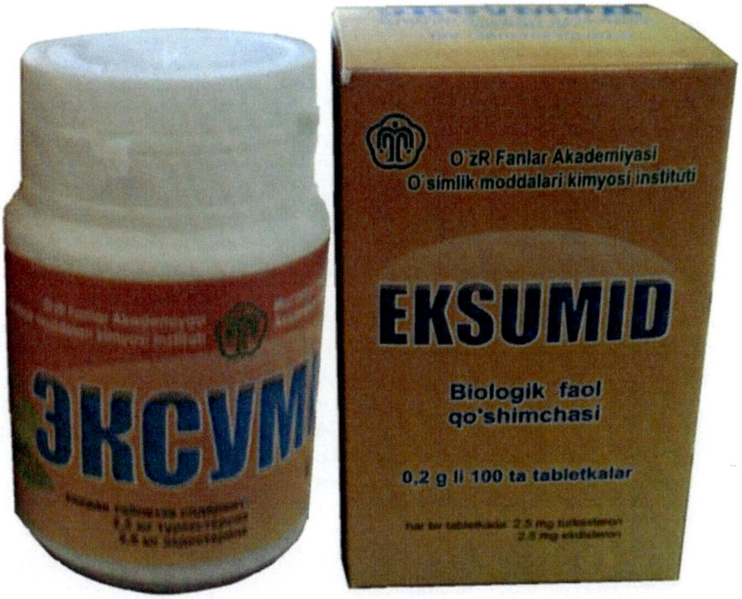

Fig. 7.7 The drug tonic action of Eksumid

References

Abdukadirov IT, Khojaeva MA, Turakhodjaev MT, Mamathanov AU (2004) Carbohydrates *Ajuga Turkestanica*. Chem Nat Compd 40:85–86. https://doi.org/10.1023/B:CONC.0000025475.31616.47

Abdukadirov IT, Turakhodjaev MT, Yakubova MR, Mamathanov AU (2005) Study of the drying process of the extract from the plant *Ajuga turkestanica*. Chem Chem Technol 2:65–66

Abdukadirov IT (2007) Development of the technology of drugs Ajustan, Eksumid, Garpachol, from plants *Ajuga Turkestanica*. Dis Cand Tech Sci Tashkent, p 105

Akhrem AA, Levina IS, Titov YuA (1973) Ecdysones are insect steroid hormones, p 232

Baltaev UA (1992) Phytoecdysteroids of *Rhaponticum carthamoides* III. Rhapisterone C. Chem Nat Compd 28:198–200. https://doi.org/10.1007/BF00630175

Baltaev UA (1991) Phytoecdysteroids of *Rhaponticum carthamoides* II. Rhapisterone B. Chem Nat Compd 27:712–713. https://doi.org/10.1007/BF00629932

Bathori M, Pongracz Z (2005) Phytoecdysteroids—from isolation to their effects on humans. Current Med Chem (Bentham Science Publishers Ltd.) 12(2):153–172. https://doi.org/10.2174/0929867053363450

Bobaev ID, Utaeva ZR, Ramazanov NSh (2009) Phytoecdysteroids of the plant *Scutellaria guttata*. Actual problems of the chemistry of natural compounds. Abstracts. Int Conf., p 126

Bondarenko ON (1971) Genus *Silene L.*-Smolevka. In the book: Key to plants of Central Asia. Tashkent: Science, pp 253–277

Borovikova EB, Shangaraeva GS, Baltaev UA (1999) Rhapisterone D 20-acetate from the seeds of *Leuzea carthamoides*. Chem Nat Compd 35:184–185. https://doi.org/10.1007/BF02234931

Boymatov IM, Bobaev ID, Ramazanov NSh (2010) Ecdysteroids of the plant *Lagochillus platycalyx*. In: VI All-Russian conference on chemistry and technology of plant substances. Thes. doc. conf., p 312

Chikov PS et al (1976) Atlas of the ranges and resources of medicinal plants of the USSR, p 263

Chueshov VI (2002a) Industrial technology of drugs. In: 2 volumes/Ed. Kharkov: MTK-Book: Publishing House of NFAU, vol 1, p 560

Chueshov VI (2002b) Industrial technology of drugs. In: 2 volumes/Ed. Kharkov, MTK-Book, Publishing House of NFAU, vol 2, p 716

Egamberdieva D, Jabborova D (2018) Medicinal plants of Uzbekistan and their traditional uses. In: Egamberdieva D, Öztürk M (eds) Vegetation of Central Asia and environs. Springer. https://doi.org/10.1007/978-3-319-99728-5_8

Ikonnikov BK, Egoyants PA, Sirotin CA, Truong TH (2010) Determination of parameters of pre- and supercritical carbon dioxide extraction of raw materials of plant origin. Super fluids. Theory Pract (moscow) 4:65–74

Ivanova SA, Weinstein VA, Kaukhova IE (2003) Features of mass transfer of lipophilic biologically active substances upon extraction of raw materials with a two-phase system of extractants. Chem Farm J (moscow) 37(8):30–33

Kaukhova IE, Demchenko YuT, Weinstein VA, Ivanova SA, Khazza IKh (2004) On the role of the polar phase in the process of extraction of medicinal plant materials with two-phase systems of extractants. Materials of international scientific and practical. In: Conference Graduate of a pharmaceutical university (faculty) in the past, present and future. St. Petersburg, pp 217–221

Kaukhova IE (2006) Features of extraction of biologically active substances by a two-phase system of extractants in the integrated processing of medicinal plant materials. Plant Resour (st. Petersburg) 42(1):82–91

Khomova TV, Gusakova SD, Glushenkova AI (1995) Lipids from Ékdysten production wastes. Chem Nat Compd 31:172–174. https://doi.org/10.1007/BF01170198

Khomova TV, Gusakova SD, Glushenkova AI, Khushbaktova ZA et al (1994) A method for producing a skin metabolism enhancing agent. No. 528. Patent Ruz

Kotenko LD, Yakubova MR, Mamatkhanov AU et al (1993) Iridoids of *Ajuga turkestanica* and their quantitative determination. Chem Nat Compd 29:606–607. https://doi.org/10.1007/BF00630207

Mamadalieva NZ, Zibareva LN, Saatov Z (2002) Phytoecdysteroids of *Silene linicola*. Chem Nat Compd 38:268–271. https://doi.org/10.1023/A:1020436128797
Mamathanov AU, Azimova SS, Turahodzhaev MT, Yakubova MR, Syrov VN (2008) The method of producing ecdysterone. No. IAP 03623. Patent Ruz
Mamathanov AU, Saatov Z, Syrov VN, Aripjanov AK, Gorovis MB, Yakubova MR, Shakirov TT, Abubakirov NK (1993) The method of obtaining ecdysten. No. 955. Patent Ruz
Mamathanov AU, Syrov VN, Soatov Z, Yakubova MR (1999) A method for producing ecdysterone. No. 5786. Patent RUz
Mamatkhanov AU, Aripov KhN, Syrov VN, Soatov Z, Aripjanov AK, Pulatov AA, Makhsumova NS, Turahodzhaev MT, Jahangirov FN (1994) A method for producing a tonic, adaptogenic, and actoprotective effect. No. 1751. Patent Ruz
Mamatkhanov AU, Shamsutdinov MR, Shakirov TT (1980) Isolation of ecdysterone from the roots of *Rhaponticum carthamoides*. Chem Nat Compd 16:381–382. https://doi.org/10.1007/BF00568372
Mamatkhanov AU (2009) Isolation of ecdysterone from the aerial part of *Ajuga turkestanica* (Rgl.) Brig. Pharm J (Tashkent) 2:33–35
Patov SA (2006) Isolation and counter synthesis of glycosides with adaptogenic properties: Abstract. Dis. Cand. Chem Sci (Moscow) p 19
Petrenko E, Parshikova V, Zaychenko E (2021) Optimization of processing branches, needles, and leaves ("woody greens") of dark-coniferous breeds of thin tree stands using a modified plant. In: IOP conference series earth and environmental science, vol 395. https://doi.org/10.1088/1755-1315/395/1/012021
Saatov Z, Gorovits MB, Abubakirov NK (1993) Phytoecdysteroids of plants of the genus Silene. Chem Nat Compd 29:551–557. https://doi.org/10.1007/BF00630195
Sagdullaev BT, Shakhidoyatov RKh, Shakhidoyatov KhM, Turakhodaev MT (2002) A method of obtaining an extract of rose hips Rosa canina No. 05273. RUz Patent
Sagdullaev BT, Turakhodaev MT, Shakhidoyatov RKh, Khojaeva MA, Ivanova IA, Rafikova RA (2001) A method for producing an extract from a plant material No. 04588. RUz Patent
Sagdullaev SS, Sultanov SA, Nigmatullaev AI et al (1999) *Rhaponticum integrifolium* as a producer of ecdysteroids. Chem Nat Compd 35:219–220. https://doi.org/10.1007/BF02234942
Sagdullaev SS (2007) The technology of complex processing of medicinal plants Aconitum leucostomum, Aconitum septentrionale, Ungernia Victoris, Rosa canina and cottonseed meal. Abstract. Dis. Doctor. Those. Sciences. ICPS AS RUz; Tashkent, p 54
Shakhidoyatov RKh (2004) Development of processing technology for plants of the genus Althaea and Capsicum. Abstract. Dis. Cand. Those. Sciences. ICPS AS RUz. Tashkent, p 22
Syrov VN, Khushbaktova ZA, Tolibaev I, Yeletskaya NV, Mamatkhanov AU (1994) The effect of lipid concentrate from the aerial part of the Turkestan survivor on metabolic processes and the dynamics of healing of wound skin defects in the experiment. J Chem Farm 11:46–48 (in Russ)
Talibaev I, Mukhamedova S, Mamatkhanov AU, Syrov VN, Khushbaktova ZA et al (1994) A method for producing a lipid concentrate from a tenacity of the Turkmenistan improving skin process metabolism. No. 1753. Patent Ruz
Vorobiev N (2008) Electrotechnologies for extraction from food plants and biomaterials. Springer, New York, p 272. https://doi.org/10.1007/978-0-387-79374-0
Yakubova MR, Mamatkhanov AU, Shamsutdinov MRI et al (1983) The monitoring of the production of ecdysterone. Chem Nat Compd 19:305–307. https://doi.org/10.1007/BF00579765